Mechanical Science

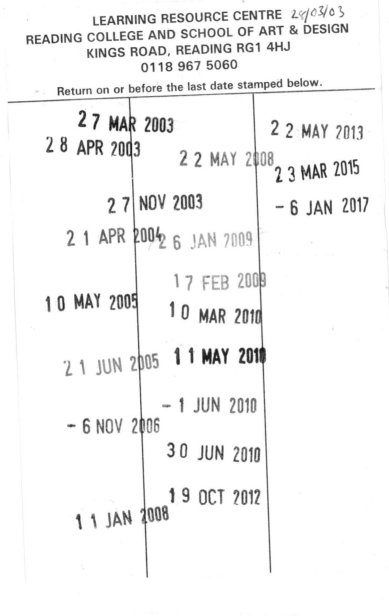

Mechanical Science

W. Bolton

Second Edition

Blackwell
Science

© 1993, 1998 W. Bolton

Blackwell Science Ltd, a Blackwell Publishing Company
Editorial Offices:
9600 Garsington Rd, OX4 2DQ, UK
 Tel: +44 (0)1865 776868
Blackwell Science, Inc., 350 Main Street, Malden, MA 02148-5018, USA
 Tel: +1 781 388 8250
Iowa State Press, a Blackwell Publishing Company, 2121 State Avenue, Ames, Iowa 50014-8300, USA
 Tel: +1 515 292 0140
Blackwell Science Asia Pty, 54 University Street, Carlton, Victoria 3053, Australia
 Tel: +61 (0)3 9347 0300
Blackwell Wissenschafts Verlag, Kurfürstendamm 57, 10707 Berlin, Germany
 Tel: +49 (0)30 32 79 060

The right of the Author to be identified as the Author of this Work has been asserted in accordance with the Copyright, Designs and Patents Act 1988.

First edition published 1993
Reprinted 1995, 1996
Second edition published 1998
Reprinted 2001, 2003

Library of Congress
Cataloging-in-Publication Data

Bolton, W. (William), 1933–
 Mechanical science/W. Bolton. – 2nd ed.
  p. cm.
 Includes index.
 ISBN 0–632–04914–6 (pb)
 1. Mechanics, Applied. I. Title.
TA350.B5385 1998
620.1′05–dc21

                                    97-34606
                                         CIP

ISBN 0-632-04914-6

A catalogue record for this title is available from the British Library

Set by Aarontype Ltd, Bristol
Printed and bound in Great Britain by
TJ International Ltd, Padstow, Cornwall

For further information on
Blackwell Publishing, visit our website:
www.blackwellpublishing.com

Contents

Preface to second edition

This book aims to give a comprehensive coverage of mechanical science, covering:

- principles of statics;
- mechanics of materials;
- principles of dynamics;
- mechanics of machines.

A background of engineering science from a National Certificate/Diploma or GNVQ Advanced Engineering, or A-level Physics or Engineering Science has been assumed, though all relevant basic principles are revised. A basic knowledge of algebra and calculus has also been assumed.

This book is aimed at giving a comprehensive coverage of mechanical science which is suitable for HNC/HND students taking Mechanical Engineering courses, including all the topics likely to be covered in both years of such courses, and other courses such as those in first year undergraduate courses in Mechanical Engineering. In the second edition the chapter on beams has been extended to consider both composite beams and plastic bending, and chapters have been added on the vibrations of multi-degree systems and matrix methods of structural analysis.

While it is recognised that analysis of mechanical science problems of any complexity is generally carried out by means of a computer, it was felt that within a book devoted to the establishment of principles the exercise of working things 'by hand' was more appropriate in this context and would enable the reader to appreciate more readily the principles involved. Almost 500 problems are included, answers being given to all. In addition each chapter includes fully worked examples, there being almost 200.

I would like to thank all those who advised me of corrections to the first print of the first edition, in particular Michael Hush, Chris Sidebotham, E. Walsh and A. Shepherd.

W. Bolton

Chapter 1
Forces and equilibrium

1.1 Scalars and vectors

There are two types of quantities in mechanics: those which have magnitude
but no directional properties and are called *scalar quantities* and those,
called *vector quantities* which are associated with a direction as well as
having magnitude.

Scalar quantities, e.g. mass and energy, can be added or subtracted by the
ordinary mathematical rules for addition and subtraction. The convention
that is used in books is that scalar quantities are represented by letters in
italic type, e.g. mass m.

Vector quantities, such as acceleration and force, cannot be added or
subtracted by the ordinary mathematical rules of addition and subtraction.
Their directions have to be taken into account. Vector quantities can be
represented by arrow-headed straight lines, the length of the line representing
the magnitude of the quantity and the direction of the arrow the direction of
the quantity. The convention that is often adopted in books is that vector
quantities are represented by letters in bold type, e.g. force **F**. When we are
referring to just the magnitude of a vector quantity then it is represented by
just the letter in italic type, e.g. the magnitude of a force F, or the letter in
italic type between vertical lines, e.g. $|F|$. Often, when the direction of a
vector is implied, or specified by some diagram, the vector is just referred to
by its magnitude.

The term *coplanar vectors* is used for vectors which lie in the same plane
and *concurrent vectors* for those which have lines of action which all pass
through the same point.

1.1.1 Vector addition and subtraction

In general, two concurrent vectors are added together by means of the
parallelogram law. The term *resultant* is used for the resulting vector. The
parallelogram law can be stated as: the resultant vector **V** obtained by
adding two vectors V_1 and V_2 is the diagonal of the parallelogram in which
V_1 and V_2 are represented by arrow-headed lines as adjacent sides. Figure
1.1 shows the parallelogram.

If a vector V_2 is drawn with its line of action unchanged but the arrow
pointing in the opposite direction then we have the vector $-V_2$. Thus the
difference between two vectors V_1 and V_2 can be obtained by using the

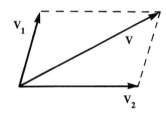

Fig. 1.1 Parallelogram law.

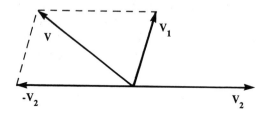

Fig. 1.2 Subtracting vectors.

parallelogram law to add the vectors $\mathbf{V}_1$ and $-\mathbf{V}_2$. Figure 1.2 illustrates this.

The result of vector addition or subtraction can be obtained by scale drawing of the parallelogram and measurement of the resulting resultant, or by calculation. With calculation useful relationships are the *cosine rule* and the *sine rule*. For the triangle shown in figure 1.3, the cosine rule gives

$$\begin{aligned}
a^2 &= b^2 + c^2 - 2bc \cos A \\
b^2 &= a^2 + c^2 - 2ac \cos B \\
c^2 &= a^2 + b^2 - 2ab \cos C
\end{aligned}$$
[1]

and the sine rule

$$\frac{a}{\sin A} = \frac{b}{\sin B} = \frac{c}{\sin C}$$
[2]

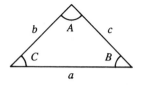

Fig. 1.3 Cosine and sine rules.

1.1.2 Resolution of vectors

It is often useful to be able to replace a single vector by two other vectors, generally at right angles to each other. The single vector is said to have been resolved into its components. It is done by using the parallelogram law in reverse, i.e. starting with the diagonal and finding the two vectors which would fit as sides of the parallelogram. Thus the magnitudes of the components of vector **V** in figure 1.4 are, in the x-direction

$$V_x = V\cos\theta \qquad [3]$$

and in the y-direction

$$V_y = V\sin\theta \qquad [4]$$

Vector addition and subtraction can often be simplified by resolving each vector into components and then adding or subtracting by simple arithmetic the vectors in each of the two directions. The resulting two components can then be recombined by means of the parallelogram law to give the resultant.

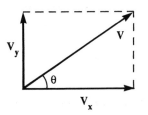

Fig. 1.4 Resolution of vectors.

1.2 Force as a vector

Forces cannot be directly observed; only their effects can be seen, these being the acceleration of an object when acted on by a resultant force or the distortion of a body acted on by a pair of forces. The concept of force arises from *Newton's laws*. These can be stated as follows.

1 A body will remain at rest, or continue to move in a straight line with a constant velocity, if there is no resultant force acting on it.
2 If a resultant force acts on a body then it will accelerate in the direction of the force with an acceleration proportional to the magnitude of the force. Or alternatively:
   If a resultant force acts on a body then it will change its momentum in the direction of the force with a rate of change of momentum proportional to the magnitude of the force.
3 If one body exerts a force on a second body then the second body will exert an equal and opposite force on the first, i.e. to every action there is an opposite and equal reaction.

The first law defines the condition for a body to be in what is termed *equilibrium*. The second law explains what happens when there is no equilibrium. The third law defines the way two bodies interact. The second law can be written as the equations

$$F = ma \qquad [5]$$

$$F = \frac{\mathrm{d}\,(mv)}{\mathrm{d}t} \qquad [6]$$

where m is called the mass of the body, a the acceleration and v the velocity, with mv being the linear momentum.

In equation [5], of the three quantities F, m and a the units of any two can be chosen arbitrarily and used to determine the unit of the third. The system of units used, the SI system, has the mass specified in kg and acceleration in m/s^2. Consequently the unit of force is defined in terms of these two units as kg m/s^2. This unit is given the name of the newton (N).

Forces, since they have both magnitude and direction, are vector quantities. They thus have to be added or subtracted by vector means. The following example illustrates this.

Example

A flag pole is held in a vertical position by two wire stays attached to the same point on the pole. If the two stays are at angles of 40° and 30° to the pole and the resultant force is to be 2 kN along the vertical axis of the pole, what are the forces acting in each stay?

Figure 1.5(a) shows the pictorial situation and figure 1.5(b) the parallelogram of forces for the concurrent forces acting at the point where the stays are attached to the flag pole. Using the sine rule with the parallelogram

$$\frac{F_1}{\sin 40°} = \frac{2}{\sin (180° - 30° - 40°)}$$

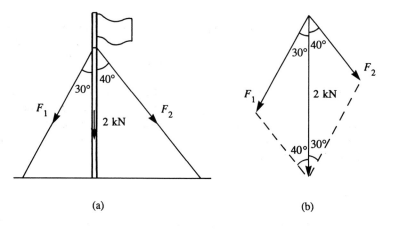

(a)   (b)

Fig. 1.5 Example 1.

Hence $F_1 = 1.4\,\text{kN}$.
Similarly

$$\frac{F_2}{\sin 30°} = \frac{2}{\sin (180° - 30° - 40°)}$$

Hence $F_2 = 1.1\,\text{kN}$.

Alternatively we could have obtained the result by resolving each of the stay forces into directions along the pole and at right angles to it. Along the pole

$$2 = F_1 \cos 30° + F_2 \cos 40°$$

and at right angles to it

$$F_1 \sin 30° = F_2 \sin 40°$$

or $F_1 = 1.29\, F_2$. Hence, substituting this in the first equation gives

$$2 = 1.29\, F_2 \cos 30° + F_2 \cos 40°$$

and hence $F_2 = 1.1\,\text{kN}$ and $F_1 = 1.4\,\text{kN}$.

1.2.1 Weight

The *weight* of a body at some location can be defined as the force that is necessary to stop it accelerating. A weightless body is thus one that is freely falling. However, it is customary to take the weight as being equal to the gravitational force. In the absence of any other force the gravitational force will cause the body to accelerate, the acceleration being known as the *acceleration due to gravity g*. At the surface of the earth, the acceleration due to gravity is about $9.8\,\text{m/s}^2$. We can thus write, for a mass m,

$$\text{weight} = W = mg \tag{7}$$

Example

A truck weighing $10\,000\,\text{N}$ stands on an incline that has a gradient of 1 in 50 (i.e. it descends by 1 m vertically for every 50 m along the hill). What is the size of the gravitational force component down the hill?

The weight of the truck is $10\,000\,\text{N}$ vertically. This can be resolved into two components, one at right angles to the hill and the other parallel to it. The parallel component for a hill which makes an angle of θ with the horizontal is $W \sin \theta$. Thus, since $\sin \theta = 1/50$, the parallel component is $10\,000/50 = 200\,\text{N}$.

1.3 Non-concurrent forces

Consider the situation where two or more forces act on a single body and the forces are not concurrent. A simple situation is a see-saw where forces, the weights of people, are applied at each end of a pivoted beam (figure 1.6). A simple experiment on such a system shows that the see-saw will balance when

$$F_1 d_1 = F_2 d_2$$

If force F_2 is removed, i.e. one person gets off the see-saw, then the see-saw begins to rotate about its pivot. The cause of the rotation is the product $F_1 d_1$. This quantity is called the *turning moment*. The turning moment of a force about some axis is thus defined as being the product of the force and its perpendicular distance from the axis.

For balance, the anticlockwise moment of one force must be balanced by the clockwise moment of the other force. This is known as the *principle of moments*.

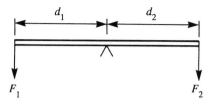

Fig. 1.6 See-saw system.

Example

For the bracket shown in figure 1.7, what is the moment of the 100 N force about the axis through point A?

AB is the perpendicular distance between the line of action of the force and the axis through A. Thus the moment is the product of the force and AB. The distance AB can be computed from the figure. However, an alternative way, and often simpler way, of considering the problem is to resolve the 100 N force into its components $100 \sin 60°$ and $100 \cos 60°$ and determine the moments for each component. These are $0.200 \times 100 \sin 60° = 17.3\,\text{N m}$ and

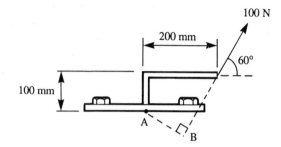

Fig. 1.7 Example.

$0.100 \times 100 \cos 60° = 5.0\,\text{N}\,\text{m}$. The $17.3\,\text{N}\,\text{m}$ moment gives an anticlockwise moment while the $5.0\,\text{N}\,\text{m}$ moment gives a clockwise moment. The total moment is thus $12.3\,\text{N}\,\text{m}$ anticlockwise.

1.3.1 Couple

A pair of equal and opposite forces that are not acting through the same point are known as a *couple*. Thus, for example, when a driver is using both hands to turn the steering wheel of a car, two non-concurrent forces are applied and hence a couple. Figure 1.8 illustrates this. The turning moment of the couple in the figure about the axis through the wheel centre is $FR + FR = 2FR$. The moment of the couple is thus the magnitude of one of the forces multiplied by their distance apart, the distance being measured perpendicular to one of the forces.

The moment of the couple does not contain any reference to the distances of the forces from an axis. Thus a couple has the same moment about any axis, indeed there is no need to specify an axis.

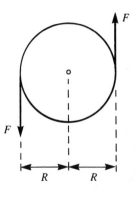

Fig. 1.8 Couple.

1.3.2 Centre of gravity

The weight of a body is an example of a distributed force in that any body can be considered to be made up of a number of particles, each having weight. However we can replace all the various weight forces of an object by a single weight force with a magnitude equal to the sum of the magnitudes of all the constituent weight forces and acting at a particular point known as the *centre of gravity*.

Consider a body made up of a large number of small elements, each having weight. The weights are non-current forces. The total moment of all the elements about some axis is

$$\text{total moment} = \delta w_1 x_1 + \delta w_2 x_2 + \delta w_3 x_3 + \dots$$

where δw_1 is the weight of element 1 which is a perpendicular distance x_1 from the axis, δw_2 the weight of element 2 and x_2 its perpendicular distance from the axis, etc. If a single weight W is to be used to replace all these distributed forces, then for it to have the same effect we must have

$$W\bar{x} = \text{sum of all the } \delta wx \text{ terms, i.e. } \Sigma \delta wx$$

where $\bar{x}$ is the distance of W from the axis. The centre of gravity is thus a distance from the axis of

$$\bar{x} = \frac{\Sigma \delta wx}{W} \qquad [8]$$

For symmetrical homogeneous objects the centre of gravity is located at the geometrical centre. For composite objects it can be obtained by considering the object to be made up of the constituent parts, each having its weight acting through its centre of gravity. For objects containing holes or cut-outs, the object can be considered to be a composite object with the hole being treated as a negative mass.

When considering a section of a constant cross-section item such as a beam the term *centroid* is often used instead of centre of gravity. This is because, if the beam is of constant cross-section, we are concerned with locating the point in the cross-section which specifies the axis throughout the length of the beam along which the centre of gravity will lie. This point is termed the centroid and is the geometrical centre of the cross-section. If the weight per unit length of a constant cross-section beam is w then equation [8] becomes

$$\bar{x} = \frac{w\Sigma \delta Ax}{wA} = \frac{\Sigma \delta Ax}{A} \qquad [9]$$

Equation [9] gives the position of the centroid within the cross-section area. The product of an element of area δA and its distance x from some axis is known as the *first moment of area*.

Example

Determine the position of the centre of gravity relative to point X for the homogeneous section shown in figure 1.9.

Consider the section as being a composite homogeneous object made up of three parts A, B and C. Part C has an area $150 \times 30 \text{ mm}^2$ with its centre of gravity 15 mm from X. Part B has an area of $100 \times 70 \text{ mm}^2$ with its centre of gravity 80 mm from X. Part A has an area of $150 \times 30 \text{ mm}^2$ with its centre of gravity 145 mm from X. The total section has an area of $2 \times 150 \times 30 + 100 \times 70 \text{ mm}^2$. Thus if the weight per unit section area is w then equation [8] gives

$$\bar{x} = \frac{(150 \times 30 \times 15 + 100 \times 70 \times 80 + 150 \times 30 \times 145)w}{(2 \times 150 \times 30 + 100 \times 70)w}$$

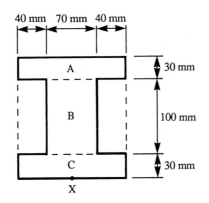

Fig. 1.9 Example.

Thus $\bar{x} = 80$ mm.

Alternatively we could have considered the object as being a rectangular section 150 mm by 160 mm, centre of gravity 80 mm from X, with two rectangular pieces 100 mm by 40 mm, centres of gravity 80 mm from X, missing. Then

$$\bar{x} = \frac{(150 \times 160 \times 80 - 100 \times 40 \times 80 - 100 \times 40 \times 80)w}{(150 \times 160 - 100 \times 40 - 100 \times 40)w}$$

Thus, as before, $\bar{x} = 80$ mm.

Alternatively we could have realised that the object is homogeneous and symmetrical and thus the centre of gravity would be at its geometrical centre which can be seen from inspection of the section to be 80 mm from X. The problem is really a determination of the position of the centroid of the section.

Example

Determine the position of the centroid for a triangular-shaped area.

Consider a small area segment of the triangle shown in figure 1.10. It has a width δy and length x, hence its area is $x\delta y$. By similar triangles

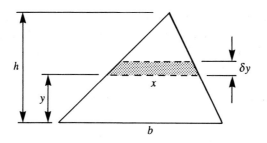

Fig. 1.10 Example.

$$\frac{x}{b} = \frac{h-y}{h}$$

Thus the area of the small segment can be written as

$$\delta A = \frac{b(h-y)\delta y}{h}$$

The sum of all the area moments is the sum of all the $y\delta A$ terms, i.e.

$$\text{sum of area moments} = \int_0^h \frac{yb(h-y)}{h}\,dy = \frac{bh^2}{6}$$

The total area of the triangle is $bh/2$ and so the position of the centroid is given by equation [9] as

$$\bar{x} = \frac{bh^2/6}{bh/2} = \frac{h}{3}$$

1.4 Free-body diagrams

The term free-body diagram is used for a diagram showing all the external forces acting on a body. The important word is 'all', in that all the active forces and the reactive forces should be included so that the forces acting on the body can be considered in isolation from the surroundings. In the case of a composite body or a structure, free-body diagrams might be drawn for the body as a whole in isolation from its surroundings and for each component part considered in isolation from the rest of the body and the surroundings.

For example, consider the structure shown in figure 1.11(a). The free-body diagram for joint C will be all the forces acting on joint C and thus will be as shown in figure 1.11(b). Such a diagram for joint C enables us to determine the resultant force on joint C or whether joint C is in equilibrium (see section 1.5.3 for an example of this).

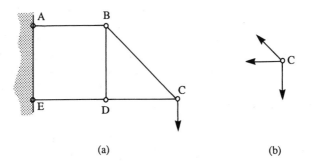

(a)   (b)

Fig. 1.11 (a) Structure, (b) free-body diagram for joint C.

1.5   Forces in equilibrium

Consider the conditions necessary for the equilibrium of a particle. A particle is an entity which is just a point, having no physical dimensions. Thus all the forces acting on a particle will be concurrent. According to Newton's first law, for equilibrium there must be no resultant force. This means that the vector sum of all the forces acting on the particle must be zero.

In many problems involving coplanar forces, it is convenient to resolve all the forces acting on a particle into two mutually perpendicular directions. Then for equilibrium the algebraic sum of the forces in each of the directions must be zero.

1.5.1   Triangle and polygon of forces

When two forces act at a point, for them to be in equilibrium the forces must be equal in size, opposite in direction and act in the same straight line.

When three forces act at a point, for them to be in equilibrium they must be all in the same plane and if the forces are represented in magnitude and direction by arrow-headed lines, then these lines when taken in the order of the forces must form a triangle. This is known as the *triangle of forces*.

When more than three forces act at a point, they will be in equilibrium if they all lie in the same plane and if the forces are represented in magnitude and direction by arrow-headed lines, then these lines when taken in the order of the forces must form a polygon, i.e. a closed shape. This is known as the *polygon of forces*.

1.5.2   Equilibrium of structures

Consider the conditions necessary for the equilibrium of a structure, i.e. a body having physical dimensions with forces being able to be applied at different points on the body. Figure 1.12(a) shows a simple body acted on by three forces of sizes F_1, F_2 and F_3 acting at different points on it in the directions shown in the figure. The force system can be simplified by inserting a pair of forces of F_3 and $-F_3$ at B (figure 1.12(b)). The force F_3 at C and $-F_3$ constitute a couple with a moment of $M_1 = F_3 d_1$, where d_1 is the perpendicular distance between the lines of action of the forces. We can thus replace these forces by the moment M_1 (figure 1.12(c)). We can perform a similar process with the force F_1, inserting additional forces of F_1 and $-F_1$ at B and then replacing the F_1 force at A and $-F_1$ force at B by a couple with a moment $M_2 = F_1 d_2$, where d_2 is the perpendicular distance between the lines of action of the forces.

Thus the force system in figure 1.12(a) can be replaced by the equivalent force system shown in figure 1.12(e). We now have three forces acting at the point B and two moments. For equilibrium the resultant force at B must be zero. This means that the algebraic sums of the vertical components and the

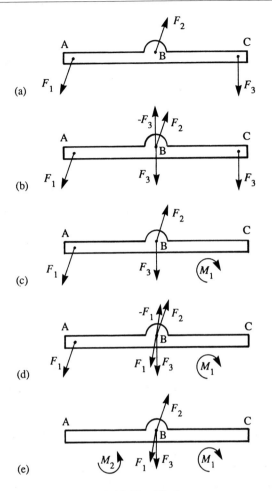

Fig. 1.12 Equilibrium.

horizontal components of the forces must both be zero. Also there must be no resultant turning moment. Both these statements when translated back to the original force system in figure 1.12(a) give the conditions for equilibrium as:

1 the algebraic sum of the vertical components of all the forces must be zero;
2 the algebraic sum of the horizontal components must be zero;
3 the moments about any axis must be zero.

1.5.3 Support reactions

Reaction forces occur at supports or points of contact between bodies (Newton's third law). Thus, for example, with a see-saw resting across a

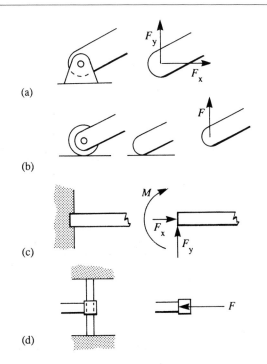

Fig. 1.13 Types of support, (a) pin connection, (b) roller or smooth contacting surface, (c) built-in or fixed support, (d) collar on smooth rod.

pivot, as in figure 1.6, the see-saw exerts a force on the pivot and the pivot exerts an opposing reaction on the see-saw. The types of reaction that occur at a support depend on the type of support concerned. A support develops a reactive force on the supported member if the support prevents a translational movement of the member, and it develops a reactive couple if it prevents rotation of the member.

Figure 1.13 shows some common forms of supports and the reactions on free-body diagrams of the members. At a pin connection, translation motion is prevented in any direction and so the reactive force can be considered to have two components at right angles to each other. The pin connection does however permit free rotation about the pin and so there is no reactive couple. A roller support, or the member resting on a smooth surface with no frictional forces involved, permits translational movement only in the direction at right angles to the surface and so there is just a reactive force at right angles to the surface. With a built-in or fixed support, both translational movement in any direction and rotational movement are not permitted. Thus there are two components of the reactive force and a reactive couple.

Example

Figure 1.14(a) shows a jib crane with a beam AB of length 5 m and mass

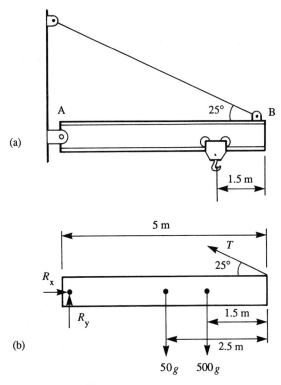

Fig. 1.14 Example.

10 kg/m, lifting a mass of 500 kg. Determine the tension in the supporting cable and the force on the pin at A.

Figure 1.14(b) shows the free-body diagram for the beam. The weight of the beam is considered to act at its centre point. Because there is a pin-joint at A there will be two reactive force components but no couple. For equilibrium, the horizontal components of all the forces will be zero, i.e.

$$R_x = T\cos 25°$$ [10]

The vertical components will also be zero, i.e.

$$R_y + T\sin 25° = 50g + 500g$$ [11]

The moments about some axis will be zero. Thus, taking moments about A, and assuming the pin is effectively on the end surface of the beam and the supporting cable is effectively connected to the centre line of the beam,

$$50g \times 2.5 + 500g \times 3.5 = T \times 5\sin 25°$$

Hence $T = 8.7$ kN. Substituting this value in equation [10] gives $R_x = 7.9$ kN, and in equation [11] gives $R_y = 1.7$ kN. These two reactive force components can be combined to give a resultant reactive force of $\sqrt{(1.7^2 + 7.9^2)} = 8.1$ kN at an angle of $\tan^{-1}(1.7/7.9) = 12.1°$ to the beam.

Example

A homogeneous smooth ball of weight 500 N rests between two smooth surfaces at 30°, as shown in Fig. 1.15(a). What are the reactive forces at the points where the sphere touches each of the surfaces?

Figure 1.15 shows the system and the free-body diagram for the sphere. We have three forces acting at a point. Thus, for equilibrium the vertical components will be zero, i.e.

$$R_A \cos 60° = 500$$

Hence $R_A = 1000\,\text{N}$. The horizontal components will also be zero, thus

$$R_B = R_A \sin 60°$$

Hence $R_B = 866\,\text{N}$.

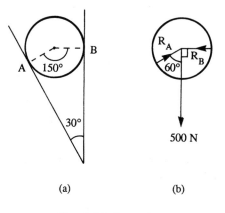

(a)   (b)

Fig. 1.15 Example.

1.5.4 *Pin-jointed structures*

The members of pin-jointed structures can only carry axial forces, no moments being transmitted at the joints. Thus it is possible to determine the equilibrium of the various parts of such structures by considering each joint individually as a free body. Such a method is called the *method of joints*. At any such joint the algebraic sum of the forces in the x-direction is zero and the sum in the y-direction is also zero. We could, alternatively, apply the triangle or polygon of forces at each joint.

The method of joints is useful when the forces in all the members of a structure are required. An alternative method, called the *method of sections*, is preferred when the forces are required in only one or a few members. The structure is considered to be cut into two pieces by a section through the member under consideration. Both parts of the structure can then be treated as structures in equilibrium and free-body diagrams drawn. The forces in the sectioned members can then be found by resolution of the forces or taking moments about a suitable point.

A member is said to be in tension if the forces applied to it have stretched it, and in compression if they have compressed it. If the forces were removed, then the member in tension would shorten, the member in compression would lengthen. Thus the internal forces in a tensile member are pulling on the pins at its ends while those in a compressive member are pushing on the pins. The convention is adopted of labelling the forces in a tensile member as being positive and those in a compressive member as negative. A member in tension is called a *tie*, a member in compression a *strut*.

Example

Determine, using the method of joints, the forces in each of the members of the plane pin-jointed bridge truss shown in figure 1.16 and the reactions at the supports, one being a pin joint and the other a roller.

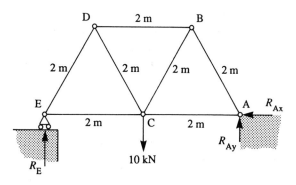

Fig. 1.16 Example.

Considering the free-body situation for the truss as an entity, the situation is essentially just a beam supported at each end and supporting a central load (figure 1.17(a)). Taking moments about end A of the beam gives

$$4R_E = 10 \times 2$$

Hence $R_E = 5\,\text{kN}$. Since we must have $R_E + R_{Ay} = 10$, then $R_{Ay} = 5\,\text{kN}$. The reaction R_{Ax} is zero, since there are no other horizontal forces acting on the beam.

Figure 1.17(b) shows the free-body diagram for the joint at E. For the vertical components of the forces

$$F_{ED} \sin 60° = 5$$

Hence $F_{ED} = 5.8\,\text{kN}$. For the horizontal components

$$F_{ED} \cos 60° = F_{EC}$$

Hence $F_{EC} = 2.9\,\text{kN}$.

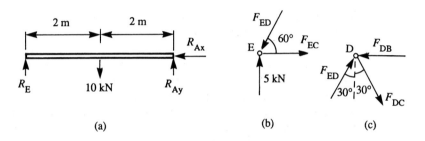

(a)   (b)   (c)

Fig. 1.17 Example.

Figure 1.17(c) shows the free-body diagram for the joint at D. For the vertical components of the forces

$$F_{ED} \cos 30° = F_{DC} \cos 30°$$

Thus $F_{DC} = 5.8\,kN$. For the horizontal components

$$F_{DB} = F_{DC} \cos 60° + F_{ED} \cos 60°$$

Thus $F_{DB} = 5.8\,kN$.

Each of the other joints can be considered in a similar way or we can recognise that the structure is symmetrical and so $F_{ED} = F_{BD} = F_{BA} = 5.8\,kN$, $F_{DC} = F_{BC} = 5.8\,kN$ and $F_{EC} = F_{AC} = 2.9\,kN$. Members ED, BD and BA are struts, DC, BC, EC and AC are ties.

Example

Determine, using the method of sections, the force acting in the member BD of figure 1.16.

Figure 1.18 shows the free-body diagrams for the structure in figure 1.16 sectioned through the member BD. In general, no more than three members with unknown forces should be cut by the section since only three equations can be generated, a moment equation and forces in the vertical and horizontal

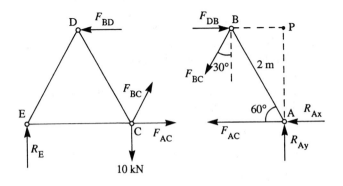

Fig. 1.18 Example.

directions, and so only three unknowns can be solved. The free-body diagrams for the two parts will then include just the external forces acting on each part and the internal forces of the cut members.

Consider the right-hand section. Taking moments about point P.

$$F_{AC} \times 2\sin 60° + R_{Ax} \times 2\sin 60° = F_{BC}\cos 30° \times 2\cos 60° \quad [12]$$

For the vertical components of the forces to be in equilibrium, then

$$R_{Ay} = F_{BC}\cos 30° \quad [13]$$

for the horizontal components to be in equilibrium

$$F_{DB} = F_{BC}\sin 30° + F_{AC} + R_{Ax} \quad [14]$$

The reaction forces can be found as in the previous example, thus $R_{Ax}=0$ and $R_{Ay}=5\,\text{kN}$. Thus equation [13] gives $F_{BC}=5.77\,\text{kN}$. Equation [12] gives

$$1.732\,F_{AC} = 0.866\,F_{BC}$$

Thus $F_{AC}=2.89\,\text{kN}$. Equation [14] then gives $F_{DB}=5.77\,\text{kN}$.

Problems

(1)   Determine the resultant force acting on the bolt shown in figure 1.19 when it is subject to the forces shown acting along the wires.

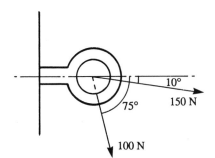

Fig. 1.19 Problem 1.

(2)   Forces of 40 N, 80 N and 100 N act at a point A in the directions AB, AC and AD respectively, where ABCD is a square. What is the resultant force acting at A?

(3)   A load of mass 12 kg is lifted by two ropes connected to the same point on the load and making angles of 20° and 30° on opposite sides of the vertical. What are the tensions in the ropes?

(4)   A bolt is acted on by a force of 120 N acting along a wire at 30° to the surface to which the bolt is attached. What are the components

of this force acting on the bolt and which are at right angles to the surface and along the surface?

(5)   A sailing boat has a force, due to the wind, of 50 N acting at right angles to its sail. If the plane of the sail is at an angle of 40° to the keel of the boat, what will be the force component parallel to the boat keel?

(6)   A loaded truck is checked at a weigh bridge and the front axle load is found to be 30 000 N and the rear axle load 40 000 N. How far is the centre of gravity of the truck from the front axle if the distance between the axles is 8.0 m?

(7)   A compound section consists of a steel I-section beam, sectional area 4750 mm², with a plate, sectional area 6000 mm², on the top flange as shown in figure 1.20. What is the position of the centroid above X?

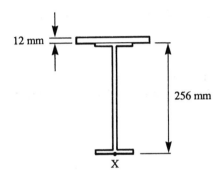

12 mm

256 mm

X

Fig. 1.20 Problem 7.

(8)   Determine the position of the centroid above the diameter for a semicircular arc of radius r.

(9)   Determine the position of the centroid of the wire shown in figure 1.21. The wire is all in the same horizontal plane.

(10)   Determine the position of the centroid of the channel section shown in figure 1.22.

(11)   A vertical telegraph pole has three horizontal cables attached to points located one above the other. The top cable is at a height of 8 m above the base of the pole and exerts a pull of 100 N. The next cable is at a height of 7 m and exerts a pull of 120 N, with the lower cable being at height 6 m and exerting a pull of 90 N. What is the resultant turning moment about the base of the pole?

(12)   Determine the tensions in the ropes supporting the objects shown in figure 1.23.

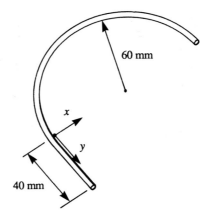

Fig. 1.21 Problem 9.

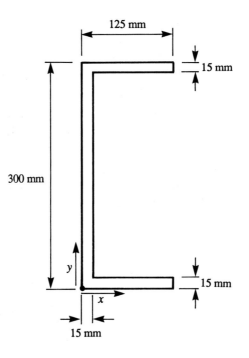

Fig. 1.22 Problem 10.

(13) A light beam AB of length 7 m is hinged with a pin joint at A and rests on a roller at B. A vertical force of 100 N acts on the beam a distance of 2 m from A and a force of 600 N inclined at 45° to the beam acts at 5 m from A. What are the reactions at A and B?

(14) A uniform light beam is pivoted halfway along its length. At one end it supports a load of 5 kN while the other end is tethered to a

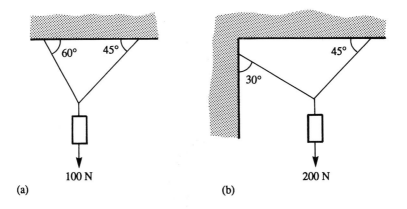

Fig. 1.23 Problem 12.

fixed point by a rope inclined at 45° to the horizontal. If the beam is in equilibrium, what is the tension in the rope?

(15)   A uniform light rod is hinged to a vertical surface and held at an angle of 30° to that surface by a light horizontal cable attached to its mid point. What will be the tension in the cable if the free end of the rod supports a vertical load of 2 kN?

(16)   A beam of length 6 m and weight 30 kN/m rests on rollers at each end. What will be the reactions at its supports if it supports a distributed load of 60 kN/m over a 3 m length from one end?

(17)   A foot operated lever is as shown in figure 1.24 and consists of a bell crank pivoted at O. What will be the tension in the horizontal cable when a vertical force of 100 N is applied to the pedal?

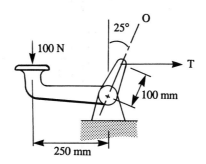

Fig. 1.24 Problem 17.

(18)   A smooth ball of weight 240 N rests between two smooth inclined planes which are at 90° to each other, as shown in figure 1.25. What are the reactive forces at the points of contact between the ball and the surfaces?

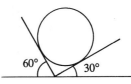

Fig. 1.25 Problem 18.

(19)   Determine the reaction at the fixed support for the loaded light frame in figure 1.26.

(20)   A uniform bar AB of weight 400 N and length 8 m is hinged at one end A to a wall and is maintained in a horizontal position by a chain attached to B and a point on the wall 5 m vertically above A. The bar carries a load of 200 N at a point along its length 6 m from A. What is the tension in the chain and the reaction force at A?

(21)   Determine the forces in each member of the pin-jointed bridge truss shown in figure 1.27 and the reactions at each end support.

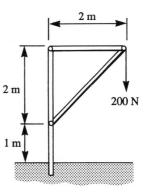

Fig. 1.26 Problem 19.

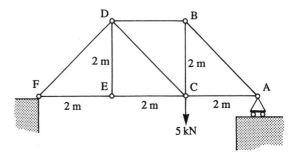

Fig. 1.27 Problem 21.

(22)   Determine the forces in each member of the frame shown in figure 1.28 and the reactions at the supports.

(23)   Determine the force in member AB of the framework shown in figure 1.29 when supporting a load of mass 200 kg.

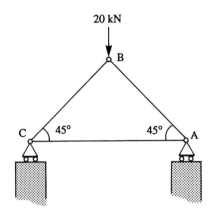

Fig. 1.28 Problem 22.

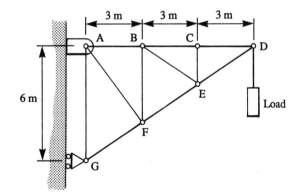

Fig. 1.29 Problem 23.

(24)   Determine the force in member BC in figure 1.30 when a load of 20 kN is being supported. All the members are the same length.

(25)   Determine the force in member DI of the roof truss shown in figure 1.31 when (a) a vertical load of 6 kN is applied at joint I, (b) wind forces on the roof result in forces at right angles to side AD of 2.5 kN at A, 5.0 kN at B, 5.0 kN at C and 2.5 kN at D.

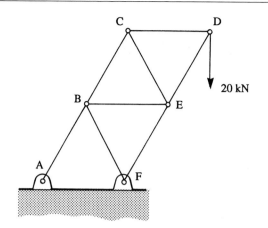

Fig. 1.30 Problem 24.

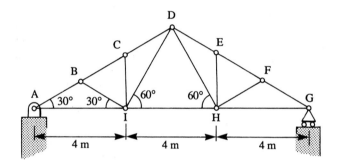

Fig. 1.31 Problem 25.

(26)   Determine, using the method of sections for the bridge truss shown in figure 1.27 (Problem 21), the force in member CD when there is (a) a vertical load of 5 kN at C (as in the figure), (b) a vertical load of 1.2 kN at C and a horizontal pull of 0.4 kN on E.

Chapter 2
Simple stress and strain

2.1 Stress and strain

When a body is subject to external forces, internal forces are set up in it which oppose the external forces. We can consider a body to be like a spring, which if stretched by external forces sets up internal forces which resist the pull extending it. The simplest case of loading is that of a straight metal bar of constant cross-section which has external forces applied at its ends along the bar axis and which either stretch the bar or compress it, as in figure 2.1. The bar is then said to be in *tension* or *compression*. If we consider a plane in the bar which is at right angles to its axis, the internal forces are at right angles to the plane. The term *direct stress* is used for the value of this force per unit area of the plane.

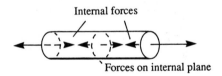

Fig. 2.1 Bar in tension.

$$\text{Direct stress } \sigma = \frac{\text{force}}{\text{area}} \qquad [1]$$

With the force in N and the area in m^2 then the stress is in N/m^2. The name pascal (Pa) is used for $1\,N/m^2$.

It is important to realise that the above equation gives the average direct stress acting on the sectional area since it is assumed that the same force acts over the entire area. With concentrated loads or where the material being loaded has changes in shape, e.g. a hole through a bar, then the stress will not generally be the same over each segment of cross-sectional area. The maximum stress will depend on the bar's geometry and the type of discontinuity. Tables and graphs are available which enable a *stress concentration factor K* to be determined, the maximum stress then being K times the average stress.

In general, when a body is subject to loading, the internal forces may be resolved into components which are at right angles to and parallel to a plane within the body. The right angle components give *tensile* or *compressive* forces and *direct stresses*, the parallel component being termed a *shear* force and giving *shear stresses*. Figure 2.2 shows a bar in shear. The shear stress is defined as being the shear force per unit area of plane.

$$\text{Shear stress } \tau = \frac{\text{force}}{\text{area}} \qquad [2]$$

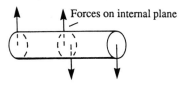

Fig. 2.2 Bar in shear.

When a bar is subject to tensile loading then it increases in length. The increase in length per unit length of bar is called the *direct tensile strain*. Similarly, when a bar is subject to compressive loading then it decreases in length, the decrease in length per unit length of bar being called the *direct compressive strain*.

$$\text{Direct strain } \varepsilon = \frac{\text{change in length}}{\text{original length}} \qquad [3]$$

Since strain is a ratio of two lengths, it has no unit.

With shear loading, rectangular shaped sections are distorted into parallelograms, as illustrated in figure 2.3. The angle ϕ by which the right angle at the corner of the section changes is called the *shear strain*. Since this angle is generally very small ϕ is approximately x/L, i.e. the change in length in the direction of the force per unit length between the sheared faces.

$$\text{Shear strain} = \phi \qquad [4]$$

The unit used for shear strain is the radian.

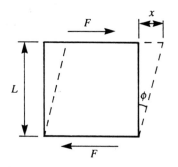

Fig. 2.3 Shear strain.

Example

A steel bolt carries a tensile load of 20 kN. What is the average tensile stress across the cross-section of the bolt at the root of the threaded section where the diameter is 16 mm?

Using equation [1] and assuming that the same force acts over the entire cross-sectional area,

$$\text{stress} = \frac{\text{force}}{\text{area}} = \frac{20 \times 10^3}{\frac{1}{4}\pi \times 0.016^2} = 99.5 \, \text{MPa}$$

Example

For the yoke and rod connection shown in figure 2.4, what is the direct stress on each rod and the average shear stress acting on the pin?

For the rods, equation [1] gives

$$\text{direct stress} = \frac{\text{force}}{\text{area}} = \frac{5 \times 10^3}{\frac{1}{4}\pi \times 0.040^2} = 3.98 \, \text{MPa}$$

The pin is in double shear, i.e. there are two planes subject to the shear. Thus the area to be sheared is double the cross-sectional area of the pin. Thus the shear stress is given by equation [2] as

$$\text{shear stress} = \frac{\text{force}}{\text{area}} = \frac{5 \times 10^3}{2 \times \frac{1}{4}\pi \times 0.020^2} = 7.96 \, \text{MPa}$$

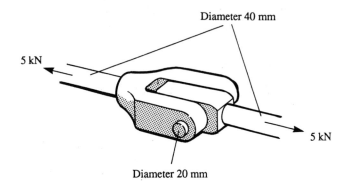

Diameter 40 mm

5 kN

5 kN

Diameter 20 mm

Fig. 2.4 Example.

Example

Two wooden beams with a square cross-section of side 50 mm, are cut at an angle of 20° and joined to form a longer beam, as illustrated in figure 2.5.

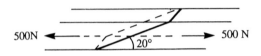

Fig. 2.5 Example.

What is the shear stress on the joint when the beam is subject to an axial tensile force of 500 N?

The force component parallel to the plane of the joint is $500 \cos 20°$ N and the area of the joint is $50 \times (50/\sin 20°)$ mm^2. Thus the shear stress is

$$\text{shear stress} = \frac{500 \cos 20°}{50 \times (50/\sin 20°) \times 10^{-6}} = 64.3 \, \text{kPa}$$

2.1.1   Tensile stress–strain graph

The tensile testing of a material involves stretching a test piece of the material in a testing machine. The machine indicates the tensile load being applied. The test piece is either a circular or rectangular cross-sectional bar which has a constant cross-section over what is termed the gauge length. This is the length for which the extension is measured. Measurements are made of the extension as the load is gradually increased. Hence the strain is determined at different stresses.

Figure 2.6 shows the type of stress–strain graph produced with mild steel. Initially the extension is proportional to the load, i.e. the strain is proportional to the stress. This relationship is called *Hooke's law* and holds up to a value of stress known as the *limit of proportionality*. Up to this point the ratio of stress/strain is a constant, this being called the *tensile modulus* (or *Young's modulus* or just *modulus of elasticity*).

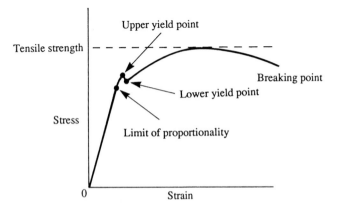

Fig. 2.6 Stress–strain graph for mild steel.

$$\text{Tensile modulus } E = \frac{\text{stress}}{\text{strain}} \qquad [5]$$

If the load is removed from a material and it returns to its original shape, it is said to be *elastic*. If it remains deformed then it is said to be *plastic*. Engineering materials are generally elastic up to a particular stress, called the *elastic limit*, and then beyond that they are partly elastic and partly plastic. The elastic limit generally coincides with the limit of proportionality or is very close to it.

When a material is stressed beyond the elastic limit, some plastic deformation occurs. With some metals, such as mild steel, a yield point occurs. At this point the material shows an appreciable increase in strain without any further increase in load. After such yielding, a further increase in strain can only be achieved by an increase in stress.

Many materials, such as aluminium alloys, do not have well-defined yield points (figure 2.7). For such materials a *proof stress* is quoted. The 0.2% proof stress is defined as that stress which results in a 0.2% offset, i.e. it is the stress given by a line drawn on the stress–strain graph which is parallel to the linear part of the graph but passes through the 0.2% strain value. Similar criteria exist for the 0.1% proof stress.

The maximum tensile stress that a material can withstand is called the *tensile strength*. The term *brittle* is used to describe materials which have small permanent elongation at the breaking point. A *ductile* material has significant permanent elongation before breaking. Cast iron is an example of a brittle material, mild steel an example of a ductile one.

When designing a structure a *factor of safety* has to be taken into account in order to ensure that working stresses keep within safe limits. For a brittle material the factor of safety is usually defined as the ratio of the tensile strength to the maximum working stress. For ductile materials it is more usually defined as the ratio of the yield or proof stress to the maximum working stress. For dead loads a factor of safety of 4 or more is often used.

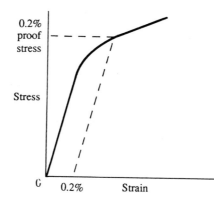

Fig. 2.7 Proof stress.

Example

What is the elongation that will be produced with a bar of aluminium alloy of length 2.0 m when stretched by a tensile force of 20 kN? The bar has a uniform cross-sectional area of 10 cm² and a tensile modulus of 70 GPa.

Assuming that the limit of proportionality is not exceeded, then the tensile modulus E is

$$E = \frac{\text{stress}}{\text{strain}} = \frac{F/A}{e/L}$$

Hence

$$\text{extension } e = \frac{FL}{AE} = \frac{20 \times 10^3 \times 2.0}{10 \times 10^{-4} \times 70 \times 10^9} = 0.57 \text{ mm}$$

Example

A flat steel bar of constant thickness 10 mm and length 1.2 m tapers from a width of 20 mm at one end to 80 mm at the other, as shown in figure 2.8. What is the change in length of the bar when it is subject to an axial tensile load of 10 kN? The material has a tensile modulus of 200 GPa.

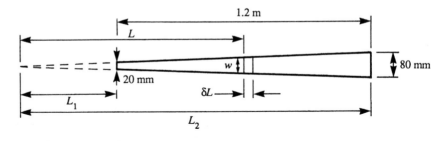

Fig. 2.8 Example.

Where there is a varying cross-sectional area it is necessary to consider the bar as being effectively made up of a number of small lengths of constant cross-sectional area elements and determine the overall extension by summing the extensions of each of the elements. Thus for this bar we can consider an element of the bar of length δL, as illustrated in figure 2.8. Then, if this element has a cross-sectional area wt, where w is the width and t the thickness, the extension δx of the element is given by

$$E = \frac{\text{stress}}{\text{strain}} = \frac{F/wt}{\delta x/\delta L}$$

$$\delta x = \frac{F}{Ewt} \delta L$$

The total extension x of the bar can be obtained by integrating over the length of the bar. Thus

$$x = \int_{L_1}^{L_2} \frac{F}{Ewt} \, dL$$

The bar has the shape shown in figure 2.8. By similar triangles $L_1/20 = L_2/80$, and since $L_2 - L_1 = 1.2$ m, then $L_1 = 0.4$ m and $L_2 = 1.6$ m. By similar triangles $w/L = 0.080/L_2 = 0.080/1.6 = 0.05$. Thus the above equation can be written as

$$x = \int_{0.4}^{1.6} \frac{10 \times 10^3}{200 \times 10^9 \times 10 \times 10^{-3} \times 0.05L} \, dL$$

$$= 1.0 \times 10^{-4} \, [\ln L]_{0.4}^{1.6}$$

$$= 1.4 \times 10^{-4} \, \text{m} = 0.14 \, \text{mm}$$

2.1.2 Shear stress–strain graph

A similar shape stress–strain graph to that occurring with tensile forces is produced with shear forces. Such a graph is obtained by subjecting test pieces in the form of thin tubes to torsional loading, i.e. twisting the tube (see chapter 6). Up to a limit of proportionality the shear strain is proportional to the shear stress and thus a *shear modulus*, or *modulus of rigidity*, G can be defined as

$$G = \frac{\text{shear stress}}{\text{shear strain}} \tag{6}$$

2.2 Thermal stresses

In general, when the temperature of a homogeneous body, such as a bar of metal, increases it expands and when the temperature drops it contracts. This change in length ΔL resulting from a temperature change ΔT is given by

$$\Delta L = L\alpha\Delta T \tag{7}$$

where α is the *linear coefficient of expansion* or *thermal expansivity*, unit $°C^{-1}$.

If the material is restrained from expanding or contracting then thermal stresses are produced. If a length L of material is prevented from expanding by ΔL then the stress produced is that due to a length $L + \Delta L$ being reduced to L, i.e. a compression. Since ΔL is small compared with L this is effectively the same as if length L had been reduced by ΔL. Thus the thermally-induced strain is

$$\text{strain} = \frac{\Delta L}{L} = \frac{L\alpha\Delta T}{L} = \alpha\Delta T \tag{8}$$

Hence the thermally-induced stress is

$$\text{stress} = E\alpha\Delta T \tag{9}$$

Example

A steel bar is constrained to just fit between two rigid supports at 20°C. What will be the direct stress induced in the bar if the temperature rises to 50°C? The linear coefficient of expansion of the material is $11 \times 10^{-6}\,°C^{-1}$ and the tensile modulus of elasticity 200 GPa.

Using equation [9],

$$\text{stress} = E\alpha\Delta T = 200 \times 10^9 \times 11 \times 10^{-6} \times (50 - 20)$$
$$= 66\,\text{MPa}$$

2.3 Composite bars

Consider a composite bar, part of the length being one material and part another or part being of one cross-sectional area and part another, with an axial force being applied, as in figure 2.9(a). For such a series arrangement, the force on each element will be the same, though the extensions of each element can differ. The total extension x will be the sum of the extensions of the constituent elements, i.e.

$$x = x_1 + x_2 \tag{10}$$

where $x_1 = FL_1/A_1E_1$ and $x_2 = FL_2/A_2E_2$, with F the axial force, L_1 and L_2 the lengths of the two materials, A_1 and A_2 their cross-sectional areas, and E_1 and E_2 the tensile modulus of the materials.

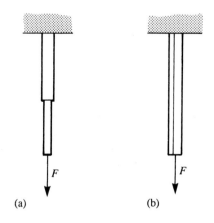

(a)   (b)

Fig. 2.9 Composite bars, (a) in series (b) in parallel.

Now consider a composite bar made up of two or more bars joined together at their ends, the bars being in parallel as in figure 2.9(b), and subject to an axial force. Because their ends are joined together they must all extend by the same amount. Thus

$$x_1 = x_2 \qquad [11]$$

where x_1 and x_2 are the extensions of two joined bars, with $x_1 = F_1L_1/A_1E_1$ and $x_2 = F_2L_2/A_2E_2$, with F_1 and F_2 being the axial forces on the two bars, L_1 and L_2 their lengths, A_1 and A_2 their cross-sectional areas, and E_1 and E_2 the tensile modulus of the materials. The force F applied to the arrangement will be the sum of the forces applied to each element, i.e.

$$F = F_1 + F_2 \qquad [12]$$

Example

A steel rod of length 0.8 m has a diameter of 25 mm for 0.2 m of its length and 50 mm for the rest. What will be the extension of the rod when it is subject to an axial force of 10 kN? The tensile modulus is 200 GPa.

Consider each element alone. For the 25 mm diameter element, the extension is given by

$$E = \frac{F/A_1}{x_1/L_1}$$

Hence

$$x_1 = \frac{FL_1}{EA_1} = \frac{10 \times 10^3 \times 0.2}{200 \times 10^9 \times \frac{1}{4}\pi \times 0.025^2} = 2.0 \times 10^{-5}\,\text{m}$$

For the 50 mm element,

$$x_2 = \frac{FL_2}{EA_2} = \frac{10 \times 10^3 \times 0.8}{200 \times 10^9 \times \frac{1}{4}\pi \times 0.050^2} = 2.0372 \times 10^{-5}\,\text{m}$$

Thus the total extension is $4.07 \times 10^{-5}\,\text{m}$.

Example

A reinforced concrete column of cross-sectional area $30\,000\,\text{mm}^2$ consists of three axial steel bars, each of cross-sectional area $100\,\text{mm}^2$, embedded in the column. What are the stresses carried by the concrete and the steel bars when the load on the column is 200 kN? The steel has a modulus of 200 GPa and the concrete 20 GPa.

We can consider the column to be essentially two columns of the same length, one being a concrete column of cross-sectional area $29\,700\,\text{mm}^2$ and one being steel of cross-sectional area $300\,\text{mm}^2$. The total axial force of 200 kN is the sum of the forces acting on each material, i.e.

$$200 \times 10^3 = F_c + F_s$$
$$= \sigma_c \times 29\,700 \times 10^{-6} + \sigma_s \times 300 \times 10^{-6} \qquad [13]$$

where σ_c is the stress on the concrete and σ_s that on the steel. Both materials will have the same extension and, since they are the same length, the same strain. Thus

$$\frac{\sigma_c}{E_c} = \frac{\sigma_s}{E_s}$$

$$\frac{\sigma_c}{20 \times 10^9} = \frac{\sigma_s}{200 \times 10^9}$$

Hence $\sigma_s = 10\sigma_c$. Substituting this into equation [13] gives

$$200 \times 10^3 = (29\,700 + 300 \times 10)10^{-6} \times \sigma_c$$

Hence $\sigma_c = 6.12\,\text{MPa}$ and $\sigma_s = 61.2\,\text{MPa}$.

2.3.1 Composite bars and thermal stresses

Consider the effect of a temperature change on a composite bar made up of two materials, with different coefficients of expansion, joined in series and held between two rigid supports at their ends (figure 2.10). Each of the materials, if free, would expand when the temperature increased. A temperature rise of ΔT would give an expansion of $L_1\alpha_1\Delta T$ (equation [7]) for bar 1 and $L_2\alpha_2\Delta T$ for bar 2, where L_1 and L_2 are the lengths of the two bars, α_1 and α_2 their coefficients of linear expansion. The expansion of the free composite bar would be the sum of the expansions of the two constituent bars.

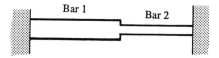

Fig. 2.10 Composite bar.

$$\text{Change in length} = L_1\alpha_1\Delta T + L_2\alpha_2\Delta T$$

The composite bar is, however, constrained and cannot expand. Thus forces are developed. We can think of the force as being that which would be necessary to squash the expanded free bar back to its contrained length. A force F would mean for bar 1 a change in length of $\sigma_1 L_1/E_1$ and for bar 2 $\sigma_2 L_2/E_2$, where σ_1 and σ_2 are the stresses developed in each bar and E_1 and E_2 the modulus of elasticity for the two materials. Thus the total change in length is

$$\text{change in length} = \frac{\sigma_1 L_1}{E_1} + \frac{\sigma_2 L_2}{E_2} = L_1\alpha_1\Delta T + L_2\alpha_2\Delta T \qquad [14]$$

In addition to the above equation, since the force is the same in both bars, we have

$$\sigma_1 A_1 = \sigma_2 A_2 \qquad [15]$$

where A_1 and A_2 are the cross-sectional areas of the two bars.

Consider the effect of a temperature change on a composite bar made up of two materials joined at both ends and having different coefficients of expansion. Though each material, if free, would expand by different amounts, they are constrained to have the same change in length. As a consequence, internal stresses are set up in the arrangement.

Consider the arrangement shown in figure 2.11(a) of three bars rigidly connected together, the outer bars being the same material. This symmetrical arrangement means that a change in temperature can result in an axial tensile force without bending. If the materials were free to expand we would have the situation shown in figure 2.11(b) when the temperature increases by ΔT. For bar 1, a temperature change of ΔT would result in an expansion of $L\alpha_1\Delta T$ (equation [7]), where α_1 is the coefficient of linear expansion for the material of bar 1. For bar 2, the expansion would be $L\alpha_2\Delta T$, where α_2 is the coefficient of expansion for the material of bar 2. But the two bars are constrained to move together, thus we have the situation shown in figure 2.11(c). Bar 1 is made to extend by an addition amount x_1 and bar 2 is compressed by an amount x_2. But $x_1 + x_2$ must equal the difference in expansion between the two bars. Thus

$$x_1 + x_2 = L\alpha_2\Delta T - L\alpha_1\Delta T = L\Delta T(\alpha_2 - \alpha_1)$$

The strain in bar 1 is x_1/L and the strain in bar 2 is x_2/L. The strain in bar 1 is σ_1/E_1 while that in bar 2 is σ_2/E_2, with σ_1 and σ_2 being the stresses in bars 1 and 2, E_1 and E_2 the tensile modulus for each material. Thus the above equation can be written as

$$\frac{\sigma_1}{E_1} + \frac{\sigma_2}{E_2} = \Delta T(\alpha_2 - \alpha_1) \qquad [16]$$

Since the composite bar is in equilibrium, then the force exerted by bar 1 on bar 2 must be equal and opposite to the force exerted by bar 2 on bar 1, i.e.

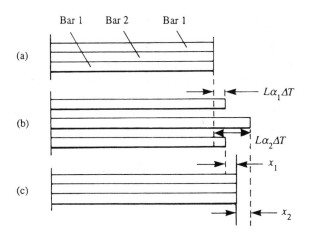

Fig. 2.11 Composite bar (a) before increase in temperature, (b) when both bars freely expand, (c) when the composite expands.

the tensile force on bar 1 has the same size as the compressive force on bar 2. Thus

$$\sigma_1 A_1 = \sigma_2 A_2 \tag{17}$$

where A_1 is the cross-sectional area of bar 1 and A_2 that of bar 2.

Example

A composite rod of length 300 mm consists of a 200 mm length of steel rod, 20 mm in diameter, joined end-to-end (as in figure 2.10) to a 100 mm length of aluminium rod, 15 mm in diameter. The composite bar is rigidly fixed between two supports. What will be the stresses developed in the two materials as a result of a temperature increase of 20 °C? The modulus of elasticity of the steel is 200 GPa and its coefficient of linear expansion is $11 \times 10^{-6} °C^{-1}$, the modulus for the aluminium being 70 GPa and its coefficient of linear expansion $23 \times 10^{-6} °C^{-1}$.

Using equation [14],

$$\frac{\sigma_s L_s}{E_s} + \frac{\sigma_a L_a}{E_a} = L_s \alpha_s \Delta T + L_a \alpha_a \Delta T$$

$$\frac{\sigma_s \times 0.200}{200 \times 10^9} + \frac{\sigma_a \times 0.100}{70 \times 10^9} = \begin{aligned} &0.200 \times 11 \times 10^{-6} \times 20 \\ &+ 0.100 \times 23 \times 10^{-6} \times 20 \end{aligned} \tag{18}$$

Hence

$$14\sigma_s + 20\sigma_a = 1.26 \times 10^9$$

But equation [15] gives

$$\sigma_s A_s = \sigma_a A_a$$
$$\sigma_s \times \tfrac{1}{4}\pi \times 20^2 = \sigma_a \times \tfrac{1}{4}\pi \times 15^2$$
$$\sigma_s = 0.563\sigma_a$$

Thus, substituting this into equation [18],

$$(14 \times 0.563 + 20)\sigma_a = 1.26 \times 10^9$$

Thus $\sigma_a = 45.2$ MPa and $\sigma_s = 25.4$ MPa.

Example

A steel tube of external diameter 30 mm and 3 mm thick has a brass rod of diameter 20 mm inside it and rigidly joined to it at each end. At 15 °C when the materials were joined there is no stress in the materials. What will be the stress in the rod and the tube when the temperature is raised to 100 °C? The steel has an elastic modulus of 200 GPa and a coefficient of linear expansion of $11 \times 10^{-6} °C^{-1}$, the brass a modulus of 120 GPa and a coefficient of linear expansion of $18 \times 10^{-6} °C^{-1}$.

Using equation [16],

$$\frac{\sigma_s}{E_s} + \frac{\sigma_b}{E_b} = \Delta T(\alpha_b - \alpha_s)$$

$$\frac{\sigma_s}{200 \times 10^9} + \frac{\sigma_b}{120 \times 10^9} = 85(18 \times 10^{-6} - 11 \times 10^{-6})$$

$$120\sigma_s + 200\sigma_b = 1.43 \times 10^{10} \qquad [19]$$

Using equation [17],

$$\sigma_s \times \tfrac{1}{4}\pi(30^2 - 24^2) = \sigma_b \times \tfrac{1}{4}\pi \times 20^2$$

$$\sigma_s = 1.23\sigma_b$$

Hence substituting into equation [19] gives

$$(120 \times 1.23 + 200)\sigma_b = 1.43 \times 10^{10}$$

Hence $\sigma_b = 41.02$ MPa and $\sigma_s = 50.66$ MPa. The brass is in compression and the steel in tension.

2.4   Poisson's ratio

When a bar is stretched by tensile forces, the axial elongation is accompanied by a lateral contraction. Figure 2.12 shows the change in shape of the bar. If Hooke's law is being obeyed, the lateral strain at a point is proportional to the axial strain at that point. The ratio of the lateral strain to the axial strain is known as *Poisson's ratio*.

$$\text{Poisson's ratio} = -\frac{\text{lateral strain}}{\text{axial strain}} \qquad [20]$$

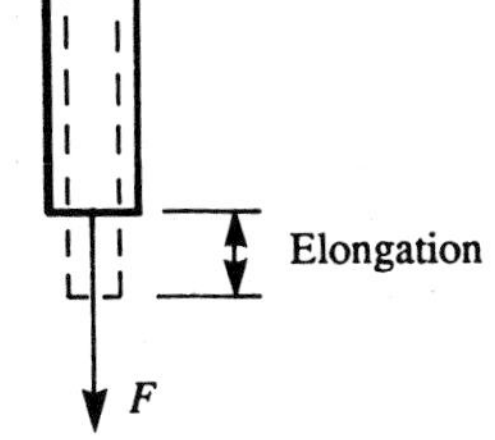

Fig. 2.12 Axial and lateral strain.

The minus sign is because when one strain is tensile, the other is compressive. Since it is a ratio it has no units. For most engineering metals, Poisson's ratio is about 0.3.

Example

A bar of length 2.0 m and constant cross-section, diameter 40 mm, is subject to axial tensile forces of 50 kN. What will be the elongation of the bar and the decrease in diameter? The bar is aluminium with a tensile modulus of 70 GPa and Poisson's ratio 0.3.

Assuming that the limit of proportionality is not exceeded, then the axial elongation x of the bar is given by

$$E = \frac{\text{stress}}{\text{strain}} = \frac{F/A}{x/L} = \frac{50 \times 10^3 \times 2.0}{x \times \frac{1}{4}\pi \times 0.040^2}$$

Thus the extension is 1.14 mm.

The axial strain is 1.14/2000. Hence, using equation [20], the lateral strain is given by

$$\text{Poisson's ratio} = 0.3 = -\frac{\text{lateral strain}}{1.14/2000}$$

Thus the lateral strain is -1.71×10^{-4}. The reduction in diameter will be $1.71 \times 10^{-4} \times 40 = 6.84 \times 10^{-3}$ mm.

2.5   Bulk modulus

If a pressure p acts on a body of volume V, such as a liquid in a pressure vessel, then the body becomes squashed and reduced in volume by δV. The term *volumetric strain* is used for the fractional change in volume, i.e. $\delta V/V$, and the term *bulk modulus K* for the stress, i.e. the pressure, divided by the strain. Thus

$$K = -\frac{p}{\delta V/V} \qquad\qquad [21]$$

The minus sign is because an increase in pressure produces a reduction in volume.

Example

Water has a bulk modulus of 2.1 GPa, what will be the reduction in volume of a volume of 1.0 m³ when subject to a pressure of 10 MPa?

Using equation [21],

$$\delta V = -\frac{pV}{K} = -\frac{10 \times 10^6 \times 1.0}{2.1 \times 10^9} = -4.8 \times 10^{-3} \, \text{m}^3$$

Problems

(1)   What is the height to which a vertical concrete wall can be built if the compressive stress must not exceed 4 MPa? Concrete has a density of 2000 kg/m^3.

(2)   A lamp of mass 50 kg is supported by two rods from the ceiling. One rod, of diameter 10 mm, is at an angle of 30° to the vertical and the other, of diameter 12 mm, is at 45° to the vertical. What are the direct stresses in the rods?

(3)   A uniform horizontal beam of length 4 m is supported at its ends by two vertical rods. If the rods have cross-sectional areas of 10 mm^2 and 15 mm^2, what will be the position of a load of 8 kN along the beam for the stresses in the two rods to be identical?

(4)   A rectangular cross-section bar has a width 100 mm and a thickness 10 mm. An axial tensile load of 100 kN is transmitted to the bar through pins passing through 25 mm diameter holes located near each end of the bar. What is the average stress at a section through the bar at (a) its midpoint and (b) through a hole?

(5)   A yoke and pin joint (as in figure 2.4) has a pin of diameter 20 mm. What will be the maximum force that can be supplied to the rods if the maximum permissible shear stress for the pin is 150 MPa?

(6)   A lap joint has two smooth plates connected by four identical bolts. What will be the maximum permissible tensile load that the joint can withstand if the maximum shear stress a bolt can withstand is 250 MPa and they each have a diameter of 25 mm?

(7)   A plate of mild steel with a shear strength of 200 MPa is to be cropped by a guillotine, i.e. sheared. What force will need to be applied by the guillotine if the plate has a width of 1.0 m and a thickness of 5 mm?

(8)   Two wires are to be used to suspend a lamp from the ceiling. Each wire has a length of 300 mm and makes an angle of 30° with the vertical. When the lamp is hung from the wires it is found that the point of suspension sinks by 3 mm under the weight of the lamp. What is the direct strain in each wire?

(9)   A tie bar has a length of 2.0 m and a diameter of 4 cm. What is the extension of the bar when subject to a tensile axial loading of 100 kN? The material has a tensile modulus of 205 GPa.

(10)   A circular metal rod of diameter 10 mm and length 1.0 m is loaded in tension with an axial tensile load of 5 kN. If the tensile modulus is 70 GPa, what will be the extension of the rod?

(11)   A surveyor's steel tape has a rectangular cross-section of 5 mm by 1 mm and a total length of 30 m. What will be the elongation of the tape when it is fully out and held taut by a force of 40 N? The

tensile modulus of the tape material may be taken as 200 GPa.

(12) Show that the total elongation produced by a vertical bar of constant cross-section subject to its own weight is the same as that which would have been produced by a load of half its weight applied to the lower end.

(13) A shaft of length L tapers from a radius of r_2 to r_1. Show that the extension of the shaft when subject to an axial tensile load F is $FL/\pi E r_1 r_2$.

(14) A brass rod of diameter 10 mm is heated to 40 °C and its ends then clamped to rigid supports. What forces will be exerted on the supports when the rod cools to 20 °C? The linear coefficient of expansion of brass is 16×10^{-6} °C^{-1} and the tensile modulus 90 GPa.

(15) A steam pipe of length 2.0 m connects two turbines. The pipe has an outer diameter of 100 mm and a wall thickness of 5 mm. If the connection was made at 20 °C with no allowance for thermal expansion, what will be the force exerted on the turbines when the pipe reaches a temperature of 100 °C and the points of attachment to the turbines are assumed to be rigid? Tensile modulus of steel = 200 GPa, coefficient of linear expansion 11×10^{-6} °C^{-1}.

(16) A bar has a total length of 1.0 m and a constant diameter of 50 mm. Three quarters of the length is steel and a quarter is aluminium. What will be the total extension of the bar when subject to an axial tensile load of 100 kN? The tensile modulus of steel is 200 GPa, and for the aluminium 70 GPa.

(17) A plastic bar has a total length of 1.2 m and is made of two parts. The first part has a length of 0.6 m and a diameter of 100 mm and the second part a length of 0.6 m and a diameter of 60 mm. What will be the shortening of the bar when it is subject to an axial force of 100 kN? The compressive modulus of the plastic is 4.0 GPa.

(18) A reinforced concrete pillar has a cross-sectional area of 250 000 mm^2. It contains 10 axial steel reinforcing bars, each having a diameter of 25 mm. What is the maximum axial compressive load that can be applied to the pillar if the maximum stress allowed for the steel is 80 MPa or for the concrete 8 MPa? The modulus of elasticity for the steel is 200 GPa and for the concrete 20 GPa. Ignore the weight of the pillar.

(19) A concrete pillar has a cross-sectional area of 160 000 mm^2 and is reinforced with axial steel rods of diameter 20 mm. How many reinforcement rods are required if, when the compressive load on the column is 4 MN, the stress on the concrete is not to exceed 20 MPa? The modulus of elasticity for the steel is 200 GPa and for the concrete 20 GPa. Ignore the weight of the pillar.

(20)   A steel pipe with an outer diameter of 80 mm and a wall thickness of 10 mm is filled with concrete and the resulting column is subject to an axial compressive load of 60 kN. What are the stresses in the concrete and the steel? The modulus of elasticity for the steel is 200 GPa and for the concrete 20 GPa.

(21)   An aluminium cylinder of diameter 75 mm is located inside a steel cylinder of internal diameter 75 mm and, wall thickness 15 mm. The assembly is compressed between two rigid cover plates by an axial force of 200 kN. What are the stresses in the steel and the aluminium? Modulus of elasticity for steel is 200 GPa, for aluminium 25 GPa.

(22)   A steel bolt of diameter 15 mm is enclosed in a copper sleeve of external diameter 32 mm and wall thickness 6 mm. The bolt is used between rigid end plates at both ends of the sleeve. A nut is screwed on to the bolt until it is just a snug fit and there are no stresses in the assembly. The nut is then rotated through 45°. What will be the stresses in the bolt and the sleeve if the screw thread has a pitch of 2 mm? The modulus of elasticity for the steel is 200 GPa and for the copper 100 GPa.

(23)   A composite bar of length 200 mm has 120 mm of its length as a steel bar of diameter 30 mm and the remaining 80 mm as a copper bar of the same diameter. At 0 °C the bar is fixed between two rigid end supports and is stress free. What will be the stresses in the two materials when the temperature rises to 100 °C? The modulus of elasticity for the copper is 100 GPa and the coefficient of linear expansion is $20 \times 10^{-6} °C^{-1}$, with the modulus for the steel being 200 GPa and the coefficient of linear expansion $11 \times 10^{-6} °C^{-1}$.

(24)   A steel bolt with a cross-sectional area of 400 mm^2 is enclosed in a sleeve of an aluminium alloy tube with a cross-sectional area of 600 mm^2 and the assembly held rigidly together at both ends by the nut. At 20 °C the nut is just snug and there are no stresses in the two materials, what will be the stresses when the temperature is raised to 85 °C? The modulus for the steel is 200 GPa and the coefficient of linear expansion $11 \times 10^{-6} °C^{-1}$, the modulus for the aluminium being 70 GPa and the coefficient of linear expansion $22 \times 10^{-6} °C^{-1}$.

(25)   An aluminium tube of external diameter 16 mm and wall thickness 2 mm has a brass rod of diameter 12 mm inside it and rigidly joined to it at both ends. At 15 °C there are no stresses in the materials, what will the stresses become when the temperature is raised to 40 °C? The aluminium has a modulus of 70 GPa and a coefficient of linear expansion of $23 \times 10^{-6} °C^{-1}$, the brass a modulus of 90 GPa and a coefficient of linear expansion of $15 \times 10^{-6} °C^{-1}$.

(26)   An aluminium rod of diameter 22 mm is enclosed in a steel tube of external diameter 28 mm and internal diameter 25 mm. The rod has

a screw thread at each end and rigid bolts are tightened up so that they just hold the rod in place inside the tube at 20 °C. What will be the stresses produced in the materials when the temperature rises by 20 °C? The aluminium has a modulus of 70 GPa and a coefficient of linear expansion of $22 \times 10^{-6}\,°C^{-1}$, the steel a modulus of 200 GPa and a coefficient of linear expansion of $11 \times 10^{-6}\,°C^{-1}$.

(27) A steel rod of diameter 30 mm is enclosed in a copper tube of external diameter 60 mm and internal diameter 35 mm. The rod and tube are attached at their ends to rigid cover plates. At 0 °C there are no stresses in the materials. What will be the stresses in the materials if an axial compressive load of 50 kN is applied at a temperature of 250 °C? The steel has a modulus of 200 GPa and a coefficient of linear expansion of $11 \times 10^{-6}\,°C^{-1}$, the copper a modulus of 100 GPa and a coefficient of linear expansion of $19 \times 10^{-6}\,°C^{-1}$.

(28) A steel bar of length 1.5 m has a rectangular cross-section of 100 mm by 50 mm. What will be the elongation of the bar and the changes in its cross-section when it is subject to an axial force of 120 kN? The steel has a tensile modulus of 200 GPa and Poisson's ratio of 0.3.

(29) A steel bar of length 1.0 m has a square cross-section of side 50 mm. What will be the elongation of the bar and the changes in its cross-section when it is subject to an axial force of 250 kN? The steel has a tensile modulus of 200 GPa and Poisson's ratio of 0.3.

(30) A pressure vessel of internal volume 2.0 m³ is filled with water at atmospheric pressure. What is the reduction in volume of the water, bulk modulus 2.1 GPa, when the gauge pressure in the vessel is 2 MPa?

Chapter 3
Beams

3.1 Types of beams

A *beam* can be defined as a structural member that is designed to resist loads applied transversely at various points along it, subjecting it to bending. When beams bend they become curved with the result that one surface of the beam becomes shorter and one longer, i.e. one surface is put in compression and one in tension. Beams can be in the following forms.

1 *A cantilever*. This is a beam which is built in or fixed at one end with the other end free to move. When a load is applied to the cantilever a reaction and a resisting moment occur at the fixed end.
2 A *simply supported* beam. This is a beam supported at its ends on rollers or smooth surfaces or with one of these combined with a pin at the other end. There are reactions at the supports but no resisting moments.
3 An *overhanging* beam. This is a simply supported beam with the supports located in some distance from the ends. There are reactions at the supports but no resisting moments.
4 A *built-in* beam. This is a beam with both ends rigidly fixed. There are reactions and resisting moments at both ends.

The term *statically determinate* is used for beams, such as the cantilever, simple beam and overhanging beam, for which the reactions at the supports may be determined by the use of just the equations of static equilibrium (see chapter 1). The reactions of such beams are independent of the deformations of the beam. A *statically indeterminate beam* is one where the equations of static equilibrium have to be supplemented by equations based on the deformation of the beam. This occurs with a beam extending over three or more supports, a beam built-in at both ends, and a beam built-in at one end and propped at the other end on a roller. This chapter is concerned with statically determinate beams.

The loads that can be carried by beams can be classified as concentrated or distributed. A *concentrated load* is one which can be considered to be applied at a point, or at least at a very small area. A *distributed load* is one that is applied over a length of the beam. This could be a uniformly distributed load as with the weight of a uniform cross-sectional beam, this being generally specified as the weight per unit length of beam. The weight of a section, with a uniformly distributed loading of a beam, can be considered to act at the centre of gravity of the section. Another kind of load that may occur is a couple.

3.2 Bending moment and shearing force

When a beam is loaded by forces or couples, internal stresses are set up in the beam. In general, both direct and shearing stresses will be set up. To determine these stresses, the internal forces and couples that act on cross-sections of the beam have to be determined.

Consider a cantilever acted on by a vertical force F at its free end (figure 3.1(a)). Now imagine the beam cut through at a distance x from the free end and consider the free-body diagrams of the two parts, as in figure 3.1(b). At the fixed end there can be a reaction force and a couple. For each part of the beam to be in equilibrium, then the free section must have a shearing force V and a bending couple of moment M and the fixed section must likewise have a shearing force V and a bending couple of moment M.

The moment of the bending couple is called the *bending moment*. Thus the bending moment (symbol M) at a transverse section of a beam is the algebraic sum of the moments about the section of all the forces acting on one (either) side of the section concerned. A bending moment is considered to be positive when the moment on the left of the section concerned is clockwise and that on the right is anticlockwise, hence producing sagging of the beam (figure 3.2(a)). A bending moment is considered to be negative when the moment on the left of the section concerned is anticlockwise and that on the right clockwise, producing hogging of the beam (figure 3.2(b)).

The shearing force at any section of a beam represents the tendency for that portion of the beam on one side of the section to slide or shear laterally relative to the other portion. The term *shearing force* at a transverse section of a beam is defined as being the algebraic sum of the external forces acting on one side of the section concerned. Thus for the beam in figure 3.1, the

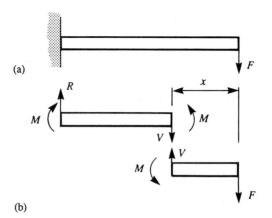

(a)

(b)

Fig. 3.1 Cantilever.

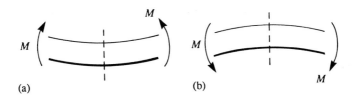

Fig. 3.2 (a) Sagging, positive bending moment; (b) hogging, negative
bending moment.

shearing force at the section is F downwards on the free segment and F
upwards (the reaction force at the fixed end) on the fixed segment. It is thus
the same size as, but in the opposite direction to, the internal shearing force
V. The convention is adopted of the shearing force being positive when the
resultant of the forces to the left is upwards, or to the right downwards, i.e.
the shearing forces are clockwise about the section. It is negative when the
resultant of the forces to the left is downwards, or to the right upwards, i.e.
anticlockwise about the section.

Example

Determine the shearing force and bending moment at the middle for the
simply supported beam shown in figure 3.3, the beam being subject to just a
concentrated force F a distance x from one end.

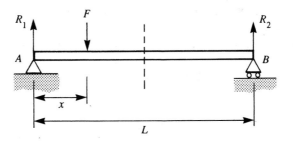

Fig. 3.3 Example.

For equilibrium we have

$$F = R_1 + R_2$$

and taking moments about A

$$Fx = R_2L$$

Hence $R_2 = Fx/L$ and $R_1 = F - Fx/L$.

Now consider the beam to be sectioned at its middle. For the right half,
the resultant external force is R_2 and is thus Fx/L. This is upwards and thus
the shearing force is $- Fx/L$. We could have considered the left half of the

beam, then the resultant force is $F - R_1$, i.e. $F - F + Fx/L$. This is downwards and thus the shearing force is $-Fx/L$, as for the other half of the beam.

For the right half of the beam, the moment about the section is $R_2 \times \frac{1}{2}L$ in an anticlockwise direction. For the left half of the beam, the moment about the section is $F(\frac{1}{2}L - x) - R_1 \times \frac{1}{2}L$, i.e. $-\frac{1}{2}Fx$ in a clockwise direction. The bending moment is thus $+\frac{1}{2}Fx$.

Example

A cantilever of negligible mass and length 2.0 m carries a load of 4 kN at its free end (figure 3.4). What is the bending moment and shearing force at (a) 0.5 m, (b) 1.0 m and (c) 2.0 m from the free end?

(a) Consider the beam cut at A. The bending moment will be $4 \times 0.5 = 2.0$ kN m and will be negative since there is hogging. The external force to the right of the section is 4 kN and so the shearing force is 4 kN, being positive since it is clockwise about the section.

(b) Consider the beam cut at B. The bending moment will be $4 \times 1.0 = 4.0$ kN m and will be negative since there is hogging. The external force to the right of the section is 4 kN and so the shearing force is 4 kN, being positive since it is clockwise about the section.

(c) Consider the beam cut at C. The bending moment will be $4 \times 2.0 = 8.0$ kN m and will be negative since there is hogging. The external force to the right of the section is 4 kN and so the shearing force is 4 kN, being positive since it is clockwise about the section.

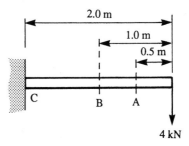

Fig. 3.4 Example.

Example

Consider the previous example of the cantilever if the beam has a uniform weight per metre length of 2 kN.

(a) Consider the beam cut at A. The external forces acting on the right-hand part of the beam are 4 kN and the weight of that part. The weight is 1.0 kN and is considered to act at a distance of 0.25 m from A. The bending moment is thus $4 \times 1.0 + 1.0 \times 0.25 = 4.25$ kN m, being negative since there is hogging. The external force to the right of the section is $4 + 1 = 5\,kN$ and so the shearing force is $5\,kN$, being positive since it is clockwise about the section.

(b) Consider the beam cut at B. The external forces acting on the right-hand part of the beam are $4\,kN$ and the weight of that part. The weight is $2.0\,kN$ and is considered to act at a distance of $0.5\,m$ from B. The bending moment is thus $4 \times 1.0 + 2.0 \times 0.5 = 5.0\,kN\,m$, being negative since there is hogging. The external force to the right of the section is $4 + 2 = 6\,kN$ and so the shearing force is $6\,kN$, being positive since it is clockwise about the section.

(c) Consider the beam cut at C. The external forces acting on the right-hand part of the beam are $4\,kN$ and the weight of that part. The weight is $4.0\,kN$ and is considered to act at a distance of $1.0\,m$ from C. The bending moment is thus $4 \times 1.0 + 4.0 \times 1.0 = 8.0\,kN\,m$, being negative since there is hogging. The external force to the right of the section is $4 + 4 = 8\,kN$ and so the shearing force is $8\,kN$, being positive since it is clockwise about the section.

3.2.1 Relationship between load, shear force and bending moment

Consider a very small segment of a beam obtained by cutting between two cross-sections a distance δx apart, as in figure 3.5. There is a distributed load of w per unit length. The segment has a weight of $w\delta x$ which can be considered to act at the centre of the segment, i.e. $\frac{1}{2}\delta x$ from the section AA. At one side, section AA, of the segment there is a bending moment of M and a shear force V, and at the other side these have both changed by some small amount to become $(M + \delta M)$ and $(V + \delta V)$.

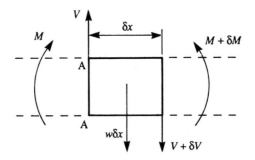

Fig. 3.5 Element of a beam.

The beam is in equilibrium, hence the small segment, length δx is in equilibrium. Thus, since there is no net force in the vertical direction

$$w\delta x + (V + \delta V) - V = 0$$

Hence

$$\frac{\delta V}{\delta x} = -w \qquad [1]$$

It should be noted that a sign convention has been assumed in the above equation, the distributed load being positive if it acts downwards, and conversely negative if it acts upwards.

Taking moments about the plane AA,

$$M + (V + \delta V)\delta x + (w\delta x)\tfrac{1}{2}\delta x - (M + \delta M) = 0$$

Ignoring the squares of small quantities

$$M + V\delta x - M - \delta M = 0$$

Hence

$$V = \frac{\delta M}{\delta x} \qquad [2]$$

In the limit as the length δx becomes infinitesimally small, equations [1] and [2] become

$$\frac{dV}{dx} = -w \qquad [3]$$

$$V = \frac{dM}{dx} \qquad [4]$$

Thus equation [3] indicates that the rate of change of shear force with distance along the beam is equal to $-w$. If there is no distributed load on part of a beam then the rate of change of shear force with distance is zero, i.e. the shear force is constant. Equation [4] indicates that, when there is a distributed load, the rate of change of bending moment with distance along a beam is equal to the shear force.

From these two equations we can obtain three other equations. We can substitute for V in equation [3] using equation [4]. Then

$$\frac{d^2 M}{dx^2} = -w \qquad [5]$$

We can integrate equation [3] along the axis of the beam. Then

$$\int_A^B dV = -\int_A^B w \, dx$$

$$V_B - V_A = -\int_A^B w \, dx \qquad [6]$$

The shear force difference between A and B is thus the area under the load–distance diagram between A and B.

We can integrate equation [4] along the axis of the beam. Then

$$\int_A^B dM = \int_A^B V \, dx$$

$$M_B - M_A = \int_A^B V\,dx \qquad [7]$$

The bending moment difference between A and B is thus the area under the shear force–distance diagram between A and B.

The above equations relate to the situation where there is just a distributed load. Now consider when there is a concentrated load acting on the beam segment, as in figure 3.6. The sign convention we will adopt is that of taking downwards loads as being positive. For equilibrium of the forces we have

$$V = F + V + \delta V$$
$$\delta V = -F \qquad [8]$$

Thus the change in the shearing force at the point where the concentrated load occurs is $-F$. For equilibrium of the moments acting on the element we have, when the considered about plane AA,

$$M + \tfrac{1}{2}F\delta x + (V + \delta V)\delta x = M + \delta M$$
$$\delta M = \tfrac{1}{2}F\delta x + (V + \delta V)\delta x = (\tfrac{1}{2}F + V + \delta V)\delta x$$

Since δx is infinitesimally small, then the bending moment does not change at the point of application of a concentrated load.

Prior to the application of the force, at plane AA, equation [2] gives $\delta M/\delta x = V$ and after the application, at plane BB, the equation gives $\delta M/\delta x = V + \delta V$. But $\delta V = -F$ (equation [8]). Thus the rate of change of M with x, i.e. dM/dx, changes abruptly at the point of application of a concentrated load by $-F$.

$$\text{Change in } dM/dx = -F \qquad [9]$$

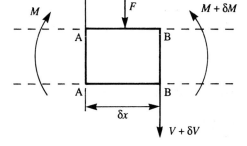

Fig. 3.6 Element of a beam.

3.2.2 Bending moment and shear force diagrams

Diagrams showing how the bending moment and the shearing force vary with distance along a beam are called *bending moment diagrams* and *shear force diagrams*. The following are some general points, arising from the equations derived in section 3.2.1, with regard to such diagrams.

1 The bending moment is a maximum when the shear force is zero (a consequence of equation [4]).
2 The shear force is a maximum when the slope of the bending moment— distance graph is a maximum and zero when the slope is zero (equation [4]).
3 For concentrated loads the shear force changes abruptly at the point of application of the load by an amount equal to the size of the load, by $-F$ where F acts downwards (equation [8]).
4 Where there are no distributed loads, the shear force is constant between concentrated load points (equation [3]).

Example

Determine the shear force and bending moment diagrams for a simply supported beam of length L with a concentrated load a distance a from one end, as in figure 3.7(a).

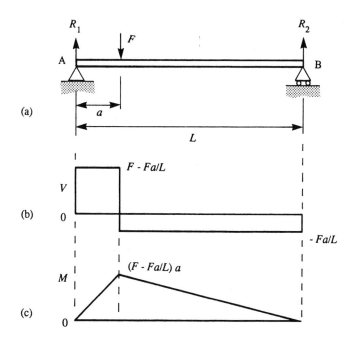

Fig. 3.7 Simply supported beam with concentrated load.

Since the beam is in equilibrium we have $R_1 + R_2 = F$ and, taking moments about end A, $R_2 L = Fa$. Thus the reactions are $R_2 = Fa/L$ and $R_1 = F - Fa/L$. Now consider the beam to be cut through a distance x from A, when

x is less than a. For the left-hand beam section we have $V = R_1 = F - Fa/L$ and this will be the same for all values of x up to a. The bending moment of this left-hand beam section is R_1x and is thus proportional to x, reaching the value of $R_1a = (F - Fa/L)a$ when $x = a$. Now consider the beam to be cut through a distance x from A when x is greater than a. For the left-hand beam section we have $V = R_1 - F = -Fa/L$ and this will be the same for all values of x between a and L. The bending moment of the left-hand beam section is $R_1x - F(x - a) = Fa - Fax/L$. The bending moment thus decreases from $(F - Fa/L)a$ at $x = a$ to 0 at $x = L$. Figures 3.7(b) and (c) show the shear force and bending moment diagrams.

Note that where the concentrated load is applied that the shear force abruptly changes by F. The slope dV/dx is zero from 0 to a and a to L, i.e. between concentrated load points, because there is no distributed load. The area under the shear force diagram up to a is $(F - FA/L)a$ and this equals the bending moment at a (see equation [7]). The area under the shear force diagram from 0 to $x = L$ is $(F - Fa/L)a - (L - a)Fa/L = 0$ and this equals the bending moment at L. The rate of change of M with x, i.e. dM/dx, prior to the concentrated load point is $(F - Fa/L)$ and changes to $-(F - FA/L)a/(L - a)$ after it. This is a change of slope of F (see equation [9]).

Example

Determine the bending moment and shear force diagrams for a simply supported beam when there is just a distributed load of w per unit length (figure 3.8(a)).

Since the beam is in equilibrium and symmetrical, then the reactions at each end are the same, with $R_1 = R_2 = \frac{1}{2}wL$. At a section a distance x from the left end, the shearing force is

$$V = R_1 - wx = \tfrac{1}{2}wL - wx \qquad [10]$$

This is a graph of a straight line which starts at $\frac{1}{2}wL$ at $x = 0$ and decreases to 0 at $x = \frac{1}{2}L$ and $-\frac{1}{2}wL$ at $x = L$, as in figure 3.8(b). The bending moment for the section is

$$M = R_1x - wx\tfrac{1}{2}x = \tfrac{1}{2}wLx - \tfrac{1}{2}wx^2 \qquad [11]$$

This is the equation of a parabolic curve, shown in figure 3.8(c). At each value of x, equation [4] gives

$$V = \frac{dM}{dx} = \tfrac{1}{2}wL - wx$$

which is in accord with the expression [10] derived earlier for the shearing force. The maximum value of the bending moment occurs when $dM/dx = 0$, i.e. when $V = 0$. This is thus when $\frac{1}{2}wL = wx$ and so $x = \frac{1}{2}L$. Putting this value in equation [11] gives the maximum bending moment as $\frac{1}{2}wL\frac{1}{2}L - \frac{1}{2}w(\frac{1}{2}L)^2 = wL^2/8$.

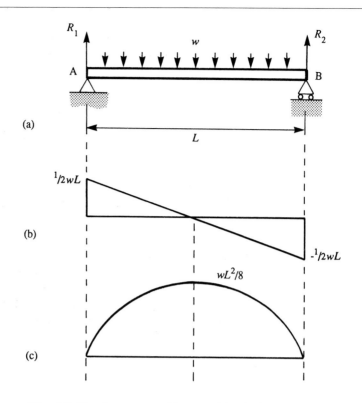

Fig. 3.8 Simply supported beam with distributed load.

Example

Determine the shear force and bending moment diagrams for the simply supported beam shown in figure 3.9 when it supports three concentrated loads.

This problem can be approached in two ways. One way is to consider the arrangement of forces as shown in the figure and take sections at different distances along the beam and so compute the shearing force and bending moment at each section. An alternative is to consider the beam with just one force acting on it and determine its shear force and bending moment diagram, then to repeat the exercise for each of the other forces. The shearing force and bending moment diagrams obtained for the three forces acting together is the sum of the diagrams obtained when each force acts alone. As an illustration, this method is used here.

For a single force F_1 acting on the simply supported beam we have the situation described in the earlier example and figure 3.7. The shear force and bending moment diagrams will thus be as shown in figure 3.10(a). With just the single force F_2 there is again just a single concentrated force acting on the simply supported beam and hence the shear force and bending

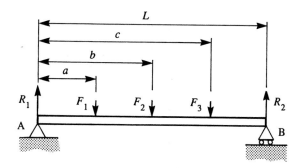

Fig. 3.9 Simply supported beam with three concentrated loads.

moment diagrams are as shown in figure 3.10(b). Likewise when there is just the single force F_3 acting on the beam the shear force and bending moment diagrams will be as shown in figure 3.10(c).

The shear force and bending moment diagrams for when the three forces are acting on the beam is the sum of the three diagrams for the forces acting alone and so is as shown in figure 3.10(d). Thus

$$V_1 = F_1 - F_1\frac{a}{L} + F_2 - F_2\frac{b}{L} + F_3 - F_3\frac{c}{L} = R_1$$

$$V_2 = -F_1\frac{a}{L} + F_2 - F_2\frac{b}{L} + F_3 - F_3\frac{c}{L}$$

V_2 is thus a drop of F_1 from V_1. Likewise V_3 is a drop of F_2 from V_2 and V_4 is a drop of F_3 from V_3 to give $V_4 = -R_2$. For the bending moment diagram, since the bending moment at distance a for a force F_1 at distance a is $F_1a - F_1a^2/L$ (see earlier example), for a force F_2 at distance b it is $F_2b - F_2ba/L$ and for F_3 at distance c the bending moment is $F_3c - F_3ca/L$,

$$M_1 = \left(F_1 - F_1\frac{a}{L}\right)a + \left(F_2 - F_2\frac{a}{L}\right)b + \left(F_3 - F_3\frac{a}{L}\right)c$$

$$= R_1a$$

In a similar way, we can also show that

$$M_2 = R_1b - F_1(b - a)$$
$$M_3 = R_2(L - c)$$

Example

Determine the equations relating the shear force and bending moment with distance between the points A and B, B and C, C and D for the cantilever beam loaded as shown in figure 3.11.

For the shear force:

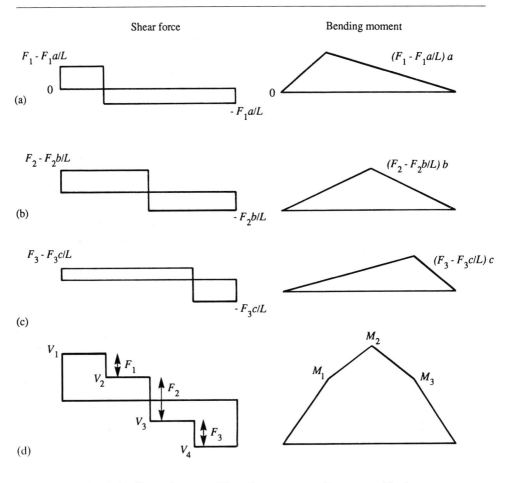

Shear force   Bending moment

Fig. 3.10 Shear force and bending moment diagrams with three concentrated loads.

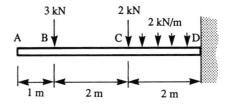

Fig. 3.11 Example.

AB	$0 < x < 1\,m$	$V = 0$
BC	$1 < x < 3\,m$	$V = -3\,kN$
CD	$3 < x < 5\,m$	$V = -3 - 2 - 2(x - 3)\,kN$

For the bending moment:

AB $0 < x < 1\,\mathrm{m}$ $M = 0$
BC $1 < x < 3\,\mathrm{m}$ $M = -3(x - 1)\,\mathrm{kN\,m}$
CD $3 < x < 5\,\mathrm{m}$ $M = -3(x - 1) - 3(x - 3)$
$$- 2(x - 1)\tfrac{1}{2}(x - 1)\,\mathrm{kN\,m}$$

3.2.3 Maximum shear force and bending moment

With cantilevers the maximum shear force and bending moment occur at the fixed support. For a cantilever of length L with just a concentrated load F at the free end the maximum shear force is F and is constant along the length of the beam, while the maximum bending moment is FL at the fixed end and decreases linearly to zero at the free end (see figure A.1(a) as the answer to problem 5). For a cantilever of length L with just a uniformly distributed load of w per unit length, the maximum shear force is wL at the fixed end and decreases linearly to zero at the free end, while the maximum bending moment is $0.5\,wL^{3}$ at the fixed end and decreases non-linearly to zero at the free end.

With a beam simply supported at its ends, the position of the maximum bending moment can be determined by finding the position at which the shear force is zero. With concentrated loads we can thus add together the downward acting loads from one reaction to the point where they equal or suddenly become greater than the reaction. This is the point of zero shear and the position of the maximum bending moment. With distributed loads we can carry out a similar process.

With a simply supported beam of length L and a central concentrated load of F, the maximum shear force is $\tfrac{1}{2}F$ at the supports and zero at the centre, while the bending moment is a maximum of $\tfrac{1}{4}FL$ at the centre and zero at the supports. With the concentrated load a distance a from one support and b from the other, with $b > a$, the maximum shear force is Fb/L and the maximum bending moment Fab/L (see figure 3.7). With a simply supported beam of length L and a uniformly distributed load along its length of w per unit length, the maximum shear force is $\tfrac{1}{2}wL$ at the supports and zero at the centre, while the bending moment is a maximum of $wL^{2}/8$ at the centre and zero at the supports (see figure 3.8).

Example

Determine the maximum bending moment and its position for the beam shown in figure 3.12.

Considering the equilibrium of the beam as a whole.

$$R_{\mathrm{A}} + R_{\mathrm{C}} = 5 + 10 + 9$$

and taking moments about A,

$$9 \times 1.5 + 10 \times 3 + 5 \times 5 = R_{\mathrm{C}} \times 4$$

Hence, $R_{\mathrm{C}} = 17.1\,\mathrm{kN}$ and $R_{\mathrm{A}} = 6.9\,\mathrm{kN}$

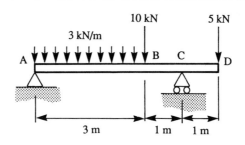

Fig. 3.12 Example.

For the shear force the equations relating it with distance x from end A are:

AB $0<x<3$ m $V = -6.9 + 3x$ kN
BC $3<x<4$ m $V = -6.9 + 9 + 10 = 12.1$ kN
CD $4<x<5$ m $V = -6.9 + 9 + 10 - 17.1 = -5$ kN

If we start at point A the shear force is -6.9 kN and by BC has become 12.1 kN. There is thus some position between A and B where the shear force becomes zero. This is when

$$-6.9 + 3x = 0$$

i.e. when $x = 2.3$ m. The bending moment at this position will have a maximum value. For section AB the bending moment is given by

AB $0<x<3$ m $M = 6.9x - 3x\tfrac{1}{2}x$ kN m

With $x = 2.3$ m then $M = 7.9$ kN m.

There is another point along the beam at which the shear force is zero, or rather abruptly changes from a positive value to a negative value, and that is C. The bending moment at C has the value

C $x = 4$ m $M = 6.9 \times 4 - 9 \times 2.5 - 10 \times 1 = -4.9$ kN m

This is a point of 'maximum negative' bending moment.

Example

A simply supported beam bridge is loaded by a point load which moves slowly across the bridge. At what position on the bridge will the load give rise to the maximum bending moment?

The beam will be of the form shown in figure 3.7. For a load F a distance x from end A the bending moment will be $R_A x$, where R_A is the reaction at A. But $R_A = F(L - x)/L$. Hence

$$M = \frac{F(L - x)}{L} x = F\left(x - \frac{x^2}{L}\right)$$

Differentiating the above equation gives

$$\frac{dM}{dx} = F\left(1 - \frac{2x}{L}\right)$$

Hence, since the maximum bending moment will occur when $dM/dx = 0$, then this will be when $x = \frac{1}{2}L$.

3.2.4 Point of contraflexure

The point on a beam where the bending moment changes sign and is zero is called the *point of contraflexure* or the *inflexion point*.

Example

For the beam in figure 3.12, i.e. the previous example, what is the position of the point of contraflexure?

For the bending moment the equations relating it with distance x from A are:

AB	$0 < x < 3$ m	$M = 6.9x - 3x\frac{1}{2}x\ kN\ m$
BC	$3 < x < 4$ m	$M = 6.9x - 9(x - 1.5) - 10(x - 3)$
CD	$4 < x < 5$ m	$M = 6.9x - 9(x - 1.5) - 10(x - 3)$
		$+ 17.1(x - 4)\ kN\ m$

The point of contraflexure occurs when $M = 0$. For AB, with the restriction $0 < x < 3$, the bending moment does not become zero. For BC the bending moment becomes zero when

$$6.9x - 9(x - 1.5) - 10(x - 3) = 0$$

i.e. when $x = 3.6$ m. This is the point of contraflexure.

3.3 Bending stresses

When a beam is loaded and bends then its longitudinal axis is bent into a curve with, say, the upper surface becoming extended and so in tension and the lower surface shortened and so in compression. Thus there is a plane intermediate between the surfaces which is unchanged in length when the beam is bent. This plane is called the *neutral plane* and the line where the plane cuts the cross-section of the beam the *neutral axis*.

A simple form of bending occurs when a beam is bent to form the arc of a circle. Consider a beam, with a cross-section that is symmetrical about the axis of bending and that was initially straight before any load is applied and is being bent by the application of a couple of moment M so that only pure bending occurs. The applied moment will be resisted by an internal moment in the beam, this moment of resistance being the same size as the applied moment but of opposite sign. Thus the beam is in equilibrium.

If the applied bending moments result in the neutral axis having a radius

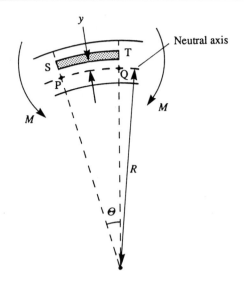

Fig. 3.13 Beam bent into the arc of a circle.

R (as in figure 3.13), then for a segment PQ (since arc length = radius ×
angle subtended)

$$PQ = R\Theta$$

The neutral axis does not change in length when the beam bends and thus
PQ is the length of the section and hence layer ST before it was bent, ST
being a section of the segment a distance y from the neutral axis. Thus

initial length of ST = PQ = $R\Theta$

Since when bent, the arc ST has a radius $(R + y)$, then

length of ST when bent = $(R + y)\Theta$

Thus the change in length on bending is $y\Theta$. Hence the strain for the layer
ST is

$$\text{strain} = \frac{y\Theta}{R\Theta} = \frac{y}{R} \qquad [12]$$

Provided the material is not extended beyond the limit of proportionality,
the stress σ acting normally to the cross-section of the beam on the layer ST
when bent will be

$$\sigma = \frac{yE}{R} \qquad [13]$$

It is assumed that the modulus of elasticity is the same in tension as in
compression.

The stress acting normally to the cross-section of the beam on any layer is
thus proportional to its distance from the neutral axis (figure 3.14). For a

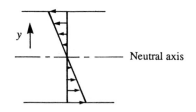

Fig. 3.14 Stress variation across the beam section.

rectangular cross-sectional beam, the neutral axis is at its centre and so the maximum distance y that is possible is half the thickness of the beam and the maximum stresses occur on the beam surfaces.

3.3.1 Position of the neutral axis

Consider a bent beam and a layer a distance y from the neutral axis, as in figure 3.15. The stress acting normally to the cross-section of this layer is given by equation [13] as yE/R, where R is the radius of the arc into which the beam is bent. Thus, if the layer has a cross-sectional area δA, the longitudinal force acting on the layer is

$$\text{force on layer} = \sigma\delta A = \frac{Ey}{R}\,\delta A$$

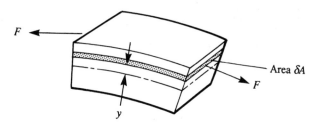

Fig. 3.15 Force acting normally to a section.

This is the force on just one layer. The total longitudinal force acting on the beam must be the sum of the forces acting on all such layers.

$$\text{Total force} = \Sigma\left(\frac{E}{R}\,y\delta A\right)$$

In the limit as δA becomes infinitesimally small,

$$\text{total force} = \frac{E}{R}\int y\,\mathrm{d}A \qquad [14]$$

But if the beam is acted on only by bending moments there is no net longitudinal force. The beam does not suffer any overall extension or contraction longitudinally. Thus we must have

$$\int y \, dA = 0$$

But this integral is the first moment of area (see section 1.3.2) about the neutral axis, as y is measured from the neutral axis. This can only be zero if the moment has been taken about the centroid. Hence the neutral axis passes through the centroid.

Example

What is the smallest radius to which a rectangular section steel strip of thickness 6 mm can be bent if the maximum stress on the material is not to exceed 100 MPa. The material has a modulus of elasticity of 200 GPa.

The neutral axis passes through the centroid, which for a rectangular cross-section is the centre. Thus the maximum distance possible from the neutral axis is half the beam thickness, i.e. 3 mm. Hence, using equation [13]

$$\sigma = 100 \times 10^6 = \frac{yE}{R} = \frac{3 \times 10^{-3} \times 200 \times 10^9}{R}$$

Hence $R = 6.0$ m. This is the radius of the neutral axis, the radius of the inner surface of the bent strip would be $6.0 - 0.003 = 5.997$ m.

3.3.2 The general bending formula

For a beam bent into the arc of a circle of radius R, as in figures 3.13 and 3.14, the force F acting on a layer a distance y from the neutral axis is

$$F = \frac{yE}{R} \delta A$$

where E is the elastic modulus and δA the cross-sectional area of the strip. The moment of the force about the neutral axis is

$$\text{moment} = Fy = \frac{yE}{R} \delta A \times y = \frac{E}{R} y^2 \delta A$$

Each layer of the beam will exert a moment about the neutral axis and thus the total moment is the sum of all these moments, i.e.

$$\text{total moment } M = \Sigma \left(\frac{E}{R} y^2 \delta A \right)$$

In the limit as δA is made infinitesimally small,

$$M = \frac{E}{R} \int y^2 \, dA$$

The quantity given by the integral is called the *second moment of area* about the neutral axis and denoted by the symbol I.

$$I = \int y^2 \, dA \tag{15}$$

Hence

$$M = \frac{EI}{R} \tag{16}$$

Since the stress σ on a layer a distance y from the neutral axis is given by equation [13] as yE/R, a general bending formula can be written as

$$\frac{M}{I} = \frac{\sigma}{y} = \frac{E}{R} \tag{17}$$

The second moment of area is sometimes referred to as the moment of inertia of the beam section. This is because it is a measure of the resistance to bending offered by a beam. The greater the second moment of area, the greater the bending moment needed to produce a particular radius of curvature.

Example

A constant cross-section I-section beam is 6.0 m long and simply supported at both ends. It has a second moment of area of $1.2 \times 10^8 \, mm^4$, a depth of 250 mm and a uniformly distributed weight of 30 kN/m. Calculate (a) the maximum bending moment, (b) the maximum bending stress and (c) the radius of curvature of the beam where the bending moment is a maximum. The elastic modulus is 200 GPa.

Figure 3.8 shows the type of beam involved. The total weight of the beam is wL, where w is the weight per unit length and L the total length, and can be considered to act at the centre of the beam. Since the beam is in equilibrium and symmetrically loaded, the reactions at the two supports of R_1 and R_2 will be the same, and thus $R_1 = R_2 = \frac{1}{2}wL$.

(a) The maximum bending moment for a simply supported beam is at the beam centre (see figure 3.8). Considering the moment about the beam centre, to the left of the centre the beam has R_1 acting upwards at $\frac{1}{2}L$ and a weight $\frac{1}{2}wL$ located $\frac{1}{4}L$ from the centre and acting downwards. Hence

$$\begin{aligned} M &= (\tfrac{1}{2}wL)\tfrac{1}{2}L - (\tfrac{1}{2}wL)\tfrac{1}{4}L \\ &= \frac{wL^2}{8} = \frac{30 \times 6^2}{8} = 135 \, \text{kN m} \end{aligned}$$

(b) Using equation [17], with the maximum stress occurring at the outer surface of the beam and a central neutral axis,

$$\sigma = \frac{My}{I} = \frac{135 \times 10^3 \times 125 \times 10^{-3}}{1.20 \times 10^8 \times 10^{-12}} = 141\,\text{MPa}$$

(c) Using equation [16],

$$R = \frac{EI}{M} = \frac{200 \times 10^9 \times 1.2 \times 10^8 \times 10^{-12}}{135 \times 10^3} = 178\,\text{m}$$

3.4 Second moment of area

The value of the second moment of area I about an axis depends on the shape of the beam section concerned and the position of the axis. Consider a rectangular cross-section, as in figure 3.16. For a layer of thickness δy a distance y from the axis XX which passes through the centroid, the second moment of area I_x about the axis XX is

second moment for strip $= y^2 \delta A = y^2 b \delta y$

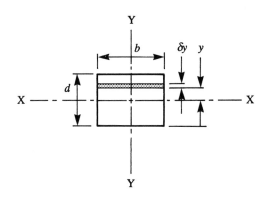

Fig. 3.16 Rectangular cross-section.

The total second moment of area about axis XX for all such strips in the section is thus

$$I_x = \int_{-d/2}^{d/2} y^2 b \, dy = \frac{bd^3}{12} \tag{18}$$

Second moments of area have the basic unit of m^4.

We could have obtained the second moment of area I_y for the rectangle about an axis YY through the centroid. This is

$$I_y = \frac{db^3}{12} \tag{19}$$

If the second moment of area of the rectangular cross-section is required about some axis X'X' which is parallel to the XX axis but a distance h from it, then since the moment of the strip would now be $(y+h)^2 \delta A$,

$$I_{x'} = \int (y + h)^2 \, dA$$
$$= \int (y^2 + 2hy + h^2) \, dA$$
$$= \int y^2 \, dA + 2h \int y \, dA + h^2 \int dA$$

The first integral is the second moment of area with respect to the XX axis, i.e. I_x. The second integral is zero because the XX axis passes through the centroid. The third integral is the area A. Thus

$$I_{x'} = I_x + Ah^2 \tag{20}$$

This equation is called the *theorem of parallel axes*.

For example, the second moment of area of the rectangle in figure 3.16 about the bottom edge, which is $\frac{1}{2}d$ displaced from the XX axis for which $I_x = bd^3/12$, is

$$I_{edge} = \frac{bd^3}{12} + bd \left(\frac{d}{2} \right)^2 = \frac{bd^3}{3} \tag{21}$$

Figure 3.17 shows the second moments of area of some other basic shapes. Many engineering sections can be considered to be composite shapes built up from basic shapes such as rectangles. The second moment of area of such a shape with respect to a particular axis is the sum of the second moments of area of its constituent parts with respect to the same axis.

Example

Determine the second moment of area about the neutral axis XX for the T-section shown in figure 3.18.

The section can be considered to be made up of two rectangular sections. The centroid of each rectangular part will be at its centre. Hence taking moments about the base of the section

$$\text{total moment} = 250 \times 50 \times 325 + 300 \times 50 \times 150$$

Hence the distance of the centroid, which is (total moment/total area), from the base is

$$\text{distance} = \frac{250 \times 50 \times 325 + 300 \times 50 \times 150}{250 \times 50 + 300 \times 50} = 230 \, \text{mm}$$

This is thus the distance of the neutral axis from the base. The second moment of area of a rectangle about its centroid is $bd^3/12$. The second moment of area of a rectangle about a different axis can be determined using the parallel axes theorem. Thus the second moment of area of a rectangle about the neutral axis if the rectangle, area bh, has a centroid a distance h away is $bd^3/12 + bdh^2$. The total second moment of the composite area about the neutral axis will be the sum of the second moments of area of the two rectangles about the neutral axis, hence

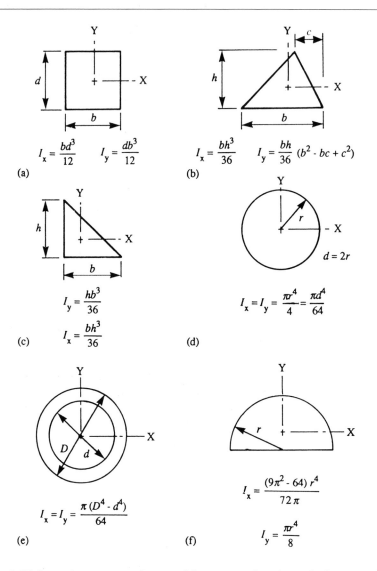

Fig. 3.17 Second moments of area with axes passing through the centroid in each case, (a) rectangle, (b) triangle, (c) right-angled triangle, (d) circle, (e) ring, (f) semicircle.

$$I_x = \frac{250 \times 50^3}{12} + 250 \times 50 \times (325 - 230)^2$$

$$+ \frac{50 \times 300^3}{12} + 50 \times 300 \times (230 - 150)^2$$

$$= 3.24 \times 10^8 \, \text{mm}^4$$

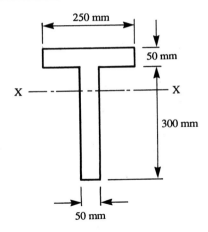

Fig. 3.18 Example.

Example

Determine the second moment of area about the neutral axis XX for the rectangular tube shown in figure 3.19.

One way of determining the second moment of area for such a section involves determining the second moment of area for the entire rectangle containing the section and then subtracting the second moment of areas for the rectangular space in that rectangle. For a rectangle $I_x = bd^3/12$. Thus

$$I_x = \frac{150 \times 200^3}{12} - \frac{100 \times 150^3}{12} = 7.19 \times 10^7 \, \text{mm}^4$$

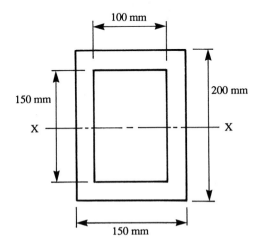

Fig. 3.19 Example.

3.5   Section modulus

For a beam the maximum bending stress σ_m will occur where the distance from the neutral axis is a maximum y_m. Hence, since $M/I = \sigma/y$ (equation [17]), we can write

$$M = \frac{I}{y_m} \sigma_m$$

The quantity I/y_m depends only on the shape of the cross-section concerned and is known as the *section modulus Z*. Hence

$$M = Z\sigma_m \qquad\qquad [22]$$

For a rectangular cross-section beam $I_x = bd^3/12$ and the maximum stress occurs at the surface which is $\frac{1}{2}d$ from the neutral axis. Hence

$$Z = \frac{I}{y_m} = \frac{(bd^3/12)}{\frac{1}{2}d} = \frac{bd^2}{6} \qquad\qquad [23]$$

Standard section handbooks give values of section modulus for different cross-section beams.

The larger the section modulus is for a beam the smaller will be the maximum stress produced by a given bending moment. With the same mass per unit length of material, different beam shapes can give different section modulus values. In general, a shape has a higher section modulus when the greater part of its area is as far as possible from the centroid. Thus a particularly good shape is the I-section or so-called *universal beam*.

Example

A 406×178 mm I-section girder has a section modulus of $1322\,\mathrm{cm}^3$. What will be the maximum bending stress produced when such a beam is subject to a bending moment of $50\,\mathrm{kN\,m}$?

Using equation [22],

$$\sigma_m = \frac{M}{Z} = \frac{50 \times 10^3}{1322 \times 10^{-6}} = 37.8\,\mathrm{MPa}$$

Example

A 305×165 mm universal beam $6.0\,\mathrm{m}$ long is supported at its ends. Loads of $40\,\mathrm{kN}$ are carried at $1.5\,\mathrm{m}$ from each end. If the section modulus is $752\,\mathrm{cm}^3$ and the weight of the beam can be ignored, what is the maximum stress?

For a simply supported beam in equilibrium

$$R_1 + R_2 = 40 + 40 = 80\,\mathrm{kN}$$

Taking moments about R_1 gives

$$6R_2 = 1.5 \times 40 + 40 \times 4.5 = 240$$

Hence, $R_2 = 40\,kN$ and $R_1 = 40\,kN$.

The bending moment for any position between the two loading positions is a constant, being for the section to the left of the centre

$$M = 40 \times 3.0 - 40 \times 1.5 = 60\,kN\,m$$

Hence

$$\sigma_m = \frac{M}{Z} = \frac{6.0 \times 10^3}{752 \times 10^{-6}} = 79.8\,MPa$$

3.6 Combined bending and direct stresses

So far in this chapter we have considered structural members as being subject to only bending loads, while in chapter 2 they were considered as subject to only axial loads. Structural members are however often subjected to the simultaneous action of both bending and axial forces. An example of this is a cantilever which is subject to a force F at its free end which is not vertical but inclined at some angle to the vertical (figure 3.20). The force can be resolved into two components, one vertical and one horizontal. The vertical component $F\sin\Theta$ produces bending stresses and the horizontal, axial, component $F\cos\Theta$ produces direct stresses. Such problems can be tackled by determining the stresses due to each form of loading acting alone and then adding the two to obtain the resultant stress.

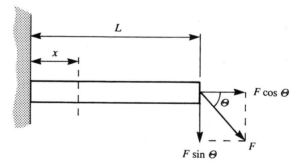

Fig. 3.20 Cantilever subject to bending and axial loads.

Consider the stress distribution at a section through the cantilever a distance x from the fixed end. The stress distribution across the section due to just the axial force component will be of the form shown in figure 3.21(a), the stress being $F\cos\theta/A$, where A is the cross-sectional area. The vertical force component will give rise to a bending moment M of $F\sin\theta$ $(L-x)$. The stress produced by this moment will vary across the section, being given by My/I, where y is the distance from the neutral axis and I the second moment of area of the section. Figure 3.21(b) shows the form of the stress distribution. The resultant stress distribution across the section is obtained by adding the two stress distributions to give the result shown in figure 3.21(c). The resultant stress is thus

$$\sigma = \frac{F\cos\theta}{A} + \frac{F\sin\theta\,(L-x)y}{I} \qquad [24]$$

It has been assumed in the above, that the beam is relatively short and stiff and so the bending is not very pronounced and does not produce any significant change in the line of action of the axial force. Typically such beams have a length-to-depth ratio of 10 or less. The situation where this is not the case is discussed in chapter 5.

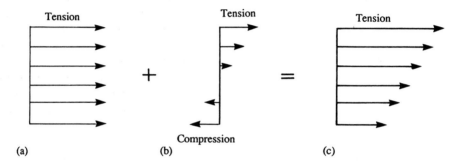

Fig. 3.21 Stress distributions due to (a) axial force, (b) bending force, (c) both forces.

Example

Figure 3.22 shows a bracket used to link two cables. What will be the maximum normal stress at the section AB? The bracket has a thickness of 40 mm.

The section AB will be subject to an axial tensile force of 20 kN and a bending moment of 20×80 kN mm (N m). The direct tensile stress will thus

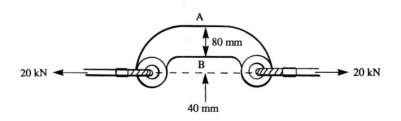

Fig. 3.22 Example.

be $20 \times 10^3/(0.080 \times 0.040) = 6.25$ MPa. The maximum bending stresses due to the bending moment will be My/I, where $y = 40$ mm and $I = db^3/12 = 0.040 \times 0.080^3/12$ m^4. Thus the maximum bending stresses are

$$\frac{20 \times 80 \times 0.040}{0.040 \times 0.080^3/12} = 37.5 \text{ MPa}$$

Thus the maximum compressive stress will be $37.5 + 6.25 = 43.75$ MPa and the maximum tensile stress $37.5 - 6.25 = 31.25$ MPa.

3.6.1   Eccentric loads

Consider a short column subject to a load F which is applied along a line parallel to the central axis of the column but displaced from it by some distance e along one of the principal axes of the column, i.e. the load is applied eccentrically (figure 3.23). The force will cause a bending moment of Fe about the YY axis and thus there will be bending stresses in the column in addition to a direct compressive stress of F/A. Note that, as shown in the figure, the force in this case is on the XX axis and is only displaced from one axis and so only produces a bending moment about one axis. The bending stress a distance x from the centroid is Mx/I_z, where I_z is the second moment of area about the Y axis. Thus the total stress at distance x is

$$\sigma = -\frac{F}{A} \pm \frac{Fex}{I_z} \tag{25}$$

A compressive stress is indicated by a minus sign, a tensile stress by a positive sign. Figure 3.24 shows the resulting stress distribution across a section of the column. The neutral axis, i.e. the unstressed axis, is when

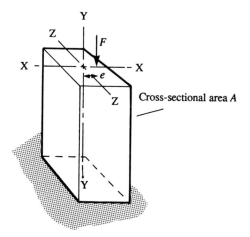

Fig. 3.23 Eccentric loading of a column.

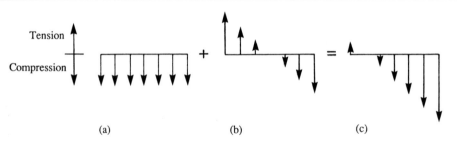

Fig. 3.24 Stress distribution.

$$\frac{F}{A} = \frac{Fex}{I_z}$$

i.e. when $x = I_z/Ae$. For a rectangular section $I_z = bd^3/12$ and $A = bd$, thus $x = d^2/12e$. If we want the section to have no tensile stresses, then we must have $x = \frac{1}{2}d$, i.e. the neutral axis must be on the surface of the section. The condition for this to occur is thus that $e = d/6$. This is known as the *middle third rule* since the eccentricity has to be kept within the middle third of the section. This condition is important if the material used for a column is one which cannot withstand significant tensile stresses, e.g. concrete or cast iron, since with this eccentricity the column is only subject to compressive forces.

When the point of application of the load is not on one of the principal axes there will be moments produced about both the centroidal principal axes. Thus if the force is applied a distance e_x in the X-direction and e_z in the Z-direction (figure 3.25) then the bending moments produced are Fe_x and Fe_z. Hence the total stress at some point in the cross-section which is defined by the coordinates x and z from the central axis is

$$\sigma = -\frac{F}{A} \pm \frac{Fe_x x}{I_z} \pm \frac{Fe_z z}{I_x} \qquad [26]$$

where I_z and I_x are the second moments of area about the ZZ and XX axes. The condition for the section to be subject to only compressive forces is when $x = \frac{1}{2}d$ and $z = \frac{1}{2}b$. This means

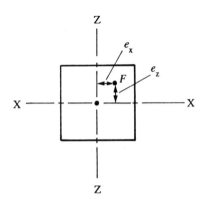

Fig. 3.25 Eccentric loading.

$$\frac{F}{bd} = \frac{Fe_x\frac{1}{2}d}{bd^3/12} + \frac{Fe_z\frac{1}{2}b}{db^3/12}$$

$$1 = \frac{6e_x}{d} + \frac{6e_z}{b} \qquad\qquad [27]$$

When $e_x = 0$ then $e_z = b/6$ and when $e_z = 0$ then $e_x = d/6$ (figure 3.26). This condition, a middle third rule, thus defines an area, called the *core*, of the section within which if the load is applied the entire section will be only subject to compressive stresses.

A similar analysis can be carried out for a circular section column and leads to a *middle quarter rule*. When the eccentricity is less than one eighth of the diameter then there will be no tensile stresses in the column. The core is thus a circle of radius one-eighth of the column diameter.

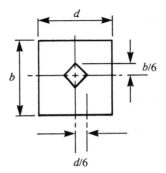

Fig. 3.26 The core.

Example

A force of 1 kN is applied to the centre of the short edge of a rectangular cross-section short column. If the column has a cross-section 100 mm by 250 mm, what will be the maximum compressive and maximum tensile stresses?

The situation is as in figure 3.23, with the eccentricity $e = 50$ mm. Thus, using equation [25], and recognising that the maximum stresses will occur at the edges, i.e. $x = \frac{1}{2}d = 50$ mm,

$$\sigma = -\frac{F}{A} \pm \frac{Fex}{I_z} = -\frac{1000}{0.100 \times 0.250} \pm \frac{1000 \times 0.050 \times 0.050}{0.250 \times 0.100^3/12}$$

$$\sigma = -40\,000 \pm 120\,000 \text{ Pa}$$

Hence the maximum compressive stress is -160 kPa and the maximum tensile stress is $+80$ kPa.

3.7   Shear stresses due to bending

Consider a beam to be made up of a number of small sections, as in figure 3.27(a). Then, when it is bent, each section is acted on by shearing forces on transverse opposite faces (see also section 3.2 for a discussion of shearing forces). There are thus transverse shearing stresses. If we consider a small element of such a section then there will be shearing stresses acting on opposite transverse faces. But the deformation of the transverse faces of a segment cannot occur without also a corresponding deformation of the longitudinal faces. They must also be acted on by shear forces and so there are longitudinal shear stresses. For equilibrium of any such segment we must have the moments of the shear forces acting on a segment equal to zero. Thus for the segment shown in figure 3.27(b) we must have, for moments about the central axis,

$$V_1 \times \tfrac{1}{2}b + V_1 \times \tfrac{1}{2}b = V_t \times \tfrac{1}{2}a + V_t \times \tfrac{1}{2}a$$

(a)   (b)

Fig. 3.27 (a) Shear forces, (b) shear forces on a segment.

The transverse shear stress for the segment is $\tau_t = V_t/bc$ and the longitudinal shear stress is $\tau_l = V_1/ac$. Thus

$$\tau_l bac = \tau_t abc$$

Thus the longitudinal shear stress must be equal to the transverse shear stress. The shear stresses are said to be *complementary*.

The existence of longitudinal shear stresses in a beam can be demonstrated by considering what happens when a longitudinally sectioned beam, as in figure 3.28, is bent. Each section can be considered as an independent beam and so each will bend about its neutral axis, the upper surface of each being in compression and the lower surfaces in tension. A consequence of this is that the bottom surface of the upper beam will have to slide over the top surface of the lower beam. Thus for a non-sectioned beam we must have longitudinal shear stresses to prevent the top half of the beam sliding over the bottom half.

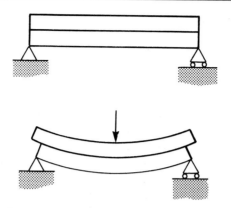

Fig. 3.28 Bending of a beam.

3.7.1   Shear stress equation

Consider a section of a rectangular cross-section beam which is being bent (figure 3.29). The section has a length δx with the bending moment on one face being M and on the other $M + \delta M$. Consider an element of the section with an area δA a distance y from the neutral axis. As a result of the bending this element will be subject to tensile stresses, given by equation [17], of $\sigma = My/I$ at one end of the section and $\sigma + \delta\sigma = (M + \delta M)y/I$ at the other. The difference in tensile forces acting on the area is thus $(\delta My/I)\delta A$. The force acting on the plane through ABC, a distance y_1 from the neutral axis, will be the sum of these elemental forces over the total area A, i.e. ABEF.

$$F = \frac{\delta M}{I} \int_{y_1}^{d/2} y \, dA$$

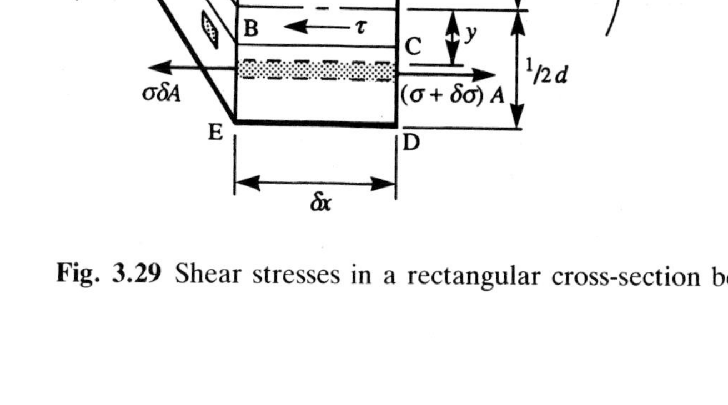

Fig. 3.29 Shear stresses in a rectangular cross-section beam.

Now if we consider the equilibrium of this plane through ABC, if the longitudinal shear stress is τ, then the longitudinal shear force of $\tau b \delta x$ acting on the plane must be balanced by the force F.

$$\tau b \delta x = \frac{\delta M}{I} \int_{y_1}^{d/2} y\, \mathrm{d}A \qquad [28]$$

The integral is the first moment of area Q of the cross-section with respect to the neutral axis. But the transverse shearing force $V = \delta M/\delta x$. Hence equation [28] can be written as

$$\tau = \frac{VQ}{Ib} \qquad [29]$$

Alternatively this equation can be written in terms of the distance of the centroid from the neutral axis.

$$\bar{y} = \frac{1}{A} \int_{y_1}^{d/2} y\, \mathrm{d}A$$

Thus equation [29] can be written as

$$\tau = \frac{VA\bar{y}}{Ib} \qquad [30]$$

This equation enables the shear stress to be determined at any point in the cross-section.

For a rectangular cross-section, the area of ABEF is $b(\frac{1}{2}d - y_1)$. The centroid of the area is at its midpoint which is $\frac{1}{2}(\frac{1}{2}d - y_1) + y_1 = \frac{1}{2}(\frac{1}{2}d + y_1)$ from the neutral axis. The second moment of area I of the section is $bd^3/12$. Thus

$$\tau = \frac{Vb(\frac{1}{2}d - y_1)\frac{1}{2}(\frac{1}{2}d + y_1)}{b \times bd^3/12}$$

$$= \frac{6V}{bd^3}\left(\frac{d^2}{4} - y_1^2\right) \qquad [31]$$

This shows there is a parabolic variation of shear stress with y_1. Figure 3.30 shows the distribution across the section. Note that the shear stress is zero when $y_1 = \frac{1}{2}d$, i.e. at the beam surface. The maximum value of the stress occurs when $y_1 = 0$, i.e. at the neutral axis, and has the value

$$\tau_{max} = \frac{3V}{2bd} \qquad [32]$$

The mean value of the shear stress across the section is $V/A = V/bd$ and thus the maximum stress is 1.5 times the mean stress.

A major assumption in the development of the above equation is that the shear stress is uniformly distributed over the breadth b. If this is not the case then errors will occur. This presents a problem when the equation is applied to determine the shear stresses in the flanges of an I-section, also at the flange−web junction where the sudden change in section gives rise to stress concentrations.

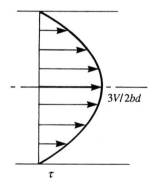

Fig. 3.30 Shear stress distribution.

Example

A simply supported beam has a width of 100 mm and a depth of 150 mm and spans a gap of 2.0 m. It carries a distributed load of 20 kN/m. What is the maximum shear stress?

The shearing force $V = \frac{1}{2} \times 20 \times 2.0 = 20$ kN. Thus, using equation [32]

$$\tau_{max} = \frac{3V}{2bd} = \frac{3 \times 20 \times 10^3}{2 \times 0.100 \times 0.150} = 2.0 \text{ MPa}$$

3.7.2 Shear stress in I-beams

Because of the shape, with an I-section beam the distribution of the shear stresses is more complicated than in the case of a rectangular beam. However, most of the transverse shear force V is carried by shear stresses in the web of the section and thus it is customary, as a reasonable approximation, to assume that the whole of the shear force is carried by the web.

Consider the I-section shown in figure 3.31 and the shear stress acting on a plane through AB, a distance y_1 from the neutral axis, and the application of equation [29]. The area is that between AB and the bottom edge of the flange and is thus made up of two parts, the area of the flange $b(\frac{1}{2}D - \frac{1}{2}d)$ and the area of part of the web $t(\frac{1}{2}d - y_1)$. The centroid of each of these areas is at its centre and thus the first moment of area of the composite area is

$$Q = b(\tfrac{1}{2}D - \tfrac{1}{2}d)(\tfrac{1}{2}d + \tfrac{1}{2}\{\tfrac{1}{2}D - \tfrac{1}{2}d\})$$
$$+ t(\tfrac{1}{2}d - y_1)(y_1 + \tfrac{1}{2}\{\tfrac{1}{2}d - y_1\})$$

This simplifies to

$$Q = \frac{b}{8}(D^2 - d^2) + \frac{t}{8}(D^2 - 4y_1^2)$$

Hence equation [29] becomes

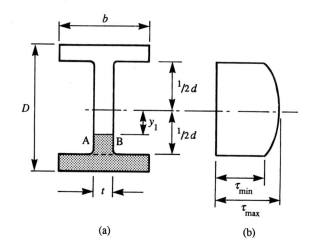

Fig. 3.31 Shear stresses in web.

$$\tau = \frac{VQ}{Ib} = \frac{V}{8It}[b(D^2 - d^2) + t(d^2 - 4y_1^2)] \tag{33}$$

The maximum shear stress is when $y_1 = 0$, i.e. at the neutral axis. Then

$$\tau_{max} = \frac{V}{8It}[b(D^2 - d^2) + td^2] \tag{34}$$

The minimum shear stress occurs at the bottom of the web when $y_1 = \frac{1}{2}d$. Then

$$\tau_{min} = \frac{V}{8It}b(D^2 - d^2) \tag{35}$$

Figure 3.31 shows the shear stress distribution in the web. The second moment of area I for the section is

$$I = \frac{bD^3}{12} - \frac{(b - t)d^3}{12} = \frac{1}{12}(bD^3 - bd^3 + td^3) \tag{36}$$

The average shear stress in the web, assuming that the entire shear force is across the web, is

$$\tau_{av} = \frac{V}{td} \tag{37}$$

For typical I-section beams, where the flanges are wide and the web thin, equation [36] approximates to

$$I \approx \frac{1}{12}b(D^3 - d^3)$$

and the maximum shear stress (equation [34]) is approximately

$$\tau_{max} \approx \frac{V}{8It}b(D^2 - d^2) \approx \frac{12Vb(D^2 - d^2)}{8tb(D^3 - d^3)} \approx \frac{12Vbd^2[(D/d)^2 - 1]}{8tbd^3[(D/d)^3 - 1]}$$

The maximum shear stress is thus of the order of V/td, i.e. the average shear stress. Generally it is within about 10% of that value and thus for design purposes the average shear stress is often used as an approximation to the maximum shear stress.

Example

An I-beam of length 3.0 m is simply supported at its ends and has a concentrated load of 20 kN at its centre. The I-beam has flanges 100 mm wide and 15 mm thick and a web 145 mm by 10 mm thick. What is the maximum shear stress in the web?

The second moment of area is given by equation [36] as

$$I = \frac{bD^3}{12} - \frac{(b - t)d^3}{12} = \frac{0.100 \times 0.175^3}{12} - \frac{0.090 \times 0.145^3}{12}$$

$$= 21.8 \times 10^{-6} \, m^4$$

The shearing force $V = 10$ kN. Thus the maximum shear stress in the web is given by equation [34] as

$$\tau_{max} = \frac{V}{8It}[b(D^2 - d^2) + td^2]$$

$$= \frac{10 \times 10^3[0.100(0.175^2 - 0.145^2) + 0.010 \times 0.145^2]}{8 \times 21.8 \times 10^{-6} \times 0.010}$$

$$= 6.7 \, \text{Mpa}$$

3.8 Composite beams

The term *composite beam* is used for a beam constructed of two or more different materials. Examples of such beams are timber joists reinforced by bolting steel plates to the sides along its full length (figure 3.32(a)), and

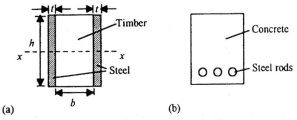

Fig. 3.32 Examples of composite beams.

reinforced concrete beams where reinforcing steel rods are embedded in the concrete (figure 3.32(b)). Concrete is weak in tension but good in compression, thus the reinforced concrete beam is designed so that the steel reinforcing rods are placed in the tension zone of the beam's cross-section.

The analysis of composite beams can be simplified by considering the behaviour of the equivalent uniform material beams. Then we can use the bending theory given earlier in this chapter.

3.8.1 Timber beams reinforced with steel plates

Consider the timber and steel beam shown in figure 3.32(a), the timber having a rectangular cross-section of breadth b and depth h and the steel rectangular cross-sections of depth h and thickness t. A bending moment M applied to the composite beam must be partly carried by the timber beam M_t and partly by the steel plates M_s, i.e.

$$M = M_t + M_s \qquad [38]$$

Because the timber and steel are fixed together, they will have the same radius of curvature R when bent. Thus, using equation [16]

$$M = \frac{E_t I_t}{R} + \frac{E_s I_s}{R} \qquad [39]$$

where E_t is the modulus of elasticity for the timber and I_t its second moment of area about xx, and E_s is the modulus of elasticity for the steel and I_s its second moment of area about xx. An equivalent uniform section beam giving the same radius of curvature for the same bending moment must have $M = EI/R$, where E is the equivalent modulus of elasticity and I its equivalent second moment of area. Thus we must have

$$EI = E_t I_t + E_s I_s \qquad [40]$$

If we consider the equivalent beam to be made entirely of timber, then equation [40] becomes

$$E_t I = E_t \left[I_t + \left(\frac{E_s}{E_t} \right) I_s \right] \qquad [41]$$

and the equivalent timber beam may be regarded as having a second moment of area of $[I_t + (E_s/E_t)I_s]$. Since $(E_s/E_t)I_s = (E_s/E_t)(td^3/12) = (tE_s/E_t)d^3/12$, this is equivalent to replacing both the steel plates in figure 3.32(a) by timber with a breadth $(E_s/E_t)t$ (figure 3.33(a)).

Alternatively we could have found the equivalent steel beam, replacing the timber by steel instead of the steel by timber. Equation [40] then becomes

$$E_s I = E_s \left[I_s + \left(\frac{E_t}{E_s} \right) I_t \right] \qquad [42]$$

and thus the equivalent steel beam may be regarded as having a second moment of area of $[I_s + (E_t/E_s)I_t]$. This is equivalent to replacing the

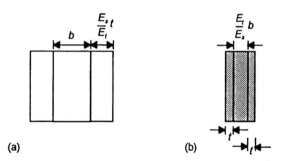

Fig. 3.33 (a) Equivalent timber, (b) equivalent steel beams.

timber in figure 3.32(a) by steel with a breadth $(E_t/E_s)b$, as illustrated in figure 3.33(b).

The bending stress σ_t in the timber of the composite beam of figure 3.32(a) a distance y from the neutral axis is given by equation [17] as

$$\sigma_t = M_t \frac{y}{I_t} \qquad [43]$$

But $M_t = E_t I_t/R$ and, for the equivalent beam, $M = EI/R = (E_t I_t + E_s I_s)/R$. Hence, eliminating R from these two equations gives

$$M_t = \frac{M}{1 + \dfrac{E_s I_s}{E_t I_t}} \qquad [44]$$

Thus substituting for M_t in equation [43] gives

$$\sigma_t = \frac{My}{I_t\left(1 + \dfrac{E_s I_s}{E_t I_t}\right)} = \frac{My}{I_t + \left(\dfrac{E_s}{E_t}\right)I_s} \qquad [45]$$

But $[I_t + (E_s/E_t)I_s]$ is the second moment of area of the equivalent timber beam. Thus, for the composite beam form of figure 3.32(a), the bending stress in the timber core is found by considering the total bending moment to be carried by the equivalent timber beam.

Because the steel and the timber are bent to the same radius of curvature, the longitudinal strains ϵ at a distance y from the neutral axis must be the same in both. Thus:

$$\epsilon = \frac{\sigma_t}{E_t} = \frac{\sigma_s}{E_s} \qquad [46]$$

and hence, at the same distances from the neutral axis:

stress in steel = modular ratio $(E_s/E_t) \times$ stress in timber   [47]

The maximum stresses will be for the maximum value of y, i.e. at $y = h/2$.

The above analysis was carried out for the bending moment about the xx axis in figure 3.32(a). We obtain the same relationships for bending moments about the yy axis of the composite beam shown in figure 3.34(a). Figure 3.34(b) shows the equivalent timber beam and figure 3.34(c) the

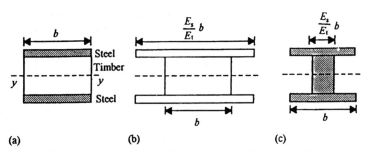

Fig. 3.34 (a) Composite section, (b) equivalent timber section, (c) equivalent steel section (not to scale).

equivalent steel beam. The maximum stresses in the timber and steel will not, however, now occur at the same distances from the y-axis. As a consequence we need to determine the stress in the equivalent beam at the appropriate distance. This is illustrated in the second of the following examples.

Example

A composite beam consists of a rectangular timber core 150 mm by 100 mm to which are fixed steel plates 150 mm by 9 mm along the entire length of the beam, the form being as in figure 3.32(a). Determine the maximum bending stress produced in the timber and the steel by a bending moment of 6 kN/m applied about the xx axis. The modulus of elasticity of the timber is 10 GPa and that of the steel 200 GPa.

Considering the equivalent timber beam

$$I = I_t + \left(\frac{E_s}{E_t}\right)I_s = \frac{0.100 \times 0.150^3}{12} + \frac{200 \times 10^9}{10 \times 10^9} \times \frac{2 \times 0.009 \times 0.150^3}{12}$$

$$= 1.29 \times 10^{-4}\,\text{m}^3$$

Using equation [45], the maximum stress in the timber is

$$\sigma_t = \frac{6 \times 10^3 \times 0.075}{1.29 \times 10^{-4}} = 3.49\,\text{MPa}$$

The maximum stress in the steel can be obtained by using equation [47] and is thus

$$\sigma_s = \frac{200 \times 10^9}{10 \times 10^9} \times 3.49 = 69.8\,\text{MPa}$$

Example

A composite beam consists of a timber core of width 75 mm and depth 150 mm with steel plates of width 75 mm and depth 12.5 mm fixed to its upper and lower surfaces. Determine the maximum bending stresses in the timber and steel when the beam is subject to a bending moment of $4\,\text{kN m}$. The modulus of elasticity of the timber is $10\,\text{GPa}$ and for the steel $200\,\text{GPa}$.

Transforming the beam to an equivalent steel beam gives an I-section of the form shown in figure 3.34(c). The second moment of area of this is

$$I = 2 \times 0.0125 \times 0.075^3 + \frac{10}{200} \times \frac{1}{12} \times 0.075 \times 0.15^3 = 11.6 \times 10^{-6}\,\text{m}^4$$

The maximum bending stress in the steel is then

$$\sigma_s = \frac{4 \times 10^3 \times 0.0875}{11.6 \times 10^{-6}} = 30.2\,\text{MPa}$$

To find the maximum bending stress in the timber we need to determine the bending stress in the equivalent steel section at the junction of the web and flange. This is

$$\sigma_s = \frac{4 \times 10^3 \times 0.0750}{11.6 \times 10^{-6}} = 25.9\,\text{MPa}$$

Hence, using equation [48], we obtain

$$\text{maximum stress in the timber} = \frac{10}{200} \times 25.9 = 1.2\,\text{MPa}$$

3.8.2 Reinforced concrete beams

The bending of a reinforced concrete beam can be treated by considering the equivalent uniform material beam. The following assumptions are made.

(1) Concrete behaves elastically so that stress is proportional to strain.
(2) It is assumed that concrete will crack in tension and so any concrete subject to tensile stresses will contribute nothing to the strength.
(3) The tensile stress in the steel reinforcing rods is uniform.
(4) There is perfect bonding between the concrete and the steel rods.

Consider the rectangular cross-section reinforced concrete beam (figure 3.35(a)) with reinforcing rods in the lower part of the section so that

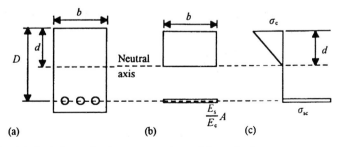

Fig. 3.35 (a) Reinforced concrete section, (b) equivalent concrete section, (c) stress diagram.

when the beam is bent the rods are in the tension. We have assumed that concrete in tension will contribute nothing to the strength and so the concrete below the neutral axis can be ignored. The bending moment of the beam is thus effectively carried by that part of the concrete in compression and the steel.

The total area of the reinforcing rods is A. If σ_s is the stress in the steel rods then the strain in them is σ_s/E_s. The strain in concrete, i.e. σ_c/E_c, is the same as the strain in the steel at that distance from neutral axis. Thus $\sigma_s/\sigma_c = E_s/E_c$. If the area A of the steel is transformed to concrete its equivalent area must be such that the bending force exerted is the same. Thus we must have $A_c/A_s = E_s/E_c$ and so the equivalent concrete area is $(E_s/E_c)A_s$ or mA, where m is the modulus ratio. Typically m is 15.

The neutral axis occurs at the centroid of the composite beam. Thus, for the equivalent concrete beam

$$bd \times \tfrac{1}{2}d = mA(D - d)$$

and hence

$$\tfrac{1}{2}bd^2 + mAd - mAD = 0$$

The distance of the neutral axis from the upper surface is the relevant root of this quadratic equation

$$d = \frac{mA}{b}\left(\sqrt{1 + \frac{2bD}{mA}} - 1\right) \tag{48}$$

The second moment of area of the equivalent beam about the neutral axis is

$$I_c = \tfrac{1}{3}bd^3 + mA(D - d)^3 \tag{49}$$

The maximum compressive stress occurs in the upper surface of the concrete, a distance d from the neutral axis, and is

$$\sigma_c = \frac{Md}{I_c} \tag{50}$$

The tensile stress σ_{sc} in the equivalent concrete for the steel rods, a distance $(D - d)$ from the neutral axis, is

$$\sigma_{sc} = \frac{M(D - d)}{I_c}$$

Hence the stress in the steel is

$$\sigma_s = \frac{M(D - d)}{I_c} \times \frac{E_s}{E_c} = \frac{mM(D - d)}{I_c} \tag{51}$$

If we divide equation [51] by equation [50] we obtain the useful relationship

$$\frac{\sigma_s}{\sigma_c} = \frac{m(D - d)}{d} \tag{52}$$

Figure 3.35(c) shows the form of the stress diagram for the beam.

Example

A rectangular cross-section reinforced concrete beam has a width of 200 mm and a depth to the steel reinforcing rods of 450 mm. It is reinforced with three steel bars, each having a cross-sectional area of 400 mm². Determine the maximum bending moment for the beam and the resulting maximum stresses in the steel and concrete. The maximum allowable stress in the concrete is 7.5 MPa and that in the steel 150 MPa. The modular ratio is 15.

Using equation [48]

$$d = \frac{mA}{b} \left(\sqrt{1 + \frac{2bD}{mA}} - 1 \right)$$

$$= \frac{15 \times 3 \times 400 \times 10^{-6}}{0.200} \left(\sqrt{1 + \frac{2 \times 0.200 \times 0.450}{15 \times 3 \times 400 \times 10^{-6}}} - 1 \right) = 0.2085\,\text{m}$$

Using equation [50] with the maximum value permissible for the stress in the concrete gives

$$M = \frac{7.5 \times 10^6 I_c}{0.2085} = 3.597 \times 10^7 I_c$$

Using equation [51] with the maximum value permissible for the stress in the steel gives

$$M = \frac{150 \times 10^6 I_c}{15(0.450 - 0.2085)} = 4.141 \times 10^7 I_c$$

Comparing these two values for the bending moment it is obvious that the maximum bending moment is set by the maximum permissible stress in the concrete, the full value of the steel strength not being used. Equation [49] gives the second moment of area I_c of the equivalent concrete beam as

$$I_c = \tfrac{1}{3} bd^3 + mA(D - d)^3$$

$$= \tfrac{1}{3} 0.200 \times 0.2085^3 + 15 \times 3 \times 400 \times 10^{-6}(0.450 - 0.2085)^3$$

$$= 8.578 \times 10^{-4}\,\text{m}^4$$

Thus

$$M = 3.597 \times 10^7 \times 8.578 \times 10^{-4} = 30.86\,\text{kN m}$$

The maximum stress in the concrete is -7.5 MPa and that in the steel is given by equation [51] as

$$\sigma_s = \frac{mM(D - d)}{I_c} = \frac{15 \times 30.86 \times 10^3 (0.450 - 0.2085)}{8.578 \times 10^{-4}} = 130.3\,\text{MPa}$$

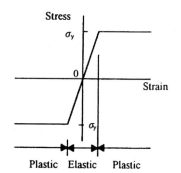

Fig. 3.36 Idealised stress–strain graph.

3.9   Plastic bending

In the bending of a beam it is usual to design it so that the maximum stress attained does not exceed the allowable working stress for the material, this being the yield stress for the material divided by some factor of safety. Consider, however, what happens with the bending of a rectangular mild-steel beam if the yield stress is exceeded.

We will assume an idealised form of stress–strain graph (figure 3.36), this being a reasonable approximation to that of mild steel. Up to the yield point the behaviour is considered to be elastic and the material reverts back to its original dimensions when the stress is removed. Beyond the yield point, the material does not revert back to its original dimensions when the applied stress is removed and the material is said to show plastic strain.

With elastic bending the maximum direct stresses occur in the outer longitudinal fibres. Initially, when the bending is such as to put the outer fibres so that they just attain the yield stress σ_y, the stress will be as shown in figure 3.37(a) with all the internal fibres elastic. With further bending, the stress will cease increasing in the outer fibres and the result will be as shown in figure 3.37(b), the region of elasticity being reduced. With yet further bending we can reach a situation where the elastic region is reduced to negligible proportions and the beam is more or less entirely plastic (figure 3.37(c)).

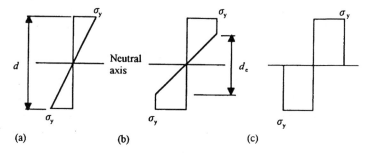

Fig. 3.37 (a) Elastic, (b) partially plastic, (c) completely plastic bending.

In figure 3.37(a), when the beam is completely elastic and the outer fibres have just attained the yield stress we have, for a rectangular cross-section beam of breadth b, depth d and second moment of area $bd^3/12$, a bending moment M_y of

$$M_y = \frac{\sigma_y I}{d/2} = \tfrac{1}{6} bd^2 \sigma_y \qquad [53]$$

As indicated in section 3.5 and equation [22], the *elastic section modulus* Z_e is M_y/σ_y and is thus $bd^2/6$ for the rectangular section.

In figure 3.37(c), when the beam is completely plastic with one half of the beam in tension and the other in compression, the resultant force in the upper half is $\tfrac{1}{2}bd\sigma_y$ and in the lower half there is the same size force but in the opposite direction. The couple, and hence bending moment M_p, due to these forces is

$$M_p = (\tfrac{1}{2}bd^2\sigma_y)(\tfrac{1}{2}d) = \tfrac{1}{4}bd^2\sigma_y \qquad [54]$$

M_p is termed the *fully-plastic moment* of the beam. We can define a *plastic section modulus* Z_p as M_p/σ_y. Thus for a rectangular section it is $\tfrac{1}{4}bd^2$. If we combine equations [53] and [54], then

$$M_p = 1.5M_y \qquad [55]$$

The ratio M_p/M_y is called the *shape factor* as it depends only on the shape of the cross-section. Thus, for a rectangular cross-section the shape factor is 1.5. For a solid circular cross-section the shape factor is 1.7.

In figure 3.37(b) when the beam is partly elastic and partly plastic, with a central region of depth d_e behaving elastically, we have for a beam bent to a radius of curvature R

$$\frac{\sigma_y}{d_e/2} = \frac{E}{R}$$

and so

$$d_e = \frac{2R\sigma_y}{E} \qquad [56]$$

One of the half core regions is in tension and acted on by a force of $\sigma_y bd_e/2$ and the other half is in compression and acted on by the same size force but in the opposite direction. The couple due to these forces, and hence the bending moment M_1 carried by the elastic core, is

$$M_1 = (\tfrac{1}{2}\sigma_y bd_e)\tfrac{1}{3}d_e = \tfrac{1}{6}bd_e^2\sigma_y$$

One of the outer plastic regions is in tension and acted on by a force of $\sigma_y b(\tfrac{1}{2}d - \tfrac{1}{2}d_e)$, the other outer plastic region being acted on by a force of the same size but in the opposite direction. The couple due to these forces, and thus the bending moment M_2 carried by the outer plastic regions, is

$$M_2 = \sigma_y b(\tfrac{1}{2}d - \tfrac{1}{2}d_e)(d_e + \tfrac{1}{2}d - \tfrac{1}{2}d_e) = \sigma_y b(\tfrac{1}{4}d^2 - \tfrac{1}{4}d_e^2)$$

The total moment M is the sum of M_1 and M_2 and is thus

$$M = \tfrac{1}{6}bd_e^2\sigma_y + \tfrac{1}{4}\sigma_y b(d^2 - d_e^2) = \sigma_y \frac{bd^2}{4}\left(1 - \frac{d_e^2}{3d^2}\right)$$ [57]

This can be written in terms of M_p (equation [54]) as

$$M = M_p\left(1 - \frac{d_e^2}{3d^2}\right)$$ [58]

Example

Determine the maximum elastic moment and the fully plastic moment for a steel rectangular cross-section beam of breadth 25 mm and depth 50 mm with a yield stress of 250 MPa.

Using equations [53] and [54]

$$M_y = \tfrac{1}{6}bd^2\sigma_y = \tfrac{1}{6} \times 0.025 \times 0.050^2 \times 250 \times 10^6 = 2.6\,\text{kN m}$$

$$M_p = \tfrac{1}{4}bd^2\sigma_y = \tfrac{1}{4} \times 0.025 \times 0.050^2 \times 250 \times 10^6 = 3.9\,\text{kN m}$$

Example

A rectangular cross-section beam of width 40 mm and depth 20 mm is used as a simply supported beam over a span of 2.0 m. Determine (a) the value of the mid-length concentrated load needed to initiate yielding in the outer fibres of the beam and (b) the depth of yielding from the upper and lower surfaces when this load is increased by 10%. The yield stress is 200 MPa.

(a) Equation [53] gives

$$M_y = \frac{\sigma_y I}{d/2} = \tfrac{1}{6}bd^2\sigma_y = \tfrac{1}{6} \times 0.040 \times 0.020^2 \times 200 \times 10^6 = 533\,\text{N m}$$

The maximum bending moment with a mid-length concentrated load is $WL/4$ at the centre and thus

$$W = \frac{4 \times 533}{2.0} = 1066\,\text{N}$$

(b) Increasing the load by 10% gives a new load of 1173 N. The maximum bending moment is then $WL/4 = (1173 \times 2)/4 = 586.5\,\text{N m}$. The yielding is then to a depth given by (equation [57])

$$M = \sigma_y \frac{bd^2}{4}\left(1 - \frac{d_e^2}{3d^2}\right)$$

$$586.5 = 200 \times 10^6 \frac{0.040 \times 0.020^2}{4}\left(1 - \frac{d_e^2}{3 \times 0.020^2}\right)$$

Hence $d_e = 17.9\,\text{mm}$. The depth of yielding is thus $\tfrac{1}{2}(20 - 17.9) = 1.05\,\text{mm}$.

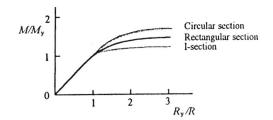

Fig. 3.38 Moment–curvature relationship.

3.9.1 Moment–curvature relationship for a rectangular section

When the outer fibres of the rectangular section just reach the yield point (figure 3.37(a)), equation [17] gives

$$\frac{\sigma_y}{d/2} = \frac{E}{R_y}$$

Hence, using equation [56]

$$\frac{d_e}{d} = \frac{R}{R_y}$$

Thus equation [58] can be written as

$$M = M_p\left[1 - \frac{1}{3}\left(\frac{R}{R_y}\right)^2\right] \tag{59}$$

and, by the use of equation [55] as

$$M = M_y\left[\frac{3}{2} - \frac{1}{2}\left(\frac{R}{R_y}\right)^2\right] \tag{60}$$

Figure 3.38 shows a graph of the moment–curvature relationship, i.e. how the values of M/M_y change with (R_y/R). The reciprocal of the radius is taken as a measure of curvature because the larger the radius the smaller is the curvature. When $(R_y/R) = 1$ then $M/M_y = 1$. For values of (R_y/R) less than 1 we have fully elastic behaviour. As (R_y/R) increases then we have a mixture of elastic and plastic behaviour and M/M_y tends to the value 1.5. For other cross-sections a similar procedure can be used, figure 3.38 showing the typical form of such results.

3.9.2 I-section beam

Consider an I-section beam (figure 3.39) in the fully plastic condition. All the tensile fibres of the beam carry the same stress σ_y and thus the total longitudinal force in the tensile flange is $\sigma_y b t_f$ and its moment about the neutral axis is $(\sigma_y b t_f)(d/2 - t_f/2)$. The longitudinal force in the tensile web is $\sigma_y t_w(d/2 - t_f)$ and its moment is $[\sigma_y t_w(d/2 - t_f)][\frac{1}{2}(d/2 - t_f)]$. The compressed

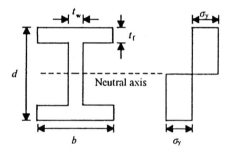

Fig. 3.39 I-section and fully plastic stress distribution.

part of the I-section contribute moments of the same magnitude. Thus the total moment carried by the beam is

$$M_p = 2\{(\sigma_y b t_f)(d/2 - t_f/2) + [\sigma_y t_w(d/2 - t_f)][\tfrac{1}{2}(d/2 - t_f)]\}$$

$$= \sigma_y(b t_f(d - t_f) + t_w(\tfrac{1}{2}d - t_f)^2] \qquad [61]$$

Using $M_p = \sigma_y Z_p$, the plastic section modulus Z_p is

$$Z_p = b t_f(d - t_f) + t_w(\tfrac{1}{2}d - t_f)^2 \qquad [62]$$

3.9.3 Plastic collapse of beams

So far we have considered the fully plastic bending moment as resulting from pure bending. This would mean that a beam would be fully plastic at all cross-sections along the entire length of the beam when M_p was reached. However, we are generally concerned with situations where the bending moment varies along the length of a beam. The point of maximum bending moment will thus be where the cross-section first becomes fully plastic. Cross-sections adjacent to the fully plastic region will then have commenced yielding to various depths. Figure 3.40(a) shows a beam simply supported at each end and carrying a central concentrated load. For such a beam the maximum bending moment is at the centre and thus it is here where yielding first starts. When a cross-section reaches the fully

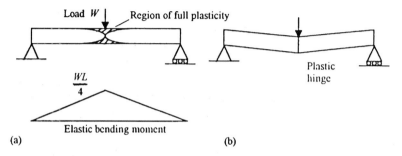

Fig. 3.40 Simply supported beam: (a) bending moment diagram and plastic zone at M_p, (b) with a plastic hinge.

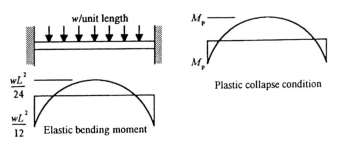

Fig. 3.41 Built-in beam.

plastic state, it cannot carry any higher loading and the beam forms a hinge at that cross-section, the term *plastic hinge* being used. Sufficient rotation can occur at this hinge to permit a redistribution of the bending moment (Figure 3.40(b)). With a beam, when sufficient plastic hinges have been formed to convert the structure into a 'mechanism' (the hinges are effectively pin joints) then collapse occurs. For the simply supported beam there is only one point of maximum bending moment and the collapse condition is reached when a plastic hinge forms at this point.

For the simply supported beam, the maximum elastic bending moment is $WL/4$ at the centre. As just one plastic hinge forms, plastic collapse occurs when $WL/4 = M_p$.

For a built-in beam (figure 3.41) of length L carrying a uniformly distributed load of w per unit length, the bending moment for elastic conditions is $wL^2/12$ at each end and $wL^2/24$ at the mid-length. For collapse, three plastic hinges must form, i.e. at each end and at the mid-length. The bending moment diagram at collapse is then obtained by making the bending moments equal to M_p at these points. By symmetry, the reactions at the ends are $wL/2$ and hence the bending moment at the mid-length point is

$$M_p = (\tfrac{1}{2}wL)(\tfrac{1}{2}L) - (\tfrac{1}{2}wL)(\tfrac{1}{4}L) - M_p$$

and thus $M_p = wL^2/16$.

Problems

(1)   A beam of negligible mass and length 4.0 m rests on supports at each end. A load of 5 kN is applied to the beam centre. What will be the bending moment and shearing force at a distance of (a) 1.0 m, (b) 2.0 m, (c) 3.0 m from one end?

(2)   A cantilever of negligible mass and length 2.0 m carries a load of 12 kN at its free end. What will be the bending moment and shearing force at a distance of (a) 0.5 m, (b) 1.0 m, (c) 2.0 m from the free end?

(3)   A cantilever has a weight per unit length of w. What will be the bending moment and shearing force at a distance x from the free end?

(4)   A beam of length 10 m is supported by a pin joint at one end and on a roller at 8.0 m from the pinned end. There is thus an overhang of 2.0 m. The beam has a weight per unit metre of 12 kN and supports a concentrated load of 28 kN at a distance of 3.0 m from the pinned end. Determine the bending moment and shearing force at 5.0 m from the pinned end.

(5)   Determine the shear and bending moment diagrams for the beams shown in figure 3.42.

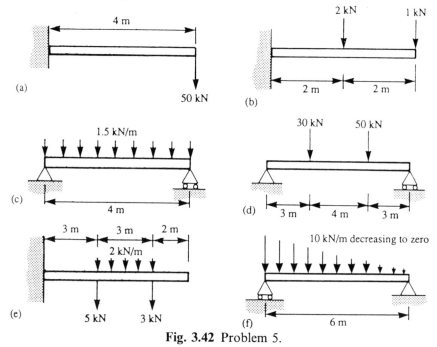

Fig. 3.42 Problem 5.

(6)   Simply supported beam AD of length L has a concentrated load of F_1 at B a distance a from end A and another concentrated load F_2 at C a distance b from A. Determine the equations relating the shear force and bending moment with distance x from A for the beam lengths AB, BC and CD.

(7)   What is (a) the maximum bending moment and (b) the maximum shear force for a cantilever having a weight of 20 kN/m and a length of 3.0 m?

(8)   A cantilever of length 3.0 m has a distributed load which varies linearly from 10 kN/m at the fixed end to 0 at the free end. What is (a) the maximum bending moment and (b) the maximum shear force?

(9)   A simply supported beam of length 5 m is supported at both ends, what will be the maximum shear force and the maximum bending moment if there is (a) a concentrated load of 50 kN at the centre,

(b) a uniformly distributed load of 10 kN/m, (c) a distributed load which varies from 0 at one support to 10 kN/m at the other?

(10)   A simply supported beam of length 2 m rests on supports at each end and supports a distributed load which varies from 30 kN/m at one end to 50 kN/m at the other. What are the shear force and the bending moment at a point half way along the beam?

(11)   A simply supported beam AB of length 12 m has a uniformly distributed load of 6 kN/m over a distance from A of 2 m to 6 m and a concentrated load of 20 kN at a distance of 9 m from A. What will be the maximum bending moment and its position?

(12)   A simply supported beam AB of length 3 m has a concentrated load of 50 kN at a distance of 1 m from A and a uniformly distributed load of 20 kN/m over the distance from A of 1 m to 3 m. What will be the maximum bending moment and its position?

(13)   Two slings are used to lift a girder of length L and uniform weight per unit length. What is the best position for the slings if the bending moment due to the weight of the girder is to be kept to a minimum?

(14)   The jib of a crane can be considered to be effectively just a cantilever. If it has a length of 10 m and a weight per unit length of 50 kN, what will be the maximum concentrated load that can be lifted at the extreme end of the jib if the maximum bending moment must not exceed 10 MN m?

(15)   A beam AB has a length of 4 m and supports at a distance of 0.5 m and 3.0 m from end A. What will be the position of the points of contraflexure if the beam has a uniformly distributed load over its entire length of 20 kN/m?

(16)   A beam AB has a length of 4 m and supports at a distance of 1 m from each end. What will be the points of contraflexure if the beam has concentrated loads of 15 kN at A, 80 kN at a distance of 2 m from A and 10 kN at B?

(17)   If a rectangular cross-section steel strip of thickness 4 mm is to be coiled on a drum of 1.0 m diameter, what will be the maximum stress produced by the coiling? The modulus of elasticity is 200 GPa.

(18)   A rectangular cross-section strip of steel 2 mm thick is to be bent round a drum. What must the diameter of the drum be if the stress in the steel strip must not exceed 100 MPa as a result of it bending to follow the circumference of the pulley wheel? The steel has a modulus of elasticity of 200 GPa.

(19)   A steel wire of 4 mm diameter is bent round a drum of diameter 1.0 m. What is the maximum bending stress? The steel has a modulus of elasticity of 200 GPa.

(20)   A steel strip of thickness 3 mm and width 35 mm is to be bent round a drum of diameter 4.0 m. What will be the maximum bending stress and the bending moment which has to be applied to bend the strip? The steel has a modulus of elasticity of 200 GPa.

(21)   What are the maximum stresses that will be set up in a rectangular cross-section beam, 200 mm by 50 mm, when a bending moment of 1 kN m is applied to it with (a) its smaller face, (b) its larger face parallel to the neutral surface?

(22)   A wooden uniform rectangular cross-section joist of depth 200 mm spans a gap of 6.0 m. What is the minimum width of joist required if the maximum tensile stress that the beam can be allowed to withstand is 14 MPa and the load to be carried is a centrally placed one of 20 kN? Neglect the mass of the joist.

(23)   A circular cross-section bar with a diameter of 200 mm and length 3.0 m is simply supported at each end. Two concentrated loads of 100 kN are applied 0.50 m from each support. What will be the maximum bending stress?

(24)   A steel cantilever beam, of rectangular cross-section 50 mm wide by 75 mm deep, has a length of 4.0 m and is subject to a distributed load of 1 kN/m. What will be the maximum bending stress?

(25)   A steel beam, of rectangular cross-section 75 mm wide by 150 mm deep, has a length of 3.0 m and is simply supported at its ends. What will be the maximum bending stress when the beam is subject to a distributed load of 20 kN/m?

(26)   Determine the second moments of area about the X and Y axes through the centroid for the sections shown in figure 3.43.

(27)   A rectangular cross-section channel to carry water is made from sheet metal 3 mm thick and has an internal cross-section 400 mm wide by 200 mm deep. The maximum depth of water that will occur in the channel is 160 mm. What is the maximum simply supported span that is possible if the maximum bending stress is not to exceed 35 MPa? The water has a density of 1,000 kg/m^3. Neglect the mass of the channel.

(28)   An L-shaped beam is made from metal 12 mm thick that has been bent to give the vertical part of the L an external length of 150 mm and the horizontal part a length of 80 mm. What is the maximum bending moment about the horizontal axis that can be applied to bend such a beam if the maximum bending stress is to be 250 MPa?

(29)   A steel scaffold tube has an external diameter of 50 mm and an internal diameter of 45 mm. What is the maximum bending moment that can be applied to such a tube if the maximum bending stress is to be 120 MPa?

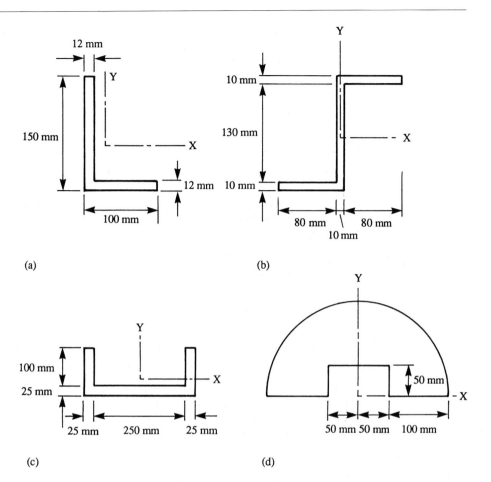

Fig. 3.43 Problem 26.

(30)   What is the required section modulus for a simply supported I-beam if it has to span 5 m, carry a concentrated load of 2 kN and the maximum permissible bending stress is to be 100 MPa?

(31)   What is the maximum uniformly distributed load that a simply supported steel I-section can carry over a span of 6.0 m if the maximum permissible stress in the beam is 50 MPa and the I-section has a section modulus of 3626 cm^3?

(32)   A timber beam has a rectangular cross-section of width 150 mm and depth 300 mm. What is (a) the section modulus and (b) the maximum bending moment that can be applied if the bending stress must not exceed 8 MPa?

(33)   A timber beam with a rectangular cross-section is required to span 4.0 m and carry a distributed load of 10 kN/m. The bending stress

must not exceed 8 MPa. What depth of beam will be required if the width is to be 120 mm?

(34)   A steel pipe has an external diameter of 150 mm and a wall thickness of 25 mm. What is the greatest distance the pipe can span if the bending stresses are not to exceed 150 MPa? The steel has a density of 7950 kg/m³.

(35)   What will be the maximum stress on the section AB of the G-clamp shown in figure 3.44 when it exerts a compressive force of 2 kN on the block? The section AB is rectangular with a cross-section 25 mm by 10 mm.

(36)   For the G-clamp shown in figure 3.44, what would be the maximum clamping force that can be applied if the section AB is an I-section beam with a second moment of area about its centroid of 5000 mm⁴ and cross-sectional area 75 mm²? The maximum allowable stress for the material is 100 MPa.

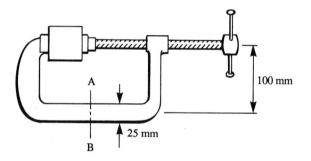

Fig. 3.44 Problem 35.

(37)   A circular cross-section vertical concrete column of diameter 0.25 m is subject to a vertical load of 500 kN applied with an eccentricity of 50 mm from the central axis of the column. What are the maximum tensile and compressive stresses in the column?

(38)   For the column described in problem 37, what would the eccentricity of the load have to be for there to be no tensile forces in the column?

(39)   A square cross-section, side 150 mm, short vertical column is subject to a vertical load of 45 kN at an eccentricity of 25 mm along an axis which bisects opposite sides of the square cross-section. What will be the stresses on at these sides of a section through the column?

(40)   A rectangular cross-section, 400 mm by 800 mm, vertical short column is subject to a vertical force of 10 kN at one corner. What will be the maximum tensile and maximum compressive stresses?

(41)   A rectangular cross-section beam of length 2.0 m is simply supported

at its ends. Determine the maximum shearing stress in the beam when it is subject to a concentrated force of 10 kN a distance of 1.5 m from one support. Beam width = 100 mm, depth = 75 mm.

(42) An I-beam has a length of 4.0 m and is simply supported at its ends. It carries a concentrated load of 30 kN at its centre. What are the maximum and minimum shear stresses in the web if the I-section has flanges of width 100 mm and thickness 25 mm, and a web of 350 mm by 10 mm?

(43) An I-beam has flanges of width 300 mm and thickness 20 mm, and a web of 600 mm by 12 mm. What will be the maximum shear stress in the web when it is subject to a shearing force of 25 kN?

(44) A composite beam consists of a timber section of width 75 mm and depth 125 mm with steel plates of thickness 20 mm and depth 125 mm fixed to the sides of the timber section. Determine the maximum bending moment that can be carried by the beam if the tensile stress in the steel must not exceed 130 MPa. The modulus of elasticity of the timber is 11 GPa and that of the steel is 200 GPa.

(45) A composite beam consists of a timber section of width 100 mm and depth 150 mm with steel plates of thickness 5 mm and depth 150 mm fixed to the sides of the timber section. Determine the maximum bending moment that can be carried by the beam if the tensile stress in the steel must not exceed 120 MPa and that in the timber must not exceed 5 MPa. The modulus of elasticity of the timber is 10 GPa and that of the steel is 200 GPa.

(46) A composite beam consists of a timber section of width 100 mm and depth 300 mm with steel plates of width 100 mm and depth 6 mm fixed to the upper and lower surfaces for the full length of the beam. The beam has a length of 4 m and is supported at each end, carrying a uniformly distributed load of 15 kN/m. Determine the maximum stresses in the timber and steel. The modulus of elasticity of the timber is 10 GPa and that of the steel is 200 GPa.

(47) A composite beam consists of a timber section of width 200 mm and depth 300 mm with steel plates of width 200 mm and depth 12 mm fixed to the upper and lower surfaces for the full length of the beam. Determine the maximum bending moment that the beam can carry if the maximum stress permitted for the timber is 8 MPa and that for the steel 120 MPa. The modulus of elasticity of the timber is 10 GPa and that of the steel is 200 GPa.

(48) A rectangular cross-section reinforced concrete beam, modular ratio 15, has a width of 300 mm and a depth to the steel reinforcing rods of 450 mm. Determine the cross-sectional area of steel required if the steel is to be stressed to 115 MPa and the concrete to 6.5 MPa.

(49) A rectangular cross-section reinforced concrete beam has a width of 200 mm and a depth to steel reinforcing rods of 420 mm. Determine the cross-sectional area of steel reinforcing rods required if the maximum stress for the steel is to be 140 MPa and that for the concrete is 8.5 MPa and both are to be fully stressed. The modular ratio is 15.

(50) A rectangular cross-section steel beam has a width of 30 mm and a depth of 72 mm and is used as a simply cupported beam on a span of 1.0 m and loaded at mid-span. If the steel has a yield stress of 250 MPa, determine the load when yield first occurs.

(51) Determine the plastic section modulus for an I-section (as in figure 3.39) with $b = 150$ mm, $d = 300$ mm, $t_f = 15$ mm and $t_w = 10$ mm.

(52) Determine the fully plastic moment of an I-section beam (as in figure 3.39) with a plastaic section modulus of 1.6×10^6 mm^3.

(53) A cantilelver carries a mid-length concentrated load W and is propped at the free end to the same height as the fixed end. Show that the collapse load is $6M_p/L$.

(54) A rectantular cross-section beam of width 50 mm and depth 20 mm is used as a simply supported beam over a span of 2.0 m. Determine the value of the mid-length concentrated load which will start yielding in the outer fibres of the beam. Take the yield stress as 200 MPa.

(55) A rectangular cross-section beam of width 30 mm and depth 80 mm is used as a simply supported beam over a span of 1.4 m and carries a uniformly distributed load. Determine the load per unit length required to give yielding to a depth of 20 mm from the upper and lower surfaces of the beam. Take the yield stress as 200 MPa.

(56) A rectangular cross-section beam of width 50 mm and depth 20 mm is used as a simply supported beam over a span of 2.0 m and carries a mid-length concentrated load. Determine (a) the load required to initiate yielding and (b) the depth to which the yielding will occur from the upper and lower surfaces if this load is then increased by 10%. The yield stress of the material is 225 MPa.

Chapter 4
Deflections of beams

4.1 The deflection curve

An effect of loading an initially straight beam is to deform it into a curved shape. The deformation can be expressed in terms of the deflection of the beam from its initially unloaded position. The convention is generally adopted of a downward deflection being positive and an upward deflection negative.

Figure 4.1 shows part of a loaded beam. At A the deflection is y, at B it is $y + \delta y$. At A the angle between the tangent to the beam and the horizontal, i.e. the unloaded position, is θ while at B it is $\theta + \delta\theta$. The angle subtended at the centre of the circle for which the curved beam between A and B is part of the circumference is thus $\delta\theta$. If δs is the arc length between A and B, then $\delta s = R\delta\theta$, where R is the radius of curvature of the arc AB. Hence we can write

$$\frac{1}{R} = \frac{\delta\theta}{\delta s}$$

Because the deflections obtained with beams are generally very small (in figure 4.1 they have been considerably exaggerated) δx is a reasonable

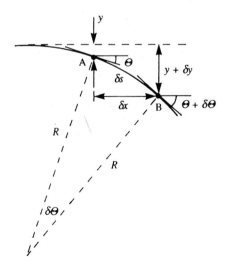

Fig. 4.1 Deflection curve of a beam.

approximation for δs. Thus the equation becomes, in the limit when $\delta\theta$ is made infinitesimally small,

$$\frac{1}{R} \approx \frac{d\theta}{dx} \qquad [1]$$

The slope of the straight line between A and B is $\delta y/\delta x$ and, to a reasonable approximation, this is $\tan\theta$. Since θ is small then $\tan\theta$ approximates to θ. Thus, in the limit, we can write

$$\theta \approx \frac{dy}{dx}$$

Hence, using equation [1],

$$\frac{1}{R} = \frac{d\theta}{dx} = \frac{d^2y}{dx^2} \qquad [2]$$

If the beam obeys Hooke's law then the bending moment $M = EI/R$ (equation [16], chapter 3), where E is the modulus of elasticity and I the second moment of area. For the beam shown the deflections y are downwards and defined as positive, thus the bending moment is negative since there is hogging. Thus equation [2] can be written as

$$\frac{d^2y}{dx^2} = -\frac{M}{EI} \qquad [3]$$

This is the basic differential equation of the deflection curve of a loaded beam. The product EI is often called the *flexural rigidity* of the beam.

The shearing force V is dM/dx (equation [4], chapter 3). Thus

$$V = \frac{dM}{dx} = -EI\frac{d^3y}{dx^3} \qquad [4]$$

If there is a distributed load of w per unit length then $dV/dx = -w$ (equation [3], chapter 3). Thus

$$w = EI\frac{d^4y}{dx^4} \qquad [5]$$

4.2   Deflections by double-integration

One method that can be used to determine the deflections of a loaded beam is by carrying out two integrations of the bending moment differential equation [3]. Thus integrating equation [3] once with respect to x gives

$$\frac{dy}{dx} = \int \frac{M}{EI}dx + A \qquad [6]$$

where A is a constant of integration. Integrating again gives

$$y = \int \left[\int \frac{M}{EI} dx + A \right] dx + B \qquad [7]$$

where B is a constant of integration.

If for a given beam there is an abrupt change in loading at some point or points, then there will be separate bending moment equations between such points and the differential equation will have to be evaluated for each.

4.2.1 Deflection of a cantilever

Consider the deflections of a cantilever beam subject to a *single concentrated load* at its free end (figure 4.2). The bending moment a distance x from the fixed end is

$$M = - F (L - x)$$

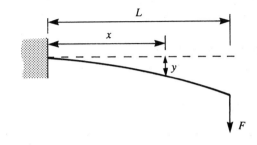

Fig. 4.2 Deflection of a cantilever.

Thus the differential equation [3] becomes

$$EI \frac{d^2y}{dx^2} = - M = FL - Fx$$

Integrating this with respect to x gives

$$EI \frac{dy}{dx} = FLx - \frac{Fx^2}{2} + A$$

The constant of integration A can be found from the condition that slope of the beam dy/dx is 0 at the fixed end, i.e. when $x = 0$. Thus $A = 0$ and so we can write for the slope of the beam, i.e. dy/dx,

$$EI \frac{dy}{dx} = FLx - \frac{Fx^2}{2} \qquad [8]$$

Integrating again gives

$$EIy = \frac{FLx^2}{2} - \frac{Fx^3}{6} + B$$

The constant of integration B can be found from the boundary condition that $y = 0$ at $x = 0$. Thus $B = 0$. Thus the deflection y at a distance x from the fixed end of the cantilever is given by

$$y = \frac{Fx^2}{6EI}(3L - x) \qquad [9]$$

The maximum deflection is at the free end when $x = L$. Thus

$$y_{max} = \frac{FL^3}{3EI} \qquad [10]$$

The maximum slope is also at the free end, equation [8] giving

$$\left(\frac{dy}{dx}\right)_{max} = \frac{FL^2}{2EI} \qquad [11]$$

For a cantilever with a *uniform distributed load* of w per unit length, the bending moment M a distance x from the fixed end is

$$M = -w(L - x)\tfrac{1}{2}(L - x) = -\tfrac{1}{2}w(L - x)^2$$

Thus the differential equation [3] becomes

$$EI\frac{d^2y}{dx^2} = -M = \tfrac{1}{2}w(L - x)^2$$

Integrating this equation with respect to x gives

$$EI\frac{dy}{dx} = -\frac{w(L - x)^3}{6} + A$$

The constant of integration A can be found from the condition that the slope $dy/dx = 0$ at $x = 0$. Thus $A = wL^3/6$. Thus

$$EI\frac{dy}{dx} = -\frac{w(L - x)^3}{6} + \frac{wL^3}{6} = \frac{w}{6}(3L^2x - 3Lx^2 + x^3) \qquad [12]$$

Integrating this equation gives

$$EIy = \frac{w}{6}\left(\frac{3L^2x^2}{2} - \frac{3Lx^3}{3} + \frac{x^4}{4}\right) + B$$

The constant of integration B can be found from the condition that $y = 0$ when $x = 0$. Thus $B = 0$. Thus the deflection is

$$y = \frac{wx^2}{24EI}(6L^2 - 4Lx + x^2) \qquad [13]$$

The maximum deflection occurs at the free end of the cantilever when $x = L$. Thus

$$y_{max} = \frac{wL^4}{8EI} \qquad [14]$$

The maximum slope of the beam is also at the free end. Thus equation [12] gives

$$\left(\frac{dy}{dx}\right)_{max} = \frac{wL^3}{6EI} \tag{15}$$

Example

A cantilever has a length of 2.0 m and a concentrated load of 30 kN applied at its free end. The beam has a second moment of area about the neutral axis of $300 \times 10^6 \, mm^4$ and an elastic modulus of 200 GPa. Determine the maximum deflection of the beam and its maximum slope.
Using equation [9],

$$y_{max} = \frac{FL^3}{3EI} = \frac{30 \times 10^3 \times 2.0^3}{3 \times 200 \times 10^9 \times 300 \times 10^6 \times 10^{-12}} = 1.33 \, mm$$

The slope is dy/dx and the maximum value is given by equation [11] as

$$\left(\frac{dy}{dx}\right)_{max} = \frac{FL^2}{2EI} = \frac{30 \times 10^3 \times 2.0^2}{2 \times 200 \times 10^9 \times 300 \times 10^6 \times 10^{-12}} = 0.001$$

This slope can be expressed as 0.001 m per m or 1 mm per m, or, since $dy/dx = \tan \theta \approx \theta$, as 0.001 rad.

4.2.2 Deflection of a simply supported beam

Consider a simply supported beam of length L with a *uniformly distributed load* of w per unit length (figure 4.3). The bending moment M a distance x from one end is

$$M = \tfrac{1}{2}wLx - wx\tfrac{1}{2}x$$

Thus the differential equation [3] becomes

$$EI \frac{d^2y}{dx^2} = -M = -\frac{wLx}{2} + \frac{wx^2}{2}$$

Integrating this equation with respect to x gives

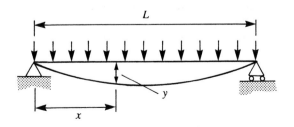

Fig. 4.3 Simply supported beam.

$$EI\frac{dy}{dx} = -\frac{wLx^2}{4} + \frac{wx^3}{6} + A$$

The constant of integration A can be obtained from the condition that the slope $dy/dx = 0$ at mid span, i.e. $x = \frac{1}{2}L$. Thus $A = wL^3/24$. Thus the slope is given by

$$EI\frac{dy}{dx} = -\frac{wLx^2}{4} + \frac{wx^3}{6} + \frac{wL^3}{24} \qquad [16]$$

Integrating this equation with respect to x gives

$$EIy = -\frac{wLx^3}{12} + \frac{wx^4}{24} + \frac{wL^3x}{24} + B$$

The constant of integration B can be obtained from the condition that $y = 0$ at $x = 0$. Thus $B = 0$. Hence the deflection is

$$y = \frac{wx}{24EI}(x^3 - 2Lx^2 + L^3) \qquad [17]$$

The maximum deflection occurs when $x = \frac{1}{2}L$, thus equation [17] gives

$$y_{max} = \frac{5wL^4}{384EI} \qquad [18]$$

The maximum slope occurs at the points of support, i.e. when $x = 0$. Thus equation [16] gives

$$\left(\frac{dy}{dx}\right)_{max} = \frac{wL^3}{24EI} \qquad [19]$$

Now consider the simply supported beam with a *concentrated load at its midpoint* (figure 4.4). Each of the reactions at the supports is $\frac{1}{2}F$. The bending moment for the left half of the beam is

$$0 \leq x \leq \frac{1}{2}L \qquad M = \frac{1}{2}Fx$$

This equation only applies to the left half of the beam, a different equation is required for the right half. Thus, for the left half, the differential equation [3] becomes

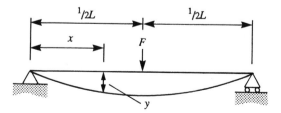

Fig. 4.4 Simply supported beam.

$$EI \frac{d^2y}{dx^2} = -M = -\frac{Fx}{2}$$

Integrating this equation with respect to x gives

$$EI\frac{dy}{dx} = -\frac{Fx^2}{4} + A$$

The constant of integration A can be obtained from the condition that, since the loading is symmetrical, the slope $dy/dx = 0$ at mid span, i.e. $x = \frac{1}{2}L$. Thus $A = FL^2/16$. Hence the slope is given by

$$EI \frac{dy}{dx} = -\frac{Fx^2}{4} + \frac{FL^2}{16} \qquad [20]$$

Integrating this equation with respect to x gives

$$EIy = -\frac{Fx^3}{12} + \frac{FL^2x}{16} + B$$

The constant of integration B can be derived from the condition that $y = 0$ at $x = 0$. Thus $B = 0$. Thus

$$y = \frac{Fx}{48EI} (3L^2 - 4x^2) \qquad [21]$$

Because of symmetry the maximum deflection will be at $x = \frac{1}{2}L$. Thus equation [21] gives

$$y_{max} = \frac{FL^3}{48EI} \qquad [22]$$

The maximum slope of the left half is when $x = 0$. Thus equation [20] gives

$$\left(\frac{dy}{dx}\right)_{max} = \frac{FL^2}{16} \qquad [23]$$

Now consider the situation where the *concentrated load is not at mid span* but a distance a from one end, as in figure 4.5. The reaction at A is Fb/L and at B is Fa/L. The bending moment equations are

$$0 \leq x \leq a \qquad M = \frac{Fb}{L} x \qquad [24]$$

$$a \leq x \leq b \qquad M = \frac{Fb}{L} x - F(x - a) \qquad [25]$$

Considering the left half of the beam, i.e. $0 \leq x \leq a$, then the differential equation [3] becomes

$$EI \frac{d^2y}{dx^2} = -M = -\frac{Fbx}{L}$$

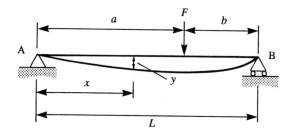

Integrating this equation with respect to x gives

$$EI \frac{dy}{dx} = -\frac{Fbx^2}{2L} + A \qquad [26]$$

Integrating this equation with respect to x gives

$$EIy = -\frac{Fbx^3}{6L} + Ax + B$$

The constant of integration B can be derived from the condition that $y = 0$ at $x = 0$. Thus $B = 0$. Thus

$$EIy = -\frac{Fbx^3}{6L} + Ax \qquad [27]$$

For the right half of the beam, i.e. $a \le x \le a + b$, equation [25] applies and thus the differential equation [3] becomes

$$EI \frac{d^2y}{dx^2} = -M = -\frac{Fbx}{L} + F(x - a)$$

Integrating this equation with respect to x gives

$$EI \frac{dy}{dx} = -\frac{Fbx^2}{2L} + \frac{F(x - a)^2}{2} + C \qquad [28]$$

Integrating this equation with respect to x gives

$$EIy = -\frac{Fbx^3}{6L} + \frac{F(x - a)^3}{6} + Cx + D \qquad [29]$$

The constants of integration are found as follows. At $x = a$ then the slopes obtained from considering the two parts of the beam must be equal. Thus equations [26] and [28] give

$$-\frac{Fba^2}{2L} + A = -\frac{Fba^2}{2L} + \frac{F(a - a)^2}{2} + C$$

Thus $A = C$. At $x = a$ then the deflections given by equations [27] and [29] must be equal.

$$-\frac{Fba^3}{6} + Aa = -\frac{Fba^3}{6L} + \frac{F(a-a)^3}{6} + Ca + D$$

Thus, since $A = C$, we have $D = 0$. At $x = L$ we have $y = 0$, thus the equation [29] which applies to this region gives, with $D = 0$ and $L = a + b$,

$$0 = -\frac{FbL^3}{6L} + \frac{Fb^3}{6} + CL$$

$$C = \frac{Fb}{6L}(L^2 - b^2)$$

Thus the deflection curves for the two parts of the beam are, for $0 \le x \le a$ equation [27] gives

$$Ely = -\frac{Fbx^3}{6L} + \frac{Fb}{6L}(L^2 - b^2)x$$

$$y = \frac{Fbx}{6EIL}(L^2 - b^2 - x^2) \qquad [30]$$

and for $a \le x \le b$ equation [29] gives

$$Ely = -\frac{Fbx^3}{6L} + \frac{F(x-a)^3}{6} + \frac{Fb}{6L}(L^2 - b^2)x$$

$$y = \frac{Fbx}{6EIL}(L^2 - b^2 - x^2) + \frac{F}{6EI}(x-a)^3 \qquad [31]$$

The maximum deflection occurs where the slope of the beam is zero, i.e. it is horizontal. Thus, using equation [26]

$$EI\frac{dy}{dx} = -\frac{Fbx^2}{2L} + \frac{Fb}{6L}(L^2 - b^2)$$

we find that the maximum deflection occurs, for $a \ge b$, at

$$x = \sqrt{\left(\frac{L^2 - b^2}{3}\right)} \qquad [32]$$

This maximum deflection can be found by substituting this value of x into equation [30]. Thus

$$y_{max} = \frac{Fb}{9\sqrt{3}EIL}(L^2 - b^2)^{3/2} \qquad [33]$$

4.2.3 Superposition

In the case of, say, a cantilever having both a distributed load and a point load then the problem can be solved by solving the bending moment differential equation for when both loads are acting on the beam. However, there is a simpler way of tackling the problem.

The bending moment M for a beam is the sum of the bending moments due to each load.

$$M = M_1 + M_2 + M_3 + \dots$$

But $EI\,d^2y/dx^2 = -M$ for the beam with all the loads. If, however, each load was considered alone then we would have for the deflections of each $EI\,d^2y_1/dx^2 = -M_1$, $EI\,d^2y_2/dx^2 = -M_2$, etc. Thus

$$EI\frac{d^2y}{dx^2} = EI\frac{d^2y_1}{dx^2} + EI\frac{d^2y_2}{dx^2} + EI\frac{d^2y_3}{dx^2} + \dots$$

Thus after two integrations

$$y = y_1 + y_2 + y_3 + \dots$$

The deflection curve for the combined loading is just the sum of the deflections due to each load alone. The deflection of a beam caused by several loads acting simultaneously can be found by superimposing the deflections caused by the loads acting separately.

Table 4.1 gives the deflections and slopes of some commonly encountered beams and loads.

Table 4.1 Deflections and slopes of beams

Symbols used: Length $= L$, distributed loads w/unit length, concentrated load F, $x =$ distance measured from fixed end of cantilever or left support of simply supported beam

Cantilevers

(1) Uniformly distributed load over entire length

$$y = \frac{wx^2}{24EI}(6L^2 - 4Lx + x^2)$$

$$\frac{dy}{dx} = \frac{wx}{6EI}(3L^2 - 3Lx + x^2)$$

(2) Uniformly distributed load over length a immediately adjacent to fixed end

For $0 \le x \le a$

$$y = \frac{wx^2}{24EI}(6a^2 - 4ax + x^2)$$

$$\frac{dy}{dx} = \frac{wx}{6EI}(3a^2 - 3ax + x^2)$$

For $a \le x \le L$

$$y = \frac{wa^3}{24EI}(4x - a)$$

$$\frac{dy}{dx} = \frac{wa^3}{6EI}$$

Table 4.1 *Continued*

(3) **Uniformly distributed load over length b immediately adjacent to free end, with $L = a + b$**

For $0 \leq x \leq a$

$$y = \frac{wbx^2}{12EI}(3L + 3a - 2x)$$

$$\frac{dy}{dx} = \frac{wbx}{2EI}(L + a - x)$$

For $a \leq x \leq L$

$$y = \frac{w}{24EI}(x^4 - 4Lx^3 + 6L^2x^2 - 4a^3x + a^4)$$

$$\frac{dy}{dx} = \frac{w}{6EI}(x^3 - 3Lx^2 + 3L^2x - a^3)$$

(4) **Concentrated load at free end**

$$y = \frac{Fx^2}{6EI}(3L - x)$$

$$\frac{dy}{dx} = \frac{Fx}{2EI}(2L - x)$$

(5) **Concentrated load a distance a from fixed end**

For $0 \leq x \leq a$

$$y = \frac{Fx^2}{6EI}(3a - x)$$

$$\frac{dy}{dx} = \frac{Fx}{2EI}(2a - x)$$

For $a \leq x \leq L$

$$y = \frac{Fa^2}{6EI}(3x - a)$$

$$\frac{dy}{dx} = \frac{Fa^2}{2EI}$$

Simply supported beams

(6) **Uniformly distributed load over entire length**

$$y = \frac{wx}{24EI}(L^3 - 2Lx^2 + x^3)$$

$$\frac{dy}{dx} = \frac{w}{24EI}(L^3 - 6Lx^2 + 4x^3)$$

(7) Uniformly distributed load over length a from left support

For $0 \leq x \leq a$

$$y = \frac{wx}{24LEI} (a^4 - 4a^3L + 4a^2L^2 + 2a^2x^2 - 4aLx^2 + Lx^3)$$

$$\frac{dy}{dx} = \frac{w}{24LEI} (a^4 - 4a^3L + 4a^2L^2 + 6a^2x^2 - 12aLx^2 + 4Lx^3)$$

For $a \leq x \leq L$

$$y = \frac{wa^2}{24LEI} (-a^2L + 4L^2x + a^2x - 6Lx^2 + 2x^3)$$

$$\frac{dy}{dx} = \frac{wa^2}{24LEI} (4L^2 + a^2 - 12Lx + 6x^2)$$

(8) Concentrated load at mid span

For $0 \leq x \leq L/2$

$$y = \frac{Fx}{48EI} (3L^2 - 4x^2)$$

$$\frac{dy}{dx} = \frac{F}{16EI} (L^2 - 4x^2)$$

(9) Concentrated load a distance a from left support and b from right support

For $0 \leq x \leq a$

$$y = \frac{Fbx}{6LEI} (L^2 - b^2 - x^2)$$

$$\frac{dy}{dx} = \frac{Fb}{6LEI} (L^2 - b^2 - 3x^2)$$

Example

Determine the equation of the deflection curve for a cantilever loaded by a uniformly distributed load over its length L of w per unit length and a concentrated load F at its free end.

For the uniformly distributed load alone, equation [13] gives

$$y = \frac{wx^2}{24EI} (6L^2 - 4Lx + x^2)$$

For the concentrated load F alone, equation [9] gives

$$y = \frac{Fx^2}{6EI}(3L - x)$$

Thus the deflection due to both loads acting simultaneously is

$$y = \frac{wx^2}{24EI}(6L^2 - 4Lx + x^2) + \frac{Fx^2}{6EI}(3L - x)$$

4.3 Moment−area method

Another method that can be used for the determination of the deflection of beams is the moment−area method. This is based on the determination of the areas under bending moment diagrams, i.e. graphs showing how the bending moment varies with distance along the beam.

Consider a beam, initially straight, which is bent under the action of some load or loads and with the slope and deflections being very small (figure 4.6). For an element dx of the beam we have (equation [3])

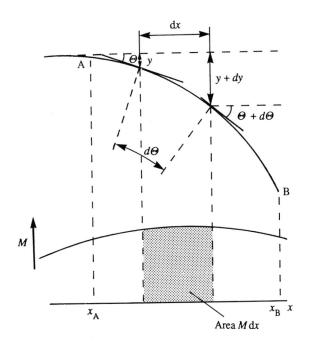

Fig. 4.6 Moment−area for a bent beam.

$$M = -EI\frac{d^2y}{dx^2} = -EI\frac{d}{dx}\left(\frac{dy}{dx}\right)$$

But dy/dx is the slope of this element dx of the beam and is approximately equal to tanθ, where θ is the angle the tangent to the beam at that point makes with the horizontal. Because the angle is small tan$\theta \approx \theta$. Thus

$$M = -EI \frac{d\theta}{dx}$$

Thus we can write

$$d\theta = -\frac{M}{EI} dx \qquad\qquad [34]$$

But $(M/EI)\,dx$ is the area under the graph of M/EI plotted against x, i.e. the bending moment diagram, over the distance dx. Thus the change in angle between the tangents to two points A and B on the beam is the negative of the area under the M/EI diagram between those points, i.e.

$$\theta_{BA} = -\int_{x_A}^{x_B} \frac{M}{EI} dx \qquad\qquad [35]$$

This relationship is known as the *first moment–area theorem*.

For the element of beam dx in figure 4.6, the contribution to the change in the vertical distance between the tangents to the slope at the beginning and end of a length of beam AB is $d\Delta$ (figure 4.7). Because the change in angle $d\theta$ is small it is a reasonable approximation to take $d\Delta = x_1\,d\theta$ (i.e. $d\Delta$ as the arc of a circle of radius x_1 subtending an angle $d\theta$ at the centre). Hence, using equation [34] we can write

$$d\Delta = x_1\,d\theta = -x_1 \frac{M}{EI} dx$$

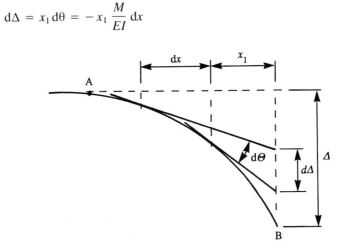

Fig. 4.7 Second moment–area theorem.

Thus the total vertical distance between the tangents at A and B for the beam is

$$\Delta = -\int_{x_A}^{x_B} x_1 \frac{M}{EI} dx \qquad\qquad [36]$$

But $(M/EI)\,dx$ is the area under the (M/EI) graph against x for the element at distance x_1. Thus the integral represents the first moment of the area of the M/EI diagram between A and B. The *second moment–area theorem* is thus: the vertical deviation of point B from the tangent at point A is equal to the negative of the first moment of the area of the M/EI diagram between A and B, when taken with respect to B.

In considering bending moment diagrams for beams with concentrated loads, the diagrams will be made up of straight line sections composed generally of triangles and rectangles and thus the areas and centroids can be determined for such composite figures. A rectangle will have its centroid at its centre, a triangle at one third the vertical height with the area being half the breadth times the height. With uniformly distributed loads the bending moment diagrams are parabolic. Figure 4.8 shows the areas and centroid positions of areas bounded by a parabola.

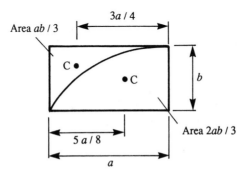

Fig. 4.8 Properties of the parabolic shape.

Example

Determine the angle of rotation, i.e. slope, and deflection at the free end of a uniform cantilever of length L with a concentrated load of F at the free end.

Figure 4.9 shows the cantilever and the bending moment diagram. Since EI is a constant the bending moment diagram is the same form as the M/EI diagram. Using the first moment–area theorem, the angle θ_{max} is equal to the negative of the area of the M/EI diagram and thus, since the area of the bending moment diagram is $\frac{1}{2}L(-FL)$ then

$$\theta_{max} = -\frac{\frac{1}{2}L(-FL)}{EI} = \frac{FL^2}{2EI}$$

The deflection at the free end can be obtained by the use of the second moment–area theorem. The vertical deviation in this case is y_{max}. The first moment of area of the M/EI diagram taken with respect to the free end is, for the triangular shape, its area multiplied by $2L/3$ (see the example in section 1.3). Thus

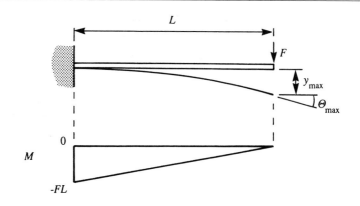

Fig. 4.9 Cantilever with concentrated load.

$$y_{max} = -\frac{\frac{1}{2}L(-FL)(2L/3)}{3EI} = \frac{FL^3}{3EI}$$

Example

Determine the maximum angle of rotation and the maximum deflection for a cantilever of length L which has a uniformly distributed load of w per unit length over a length a from the fixed end.

Figure 4.10 shows the cantilever and its bending moment diagram. For the length a the bending moment is $-\frac{1}{2}w(a-x)^2$, when x is measured from the fixed end. The bending moment has thus the value $-\frac{1}{2}wa^2$ at $x=0$, i.e. the fixed end, and 0 at $x=a$. For $x \geq a$ the bending moment is zero. The

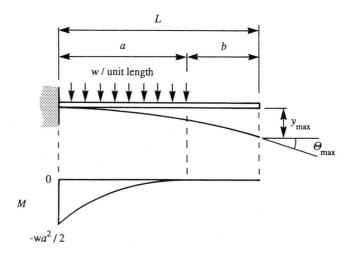

Fig. 4.10 Cantilever with part distributed load.

area of the bending moment diagram up to $x = L$ is $-a(wa^2/2)/3$. Thus, using the first moment area theorem

$$\theta_{max} = -\frac{-a\,(wa^2/2)/3}{EI} = \frac{wa^3}{6EI}$$

The centroid of the area is $\frac{1}{4}a$ from the fixed end and so $(b + 3a/4)$ from the free end. Thus, using the second moment–area theorem

$$y_{max} = -\frac{[-a\,(wa^2/2)/3]\,(b + 3a/4)}{EI} = \frac{wa^3(b + 3a/4)}{6EI}$$

and, since $b = L - a$,

$$y_{max} = \frac{wa^3}{24EI}\,(4L - a)$$

Example

A cantilever of length L has the half of its length nearest to the fixed end with a first moment of area of $2I$ and the half nearest the free end with a first moment of area of I. What will be the deflection at the free end when a concentrated load of F is applied there?

Figure 4.11(a) shows the cantilever, figure 4.11(b) its bending moment M diagram and figure 4.11(c) the M/I diagram. The M/I diagram can conveniently be regarded as made up of two triangular areas (figure 4.11(d)). Thus, applying the second moment–area theorem

$$
\begin{aligned}
y_{max} &= -\frac{1}{E}\left(-\frac{FL}{2I}\,L \times \frac{2L}{3} - \frac{FL}{2 \times 4I} \times \frac{L}{2} \times \frac{2L}{3 \times 2}\right) \\
&= \frac{3FL^3}{16EI}
\end{aligned}
$$

4.4   Macaulay's method

For a beam with a number of concentrated loads, separate bending moment equations have had to be written for each part of the beam between adjacent loads. Integration of each expression then gives the slope and deflection relationships for each part of the beam. Each slope and deflection relationship includes constants of integration and these have to be determined for each part of the beam. The constants of integration are then found by equating slopes and deflections given by the expressions on each side of each load. This can be very laborious if there are many loads. A much less laborious way of tackling the problem is to use the method known as *Macaulay's method*. This involves writing just one equation for the bending moment of the entire beam and enables boundary conditions for any part of the beam to be used to obtain the constants of integration.

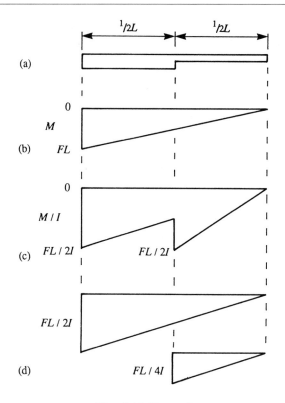

Fig. 4.11 Example.

Consider a concentrated force F applied some distance a along a beam, as in figure 4.12. For distances x up to a we would conventionally write for the bending moment when we consider the value at x due to forces to the left of x,

$$0 \leq x < a \qquad M = 0$$

For x greater than a we would write

$$a \leq x \qquad M = F(x - a)$$

The Macaulay expression for the bending moment for all values of x is

$$M = F\{x - a\} \qquad [37]$$

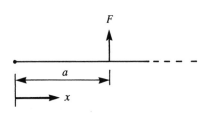

Fig. 4.12 Concentrated force.

This equation is subject to the condition that all terms for which the quantity inside the { } brackets is negative are given the value zero. Note that in some books ⟨ ⟩ brackets and in others [ ] are used for terms which have to be treated in this special way.

Consider a uniformly distributed load of w per unit length which starts a distance a along a beam, as in figure 4.13. The bending moment at a distance x along the beam, when we consider forces to the left of x, is

$$0 \leq x < a \qquad M = 0$$
$$a \leq x \qquad M = \tfrac{1}{2}w(x - a)^2$$

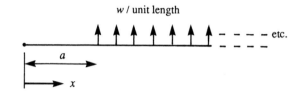

Fig. 4.13 Uniformly distributed load.

The Macaulay expression for the bending moment for all values of x is

$$M = \tfrac{1}{2}w \{x - a\}^2 \tag{38}$$

The quantity inside the { } brackets is zero when it has a negative value.

A beam with a number of concentrated loads and/or distributed loads can thus have a Macaulay expression written for its bending moment. A procedure that can be adopted to obtain such an expression for an entire beam is to: take the origin at the left-hand end of the beam, then write down the bending moment expression at a section at the extreme right-hand end of the beam in terms of the loads to the left of that section. The resulting expression must not be simplified by expanding any of the terms in the Macaulay brackets { }. When integrating, all the terms inside such brackets must be kept within the brackets and each integrated as a whole. The resulting equation can then have the constants of integration evaluated for boundary conditions from any point on the beam. The equation can thus be written for any particular value of x along the beam if all terms within Macaulay brackets are zero whenever they are negative.

Example

Determine the Macaulay expression for the deflection y of the simply supported beam shown in figure 4.14 and which is subject to three concentrated loads.

Consider the bending moment at the section XX due to all the forces to the left of the section.

$$M = R_1x - F_1\{x - a\} - F_2\{x - b\} - F_3\{x - c\}$$

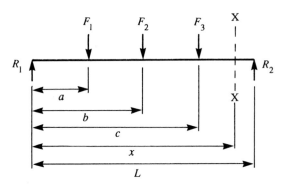

Fig. 4.14 Example.

Thus applying equation [3],

$$\frac{d^2y}{dx^2} = -\frac{M}{EI}$$

$$= -\frac{1}{EI}\left(R_1x - F_1\{x - a\} - F_2\{x - b\} - F_3\{x - c\}\right)$$

Integrating this expression gives

$$\frac{dy}{dx} = -\frac{1}{EI}\left(\frac{R_1x^2}{2} - \frac{F_1\{x - a\}^2}{2} \cdot - \frac{F_2\{x - b\}^2}{2} - \frac{F_3\{x - c\}^2}{2} + A\right)$$

Integrating again gives

$$y = -\frac{1}{EI}\left(\frac{R_1x^3}{6} - \frac{F_1\{x - a\}^3}{6} - \frac{F_2\{x - b\}^3}{6}\right.$$
$$\left. - \frac{F_3\{x - c\}^3}{6} + Ax + B\right)$$

The constants of integration can be evaluated by taking the boundary conditions $y = 0$ when $x = 0$, and $y = 0$ at $x = L$. The first condition gives $B = 0$. The second condition gives, since L is greater than a, b or c,

$$A = -\frac{R_1L^2}{6} + \frac{F_1(L - a)^3}{6L} + \frac{F_2(L - b)^3}{6L} + \frac{F_3(L - c)^3}{6L}$$

The slope and deflection can be found for any point along the beam, provided that all the Macaulay brackets which come out negative are made zero.

Example

Determine the Macaulay expression for the deflection of the simply supported beam shown in figure 4.15(a), it carrying a uniformly distributed load over part of its length.

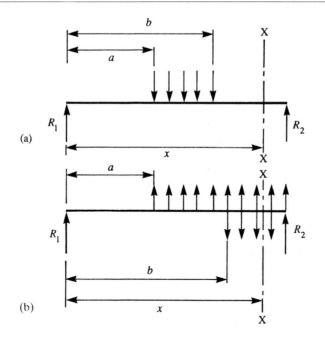

Fig. 4.15 Example.

If a distributed load is not continuous to the right end of the beam then it is necessary to introduce two equal but opposite dummy loads, as shown in figure 4.15(b). These two loads will have no net effect on the behaviour of the beam but enable a Macaulay equation to be written which extends over the entire beam. If the loading is w per unit length, then the bending moment equation for the section XX, due to all the forces to the left, is

$$M = R_1 x - \tfrac{1}{2}w\{x - a\}^2 + \tfrac{1}{2}w\{x - b\}^2$$

Thus applying equation [3],

$$\frac{d^2 y}{dx^2} = -\frac{M}{EI}$$

$$= -\frac{1}{EI}\left(R_1 x - \tfrac{1}{2}w\{x - a\}^2 + \tfrac{1}{2}w\{x - b\}^2\right)$$

Integrating this equation gives

$$\frac{dy}{dx} = -\frac{1}{EI}\left(\frac{R_1 x^2}{2} - \frac{w\{x - a\}^3}{6} + \frac{w\{x - b\}^3}{6} + A\right)$$

Integrating again gives

$$y = -\frac{1}{EI}\left(\frac{R_1 x^3}{6} - \frac{w\{x - a\}^4}{24} + \frac{w\{x - b\}^4}{24} + Ax + B\right)$$

A and B can be evaluated in the same way as in the previous example.

Example

A simply supported beam of length 3.0 m has concentrated loads of 40 kN at 1.0 m from one end and 60 kN at 2.0 m from the same end (figure 4.16). What is the deflection and slope of the beam under the 40 kN load? What is the maximum deflection and at what position along the beam does it occur? The beam has $EI = 20\,\text{MN}\,\text{m}^2$.

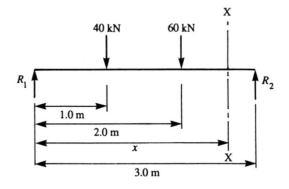

Fig. 4.16 Example.

Since the beam is in equilibrium we have $R_1 + R_2 = 100\,\text{kN}$ and for moments taken about the R_1 end we have

$$40 \times 1.0 + 60 \times 2.0 = 3.0R_2$$

Thus $R_2 = 53.3\,\text{kN}$ and $R_1 = 46.7\,\text{kN}$.

Considering the bending moment about section XX due to the forces to the left of the section, then

$$M = R_1x - 40\{x - 1\} - 60\{x - 2\}$$

Thus applying equation [3],

$$\frac{\mathrm{d}^2y}{\mathrm{d}x^2} = -\frac{M}{EI}$$

$$= -\frac{1}{EI}(R_1x - 40\{x - 1\} - 60\{x - 2\})$$

Integrating this expression gives

$$\frac{\mathrm{d}y}{\mathrm{d}x} = -\frac{1}{EI}\left(\frac{R_1x^2}{2} - \frac{40\{x - 1\}^2}{2} - \frac{60\{x - 2\}^2}{2} + A\right) \qquad [39]$$

Integrating again gives

$$y = -\frac{1}{EI}\left(\frac{R_1x^3}{6} - \frac{40\{x - 1\}^3}{6} - \frac{60\{x - 2\}^3}{6} + Ax + B\right) \qquad [40]$$

When $x = 0$ we have $y = 0$, thus $B = 0$. When $x = 3\,\text{m}$ we have $y = 0$. Thus

$$0 = -\frac{1}{EI}\left(\frac{46.7 \times 3^3}{6} - \frac{40 \times 2^3}{6} - \frac{60 \times 1^3}{6} + 3A\right)$$

Hence $A = -48.9$. Thus equation [40] becomes

$$y = -\frac{1}{EI}\left(\frac{R_1 x^3}{6} - \frac{40\{x-1\}^3}{6} - \frac{60\{x-2\}^3}{6} - 48.9x\right) \qquad [41]$$

The deflection at $x = 1.0\,\text{m}$ is thus

$$y = -\frac{10^3}{20 \times 10^6}\left(\frac{46.7 \times 1.0^3}{6} - 0 - 0 - 48.9 \times 1.0\right)$$

The 10^3 is included because all the forces have until now been in kN. The $\{x-2\}$ term is zero because it has a negative value when $x = 1.0\,\text{m}$. The $\{x-1\}$ term is also zero. Thus $y = 2.06\,\text{mm}$.

The slope of the beam at $x = 1.0\,\text{m}$ is given by equation [39] as

$$\frac{dy}{dx} = -\frac{10^3}{20 \times 10^6}\left(\frac{46.7 \times 1.0^2}{2} - 0 - 0 - 48.9\right)$$

Thus the slope is $1.28 \times 10^{-3}\,\text{rad}$.

Maximum deflection occurs when $dy/dx = 0$. Thus we have for equation [39]

$$\frac{dy}{dx} = 0 = -\frac{1}{EI}\left(\frac{46.7x^2}{2} - \frac{40\{x-1\}^2}{2} - \frac{60\{x-2\}^2}{2} - 48.9\right)$$

By inspection of the problem it looks likely that the maximum deflection will occur between the two load positions. Hence the solution of the above equation will be for x having a value between 1 and 2 m. With x having such a value the $\{x-2\}$ bracket will be zero and the $\{x-1\}$ bracket can be written as $(x-1)$. Thus

$$46.7x^2 - 40(x-1)^2 - 0 - 2 \times 48.9 = 0$$
$$6.7x^2 - 80x - 97.8 = 0$$

The roots of a quadratic equation of the form $ax^2 + bx + c = 0$ are given by

$$x = \frac{-b \pm \sqrt{(b^2 - 4ac)}}{2a}$$

Thus for our equation

$$x = \frac{-80 \pm \sqrt{(80^2 + 4 \times 6.7 \times 97.8)}}{2 \times 6.7} = -5.95 \pm 7.09$$

Thus $x = 1.12\,\text{m}$ (only the positive value has any significance). The deflection at this value of x is obtained using equation [41].

$$y = -\frac{1}{EI}\left(\frac{R_1 x^3}{6} - \frac{40\{x-1\}^3}{6} - \frac{60\{x-2\}^3}{6} - 48.9x\right)$$

$$= -\frac{10^3}{20 \times 10^6}\left(\frac{46.7 \times 1.12^3}{6} - \frac{40 \times 0.12^3}{6} - 48.9 \times 1.12\right)$$

Hence $y = 2.19$ mm.

4.5 Statically-indeterminate beams

In the case of statically-determinate beams, such as the cantilevers or simply supported beams so far considered in this chapter, there are only two reaction forces and these can be found from the two equilibrium equations, i.e. the sum of the forces in any direction is zero and the sum of the moments about any axis is zero. A beam which is supported in such a way as to produce three or more reaction forces or moments is said to be *statically indeterminate* since the two equilibrium equations are insufficient to enable them to be determined. Examples of such beams are beams fixed at both ends, beams fixed at one end and supported at the other, and beams supported on three or more supports at the same level.

Statically-indeterminate beams thus require more than just the equilibrium equations for the reaction forces or moments to be determined. These can be provided by the equations arising from the deflection of the beam. Such equations can be obtained by the use of the methods described earlier in this chapter, namely the double integration method, superposition, the moment−area method or Macaulay's method.

4.5.1 Built-in beams

A beam is said to be *built-in* or *encastre* if both its ends are rigidly fixed so that the slope of the beam cannot change when the beam is subject to loads. Usually the slope at the ends is fixed to be zero, i.e. the ends are horizontal, and usually the ends are also at the same level.

Consider a simply supported beam (figure 4.17). The ends of such a beam are free to move and the effect of loading the beam is to make the ends rotate upwards and assume some slope to the horizontal. In order to make the ends of the beam lie flat and not rotate it is necessary to apply couples at each end. Thus for a built-in beam, loading will result in couples, called *fixed-end moments*, appearing at each end.

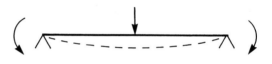

Fig. 4.17 Beam with end couples.

Derive, using the double integration method, the fixed-end moments and the maximum deflection for a beam of length L which is built-in at both ends and supports a uniformly distributed load of w per unit length over its full length.

Figure 4.18 shows the free body diagram for the beam, there being a fixed-end moment and a reaction force at each end. Because of the symmetry we have $M_1 = M_2$ and $R_1 = R_2 = \frac{1}{2}wL$. The bending moment at a section a distance x from the left end of the beam is thus

$$M = -M_1 + \tfrac{1}{2}wLx - \tfrac{1}{2}wx^2 \tag{42}$$

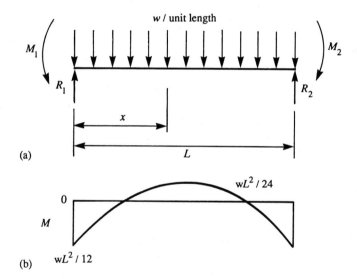

Fig. 4.18 Built-in beam with distributed load.

Thus

$$\frac{d^2y}{dx^2} = -\frac{M}{EI} = \frac{1}{EI}\left(M_1 - \tfrac{1}{2}wLx + \tfrac{1}{2}wx^2\right)$$

Integrating this equation gives

$$\frac{dy}{dx} = \frac{1}{EI}\left(M_1x - \frac{wLx^2}{4} + \frac{wx^3}{6} + A\right) \tag{43}$$

Integrating again gives

$$y = \frac{1}{EI}\left(\frac{M_1x^2}{2} - \frac{wLx^3}{12} + \frac{wx^4}{24} + Ax + B\right) \tag{44}$$

The boundary conditions are that when $x = 0$, $y = 0$ and $dy/dx = 0$, and when $x = L$, $y = 0$ and $dy/dx = 0$. Applying the first set of conditions to equations [43] and [44] gives $A = 0$ and $B = 0$. Now applying the $x = L$, $y = 0$ condition to equation [44] gives

$$0 = \frac{M_1L^2}{2} - \frac{wL^4}{12} + \frac{wL^4}{24}$$

Thus $M_1 = wL^2/12$. This value can be substituted into equation [42] to enable the bending moment to be determined for all values of x and so obtain the bending moment diagram (figure 4.18(b)). Two points of contra-flexure occur, at $x = 0.211L$ and $0.789L$.

Because the beam is symmetrical the maximum deflection will be at mid span. Thus equation [44] gives

$$y_{max} = \frac{1}{EI}\left(\frac{wL^2}{24} \times \frac{L^2}{4} - \frac{wL}{12} \times \frac{L^3}{8} + \frac{w}{24} \times \frac{L^4}{16}\right) = \frac{wL^4}{384EI}$$

Example

For the built-in beam considered in the previous example, use the principle of superposition to determine the fixed-end moments.

The slopes and deflections of the built-in beam can be considered to be the sum of those occurring when we consider the reaction forces alone acting on the beam and the fixed-end moments alone acting on the beam, as in figure 4.19. Because of the fixed ends, the sum of the slopes at the ends must be zero. The slope at the end of a simply supported beam subject to a uniformly distributed load is $+ wL^3/24EI$. The slope at the end of a beam subject to just the fixed-end couples is $- M_1L/2EI$, with $M_1 = M_2$. Thus, since these two slopes must add together to give zero slope, $M_1 = wL^2/12$.

Example

Derive, using the moment−area method, the fixed-end moment and the maximum deflection for a built-in beam of length L and carrying a concentrated load of F at mid span (figure 4.20(a)).

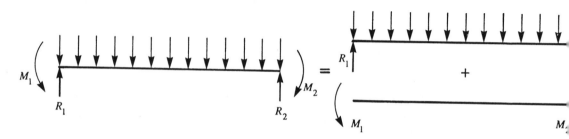

Fig. 4.19 Built-in beam with distributed load.

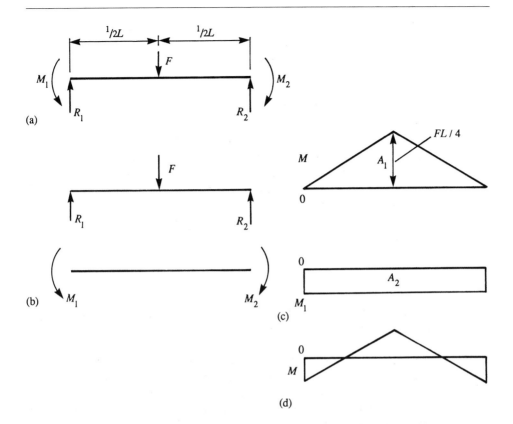

Fig. 4.20 Built-in beam with concentrated load.

We can obtain the bending moment diagram by using the superposition method of considering the built-in beam as being equivalent to a simply supported beam with a central load plus a beam subject to just the fixed-end moments, as in figure 4.20(b). Since the beam is symmetrical, the fixed-end moments at each end are the same, i.e. $M_1 = M_2$, and so the bending moment diagram for the beam subject to just the moments is a rectangle of height $-M_1$ (figure 4.20(c)). Since the beam is symmetrical, the bending moment diagram for the simply supported beam will be a triangle with a maximum ordinate of $FL/4$ (figure 4.20(c)). The composite bending moment diagram is thus as shown in figure 4.20(d).

For a horizontal built-in beam the angles of the beam at each end will remain the same when loading occurs. Thus the difference in angle between the tangents at each end of the beam will be zero. Thus the first moment–area theorem must give a total area under the M/EI graph of zero. For a beam of constant cross-section and material, this means that the total area under the bending moment graph is zero. Therefore, for figure 4.20(c) we must have area $A_1 =$ area A_2. Thus

$$\tfrac{1}{2}(FL/4)L = M_1 L$$

Hence $M_1 = FL/8$.

Because the beam is symmetrical, the maximum deflection will be at the mid span. Thus, taking the first moments of the areas about one end for half the beam gives, for each of the areas in the bending moment diagrams in figure 4.20(c) that add together to give the composite bending moment diagram in figure 4.20(d),

$$y_{max} = \Delta = \frac{1}{EI} \left(\frac{1}{2} \times \frac{FL}{4} \times \frac{L}{2} \times \frac{2}{3} \times \frac{L}{2} - M_1 \times \frac{L}{2} \times \frac{L}{2} \right)$$

Thus $y_{max} = FL^3/192EI$.

Note that because there is no change in the displacement between the two ends of the beam, the second moment−area theorem must give a total first moment of area of the total area under the M/EI graph of zero. For a beam of constant cross-section and material, this means that the first moment of the total area under the bending moment diagram must be zero.

Example

Derive, using Macaulay's method, the fixed-end moment and the maximum deflection for a built-in beam of length L and carrying a concentrated load of F at mid span (as in the previous example in figure 4.20(a)).

Taking the origin as the left end of the beam and considering the moments at a section XX a distance x from that end and near the right end of the beam, then

$$M = R_1 x - M_1 - F\{x - \tfrac{1}{2}L\}$$

Thus

$$\frac{d^2 y}{dx^2} = -\frac{M}{EI} = \frac{1}{EI}(-R_1 x + M_1 + F\{x - \tfrac{1}{2}L\})$$

Integrating gives

$$\frac{dy}{dx} = \frac{1}{EI}\left(-\frac{R_1 x^2}{2} + M_1 x + \frac{F\{x - \tfrac{1}{2}L\}^2}{2} + A \right) \qquad [45]$$

Integrating again gives

$$y = \frac{1}{EI}\left(-\frac{R_1 x^3}{6} + \frac{M_1 x^2}{2} + \frac{F\{x - \tfrac{1}{2}L\}^3}{6} + Ax + B \right) \qquad [46]$$

The boundary conditions are $y = 0$ when $x = 0$, hence $B = 0$; $dy/dx = 0$ when $x = 0$, hence $A = 0$; and when $x = L$ then $y = 0$ and $dy/dx = 0$. Using the last condition in equation [45] gives

$$0 = -\frac{R_1 L^2}{2} + M_1 L + \frac{F\tfrac{1}{4}L^2}{2}$$

Because the beam is symmetrical we also have, when the equations of equilibrium are applied, $R_1 = R_2 = \frac{1}{2}F$ and $M_1 = M_2$. Thus the above equation gives $M_1 = FL/8$.

Because the beam is symmetrical, the maximum deflection will occur when $x = \frac{1}{2}L$. Thus equation [46] gives

$$y_{max} = \frac{1}{EI}\left(-\frac{R_1}{6} \times \frac{L^3}{8} + \frac{M_1}{2} \times \frac{L^2}{4} + 0\right)$$

Hence, using the values of R_1 and M_1 obtained above, $y_{max} = FL^3/192EI$.

4.5.2 Propped cantilever

For a propped cantilever, one end of the beam is fixed and the beam, though supported at some other point, is free to rotate at that support. Thus, at the fixed end there can be a fixed-end moment and a reaction force, while at the other support the beam is subject to just a reaction force. When the beam is supported in a horizontal position, then the deflections at the support points are zero, with the slope being only necessarily zero at the fixed end.

Example

A uniform cantilever of length L is fixed at one end and freely supported at the same level at the other end. If the beam supports a uniformly distributed load of w per unit length, determine the reactions, the fixed-end moment and the deflection curve.

Figure 4.21 shows the beam and the forces acting on it. The bending moment M at a section a distance x from the fixed end is

$$M = R_1 x - M_1 - \tfrac{1}{2}wx^2$$

Hence

$$\frac{d^2y}{dx^2} = -\frac{M}{EI} = -\frac{1}{EI}(R_1 x - M_1 - \tfrac{1}{2}wx^2)$$

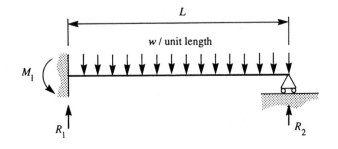

Fig. 4.21 Propped cantilever.

Integrating this equation gives

$$\frac{dy}{dx} = -\frac{1}{EI}\left(\frac{R_1x^2}{2} - M_1x - \frac{wx^3}{6} + A\right) \tag{47}$$

Integrating again gives

$$y = -\frac{1}{EI}\left(\frac{R_1x^3}{6} - \frac{M_1x^2}{2} - \frac{wx^4}{24} + Ax + B\right) \tag{48}$$

The boundary conditions are that $y = 0$ and $dy/dx = 0$ at $x = 0$. Thus $A = 0$ and $B = 0$. We also have $y = 0$ at $x = L$. Thus equation [48] gives

$$\frac{R_1L^3}{6} - \frac{M_1L^2}{2} - \frac{wL^4}{24} = 0$$

We also have the equilibrium condition that, for moments about the propped end,

$$M_1 = -wL \times \tfrac{1}{2}L + R_1L$$

Hence

$$\frac{R_1L^3}{6} - \left(-\frac{wL^2}{2} + R_1L\right)\frac{L^2}{2} - \frac{wL^4}{24} = 0$$

Thus $R_1 = 5wL/8$ and so $M_1 = wL^2/8$. The deflection curve for the beam can be found by substituting these values in equation [48]. Thus

$$y = -\frac{1}{EI}\left(\frac{5wLx^3}{48} - \frac{wL^2x^2}{16} - \frac{wx^4}{24}\right)$$

The reaction R_2 can be found from the equilibrium condition that

$$R_1 + R_2 = wL$$

Hence $R_2 = 3wL/8$.

Example

Determine, using superposition, the reaction forces and the fixed-end moment for the cantilever shown in figure 4.22.

Using the principle of superposition, we can consider the deflection curve to be the sum of the deflection curves due to a cantilever with a concentrated force F a distance a from the fixed end plus that of a cantilever with an upwards directed force of R_2 at the free end. Thus, for $a \le x \le L$,

$$y = \frac{Fa^2}{6EI}(3x - a) - \frac{R_2x^2}{6EI}(3L - x)$$

We have the condition that at $x = L$ that $y = 0$. Thus

$$Fa^2(3L - a) = R_2L^2 \times 2L$$

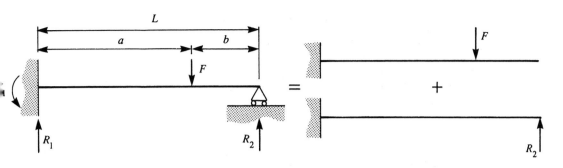

Fig. 4.22 Propped cantilever.

$$R_2 = \frac{Fa^2}{2L^3}(3L - a)$$

For equilibrium we must have $R_1 + R_2 = F$. Hence

$$R_1 = F - \frac{Fa^2}{2L^3}(3L - a)$$

If moments are taken about the propped end, then for equilibrium we have

$$M_1 = R_1L - Fb = \frac{Fa^2}{2L^3}(3L - a) - Fb$$

Example

Determine, using the moment–area method, the reaction forces and the fixed-end moment for the cantilever shown in figure 4.23.

We can consider the propped cantilever to give a bending moment diagram which is the sum of that due to a cantilever with a concentrated load F a distance a from the fixed end and a cantilever with a concentrated upwards load of R_2 at a distance of L from the fixed end (see the previous example and figure 4.22). Figure 4.23 shows the bending moment diagrams. Since both the ends of the beam suffer no deflections, then the second moment–area theorem indicates that the total of the first moments of area of the bending moment diagram must be zero. Thus

$$\tfrac{1}{2}R_1L \times L \times 2L/3 = \tfrac{1}{2}Fa \times a(b + 2a/3)$$

Hence

$$R_1 = \frac{3Fa^2}{2L^3}\left(b + \frac{2a}{3}\right) = \frac{Fa^2}{2L^3}(2L + b) = \frac{Fa^2}{2L^3}(3L - a)$$

Since for equilibrium we have $R_1 + R_2 = F$, then

$$R_2 = F - \frac{Fa^2}{2L^3}(3L - a)$$

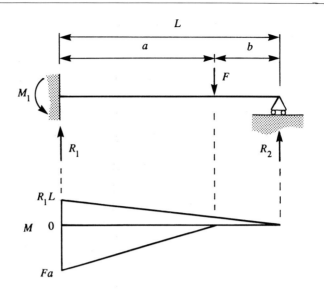

Fig. 4.23 Propped cantilever.

If moments are taken about the propped end, then for equilibrium we have

$$M_1 = R_1L - Fb = \frac{Fa^2}{2L^3}(3L - a) - Fb$$

Problems

(1) A cantilever has a length of 3.0 m and a concentrated load of 40 kN applied at its free end. If the beam has a second moment of area about the neutral axis of $300 \times 10^6 \, \text{mm}^4$ and a tensile modulus of 200 GPa, what will be the maximum deflection and maximum slope?

(2) A cantilever has a length of 2.0 m and carries a uniformly distributed load of 20 kN/m. If the beam has a second moment of area of $20 \times 10^6 \, \text{mm}^4$ and an elastic modulus of 200 GPa, what will be the slope and deflection at the free end?

(3) A simply supported beam of length 5.0 m carries a concentrated load of 50 kN at mid span. What will be the maximum deflection of the beam if it has a second moment of area of $150 \times 10^6 \, \text{mm}^4$ and an elastic modulus of 200 GPa?

(4) A simply supported beam of length 5.0 m carries a concentrated load of 40 kN at mid span. What will be the deflections at 1.0 m and 3.0 m from one end? The beam has a second moment of area of $100 \times 10^6 \, \text{mm}^4$ and an elastic modulus of 200 GPa.

(5) A simply supported beam has a length of 5.0 m and has a circular cross-section of diameter 100 mm. What will be the maximum de-

flection when a concentrated load of 1 kN is applied mid span? The material has a modulus of 200 GPa.

(6)   What is the maximum deflection of a simply supported beam of length 1.5 m when a concentrated load of 20 kN is applied 1.0 m from one support? The beam has a tensile modulus of 200 GPa and a second moment of area of 10×10^6 mm^4.

(7)   Derive the equation for the deflection curve of a cantilever which supports a distributed load which linearly decreases from w_0 at the fixed end to zero at the free end.

(8)   Derive the equations for the deflection curve of a beam of length $1.5L$ and which has a distance between supports of L and $0.5L$ overhang and a concentrated load of F at the free end.

(9)   Derive the equations for the deflection curve of a cantilever which has a uniformly distributed load of w per unit length over a distance a from the fixed end and no load over the remaining part of the beam.

(10)   Derive the equation for the deflection at the free end of a cantilever, length L, which has a uniformly distributed load of w per unit length over a distance a from the fixed end and a concentrated load of F at the free end. (Hint: the answer to problem 9 can help.)

(11)   Determine the maximum deflection of a simply supported beam which has a uniformly distributed load of 10 kN/m and a mid-span concentrated load of 70 kN. The beam has a span of 2.0 m, an elastic modulus of 200 GPa and a second moment of area of 100×10^6 mm^4.

(12)   Determine the equations for the deflection curve of a simply supported beam of length L when it is subject to two concentrated loads, each being F, with one load being a distance a and the other $(a + b)$ from one end. $L = 2a + b$.

(13)   Determine the maximum deflection of a cantilever of length L carrying two concentrated loads of F at distances of $L/3$ and $2L/3$ from the fixed end.

(14)   A simply supported uniform beam AB of length L has a concentrated load at mid span. Use the moment–area method to determine the slope of the beam at a point $3L/4$ from A.

(15)   A uniform cantilever of length L is subject to a uniformly distributed load over its entire length of w per unit length. Use the moment–area method to determine the deflection at the free end.

(16)   A cantilever of length $3L$ has a concentrated load of F at the free end. The length $2L$ nearest to the fixed end has a first moment of area of $2I$ and the length L nearest to the free end a first moment of

area of I. Use the moment–area method to determine the deflection at the free end.

(17) A simply supported shaft has a span of L. The central half of the shaft has a first moment of area of $2I$ with the quarters at each end having a first moment of area of I. Use the moment–area method to determine the deflection at mid span.

(18) A uniform simply supported beam AB has a length of 8.0 m and has a distributed load of 20 kN/m over a 4.0 m length from end A and a concentrated load of 120 kN at 6.0 m from the same end. The beam has a value of EI of 100 MN m^2. Use Macaulay's method to obtain an expression for the deflection at any distance x from end A and hence the deflection at mid span.

(19) A uniform cantilever of length L has no load over a from the fixed end and a uniformly distributed load of w over the remaining length b, i.e. $L = a + b$. Determine the Macaulay expression for the deflection curve of the beam.

(20) A uniform simply supported beam of length L has a concentrated load of F a distance of a from one end and b from the other. Determine the Macaulay expression for the deflection curve of the beam.

(21) A uniform simply supported beam of length L has a concentrated load of F a distance a from one end and another concentrated load of F a distance a from the other end. Determine the Macaulay expression for the deflection curve of the beam.

(22) A uniform cantilever of length 4.0 m has a uniformly distributed load of 10 kN/m over a length of 2.0 m from the fixed end. The beam has a value of EI of 20 MN m^2. Use Macaulay's method to obtain the deflection of the free end.

(23) A uniform simply supported beam AB has a length of 8.0 m and carries a concentrated load of 20 kN at 2.0 m from A and a uniformly distributed load of 5 kN/m over the length from the mid-span point to B. The beam has a value of EI of 20 MN m^2. Use Macaulay's method to obtain the deflection at mid span.

(24) A uniform beam AB of length 8.0 m rests symmetrically on two supports, the supports being 2.0 m from each end of the beam. Concentrated loads of 40 kN are applied at each end, A and B, of the beam. The beam has a value of EI of 20 MN m^2. Use Macaulay's method to determine the deflections of the ends A and B.

(25) A uniform simply supported beam AB has a span of 8.0 m and has concentrated loads of 15 kN at 1.5 m from A and 20 kN at 3.5 m from A. At what position along the beam will the maximum deflection occur and what is its value? The beam has $EI = 20$ MN m^2.

(26) A uniform simply supported beam AB has a span of 5.0 m and is subject to a uniformly distributed load of 5 kN/m over a 2.0 m length from A and a concentrated load of 10 kN at 4.0 m from A. At what position along the beam will the maximum deflection occur and what is its value? The beam has $EI = 20\,\text{MN m}^2$.

(27) A uniform beam AB of length 8.0 m is supported at A and a point 6.0 m from A. The beam is subject to concentrated loads of 50 kN at a point 2.0 m from A and 40 kN at the overhanging end B. At what position along the beam will the maximum deflection occur and what is its value? The beam has $EI = 20\,\text{MN m}^2$.

(28) Derive, using (a) the moment−area method and (b) the Macaulay method, the fixed-end moments and the deflection under the load for a built-in beam, of length L, supporting a concentrated load F a distance a from one end.

(29) A built-in beam of length L has a uniformly distributed load of w/unit length over the left half. Determine, using (a) the moment−area method and (b) the Macaulay method, the mid-span deflection of the beam.

(30) A horizontal uniform built-in beam of length 5.0 m supports two concentrated loads of 50 kN at 0.75 m from each end. Determine the reactions, the fixed-end moments and the mid span deflection. The beam has $EI = 20\,\text{MN m}^2$.

(31) A horizontal uniform built-in beam of length 5.0 m carries a concentrated load of 100 kN at mid span. Determine the reactions, the fixed-end moments and the mid span deflection. The beam has $EI = 20\,\text{MPa}$.

(32) A horizontal uniform beam is built-in at one end and is propped at the same level a distance a from the fixed end and b from the free end. Determine the reaction force in the prop support if the beam has a distributed load over its entire length of w/unit length?

(33) A horizontal uniform beam of length $2L$ is built-in at one end and supported at the same level on a prop a distance L from the fixed end. What are the reaction forces and the fixed-end moment of the beam when it is subject to a concentrated load of F at the free end?

(34) A horizontal uniform beam of length $3L$ is built-in at one end and supported at the same level on a prop at the far end. Concentrated loads of F are supported at distances of L and $2L$ from the fixed end. What are the reaction forces and the fixed-end moment of the beam?

Chapter 5
Struts and columns

5.1 Buckling

A structural member which is long and slender and subject to compressive forces is called a *strut*. The term *column* is however often used when they are vertical members. Examples of these types of members are the connecting rod of an internal combustion engine or the vertical structural steel work columns used in buildings.

In chapter 1 struts were considered as structural members and in chapter 2 the effect of axial compressive loads on such members was considered. If such a member is subject to only axial compressive forces then the effect is just to reduce the length of the strut and yielding will occur when the compressive stress reaches the yield stress. However with long and slender members the loading may cause the member to bend and deflect in a transverse direction. Under an increasing compressive axial load the transverse deflections increase and eventually the strut can collapse completely. This is called *buckling* and can occur at compressive loads which are below that at which direct yielding of the strut would occur. For example, if you step on an empty vertical aluminium drink can it collapses by buckling.

Buckling leads to a sudden dramatic collapse of a structure and care has to be taken in the design of structures involving struts that they can safely support their loads without buckling. The maximum load that a strut can support without buckling is called the *critical load*.

5.1.1 *Equilibrium*

To illustrate the phenomena of buckling, consider the idealised structure shown in figure 5.1. It consists of two weightless bars AB and BC that are rigid and pinned at their ends. The middle joint at B is tethered by a horizontal spring. When the two bars are in the same straight line and vertical the spring is in the unstretched state. A force F is applied to compress the structure. Now suppose the joint at B is displaced sideways by a small amount y. The applied load F is still acting in a vertical line and so there is a force F acting vertically at B. This produces a clockwise moment about C of Fy. The displacement y results in the spring being compressed and so giving a restoring force of ky, where k is the spring constant. This force produces an anticlockwise moment about C of $ky \times \frac{1}{2}kL$. If this restoring moment is greater than the displacement moment, i.e. $\frac{1}{2}kyL > Fy$

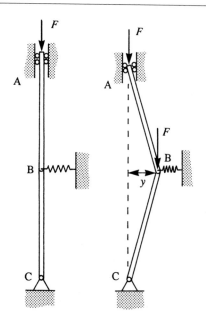

Fig. 5.1 Idealised model of a strut.

or $F < \frac{1}{2}kL$ then the system will return to its initial stable, straight-line, position. If, however, the restoring moment is less than the displacement moment, i.e. $\frac{1}{2}kyL < Fy$ or $F > \frac{1}{2}kL$, then the system is unstable and bars will continue to rotate and collapse result. The intermediate value, when the restoring moment is equal to the displacement moment, i.e. $\frac{1}{2}kyL = Fy$ or $F = \frac{1}{2}kL$, will result in the bars remaining in their deflected position and the system can be said to be in neutral equilibrium. This force of $\frac{1}{2}kL$ marks the transition from instability to stability and is called the critical load.

The above can be considered to be an idealised model of a strut which bends under the action of an axial force. The restoring force in this case is provided by the elastic nature of the beam. Up to the critical load the strut is stable. Above the critical load it becomes unstable and buckling occurs.

5.2 Euler's equation

In considering the buckling of columns we will begin by considering a slender strut with pinned ends (figure 5.2). The strut, of length L, is homogeneous and initially perfectly straight. Axial compressive forces are applied, the axis passing through the centroid of the strut's cross-section. The strut material is assumed to obey Hooke's law.

When bending occurs, bending moments are developed in the column. If the transverse deflection of the strut is y at a distance x from one end, then using equation [3] of chapter 4,

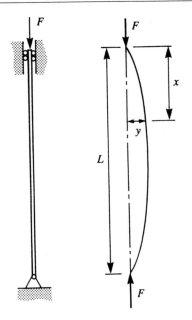

Fig. 5.2 Strut with pinned ends.

$$\frac{d^2y}{dx^2} = -\frac{M}{EI}$$

But, for the bent column, $M = Fy$. Thus

$$\frac{d^2y}{dx^2} = -\frac{Fy}{EI} \qquad [1]$$

For convenience in the writing of the equations that follow, let $K^2 = F/EI$. Then the equation can be written as

$$\frac{d^2y}{dx^2} + K^2y = 0 \qquad [2]$$

Such a second-order differential equation has the general solution

$$y = A\cos Kx + B\sin Kx$$

where A and B are the constants of integration. The boundary conditions are $y = 0$ when $x = 0$, and $y = 0$ when $x = L$. Using the first condition in equation [3] gives, since $\sin 0 = 0$, that $A\cos 0 = 0$. Since $\cos 0 = 1$, then we must have $A = 0$. Thus

$$y = B\sin Kx \qquad [3]$$

Using the second boundary condition, then $B\sin KL = 0$. Thus, for $\sin KL$ to be 0, we must have

$$KL = 0, \pi, 2\pi, \ldots n\pi$$

Hence $K = n\pi/L$. Thus, since $K^2 = F/EI$,

$$F = \frac{n^2\pi^2 EI}{L^2}$$
[4]

where $n = 0, 1, 2, 3, \dots$. Only when the force F has one of the values given by equation [4] is it possible for the strut to have the bent shape. If $n = 0$ then $K = 0$ and then equation [3] indicates that $y = 0$. The strut will not buckle. The smallest value of F which will cause buckling is when $n = 1$, thus the critical load F_c for the column is

$$F_c = \frac{\pi^2 EI}{L^2}$$
[5]

This critical load is sometimes referred to as the *Euler load* and the equation as the *Euler equation*.

The critical load of a strut is proportional to the flexural rigidity EI. Thus the critical load can be increased by selecting a stiffer material, i.e. one with a higher modulus of elasticity E. Note that the critical load is not determined by the strength of the material, a stronger material will not necessarily buckle at higher loads. Since the critical load is proportional to the second moment of area I, then increasing I will increase the critical load. The second moment of area is increased by distributing the material away from the centroid of the cross-section. Hence, as struts, tubes are more economical than solid rods. If the cross-section has second moments of area which differ in the two directions at right angles to the longitudinal axis of the strut, then the strut will buckle in the direction for which the second moment of area is the least.

The equation of the deflection curve, i.e. equation [3], is

$$y = B \sin\frac{n\pi x}{L}$$
[6]

The deflection curve at the smallest critical load is when $n = 1$ and corresponds to a deflection curve with a shape of half a sine wave (figure 5.3). With $n = 2$ the deflection curve is that of a full sine wave. With $n = 3$ the deflection curve is that of $1\frac{1}{2}$ sine waves. Unless the strut is restrained it will not go beyond the lowest critical load value, i.e. when $n = 1$, since it will have buckled.

For a strut which obeys Euler's equation, as the force is increased from zero there is no deflection until the critical load is reached. Up to that point $n = 0$ and the beam is in stable equilibrium. At the critical load buckling starts and there is a deflection. The beam is in neutral equilibrium and the deflection can take any value. The load−deflection graph is of the form shown in figure 5.4.

Example

A 8.0 m length of straight steel tubing has an inner diameter of 55 mm and

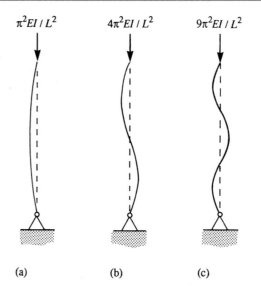

$\pi^2 EI / L^2$ $4\pi^2 EI / L^2$ $9\pi^2 EI / L^2$

(a)   (b)   (c)

Fig. 5.3 Buckled shapes for (a) $n = 1$, (b) $n = 2$, (c) $n = 3$.

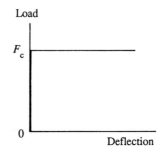

Fig. 5.4 Load–deflection diagram for a strut obeying Euler's equation.

an outer diameter of 60 mm. What will be the maximum allowable load the tube can support without buckling when used as a pin-ended column? The material has an elastic modulus of 200 GPa.

The second moment of area I of the tube is $\pi(D^4 - d^4)/64$ (see figure 3.17). Thus, assuming Euler's equation is valid, equation [5] gives

$$F_c = \frac{\pi^2 EI}{L^2} = \frac{\pi^2 \times 200 \times 10^9 \times \pi(0.060^4 - 0.055^4)}{64 \times 8.0^2}$$

$$= 5.77 \, \text{kN}$$

Example

A straight strut has a length of 2.0 m and a rectangular cross-section of 12 mm by 5.0 mm and is subject to an axial load. Assuming that Euler's equation applies, what will be the maximum central deflection of the strut before it reaches the yield point of 280 MPa? The modulus of elasticity is 72 GPa.

According to the Euler equation, there will be no deflection until the critical load is reached. Buckling will occur in the direction determined by the smallest second moment of area. Thus the critical load is (equation [5])

$$F_c = \frac{\pi^2 EI}{L^2} = \frac{\pi^2 \times 72 \times 10^9 \times 0.012 \times 0.005^3}{2.0^2 \times 12} = 22.2\,\text{N}$$

The maximum bending moment occurs at the centre of the strut and is $Fy = 22.2\delta$, where δ is the central deflection. The maximum stress is the sum of the direct stress (F_c/A) and bending stress (My/I) at the centre. Thus

$$280 \times 10^6 = \frac{22.2}{0.012 \times 0.005} + \frac{22.2\delta \times 0.0025 \times 12}{0.012 \times 0.005^3}$$

Hence $\delta = 0.631\,\text{m}$.

5.2.1 *Support conditions*

So far only struts or columns with both ends pinned has been considered. Other support conditions are however possible. Basically the four possible conditions can be considered to be:

1 both ends pinned;
2 one end fixed, one end free;
3 both ends fixed;
4 one end fixed, one end pinned.

The critical loads for such struts can be determined from the differential equation $d^2y/dx^2 = -M/EI$ in the same way as with the pinned joint. The critical loads so derived can then be related to the critical load of a strut pinned at both ends through the concept of an *effective length*.

 Consider the deflected shape of a strut which has one end fixed and one free (figure 5.5(a)). It buckles into a shape which is one quarter of a full sine wave. The strut pinned at both ends buckles into a shape which is one half of a full sine wave. Thus the effective length L_e for the fixed–free strut is twice the length L of the equivalent pin-ended strut. The deflected shape of a strut which has both ends fixed (figure 5.5(b)) is that of a full sine wave. Thus the effective length is half the length of the equivalent pin-ended strut. The deflected shape of a strut with one end fixed and the other pinned (figure 5.5(c)) is such that the effective length is $0.899L$.

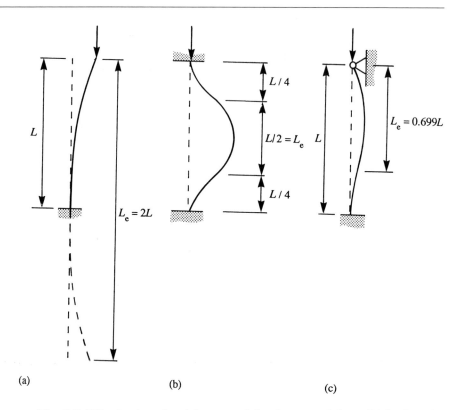

(a) (b) (c)

Fig. 5.5 Effective lengths, (a) one end fixed, one end free, (b) both ends fixed, (c) one end fixed, one end pinned.

Since the effective length is the length of the equivalent pin-ended strut, then the critical load is, in general,

$$F_c = \frac{\pi^2 EI}{L_e^2} \qquad [7]$$

Example

What is the critical load for a strut of length 2.0 m and flexural rigidity 20 MPa if both ends are fixed?

Using equation [7] with the effective length as $\frac{1}{2}L = 1.0$ m,

$$F_c = \frac{\pi^2 EI}{L_e^2} = \frac{\pi^2 \times 20 \times 10^6}{2.0^2} = 49.3 \, \text{MN}$$

Example

A pipe column of external diameter 100 mm supports a beam carrying a concentrated load of 100 kN as shown in figure 5.6. The pipe is fixed at its

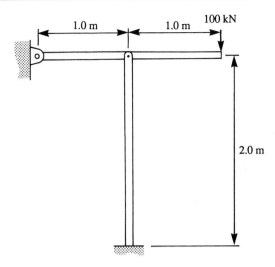

Fig. 5.6 Example.

lower end and pinned at the beam. What thickness wall should the pipe have if buckling is not to occur? The pipe material has an elastic modulus of 70 GPa.

Taking moments about the fixed end of the beam gives

$$1.0 \times F = 2.0 \times 100$$

Thus F, the reaction force at the column and hence the compressive force acting internally on it, is 200 kN. The second moment of area of the pipe is $\pi(D^4 - d^4)/64$ and its effective length is $0.699 \times 2.0 = 1.398$ m. Thus, assuming Euler's equation holds

$$F_c = 200 \times 10^3 = \frac{\pi^2 EI}{L_e^2} = \frac{\pi^2 \times 70 \times 10^9 \times \pi(0.100^4 - d^4)}{1.398^2 \times 64}$$

Thus $d = 0.0970$ m. Hence the minimum wall thickness is 0.0030 m.

5.3 Limitations of Euler's equation

The Euler equation for the critical load can be written in terms of the stress σ acting over the cross-sectional area A of a strut as

$$F_c = \sigma_c A = \frac{\pi^2 EI}{L_e^2} \tag{8}$$

where σ_c is the critical stress which will cause buckling.

The Euler equation only gives the maximum force that a strut can withstand provided the yield stress is not exceeded. Up to the yield stress the strut will fail by buckling, at the yield stress the strut will fail by crushing. Thus the maximum buckling loading F_c will be that which gives the yield stress σ_y.

The second moment of area of any section can be expressed in the form

$$I = Ak^2 \tag{9}$$

where A is the cross-sectional area and k is called the *radius of gyration*. This radius of gyration can be considered to be the distance from the axis at which the entire area could be concentrated and still have the same second moment of area as the original section area. The term *slenderness ratio* is used for the ratio of the length L of a strut to its radius of gyration k, i.e.

$$\text{slenderness ratio} = \frac{L}{k} \tag{10}$$

The term *effective slenderness ratio* is used for the ratio of the effective length L_e to its radius of gyration k.

Equation [8] can thus be written as

$$F_c = \sigma_c A = \frac{\pi^2 EI}{L_e^2} = \frac{\pi^2 EAk^2}{L_e^2} = \frac{\pi^2 EA}{(L_e/k)^2} \tag{11}$$

Figure 5.7 shows a graph of the stress at failure plotted against the effective slenderness ratio (L_e/k). Buckling is the failure mode up to the yield stress when crushing becomes the mode.

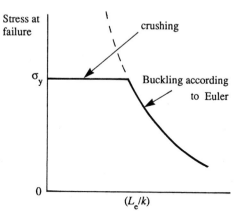

Fig. 5.7 Stress at failure.

The slenderness ratio is thus

$$\text{slenderness ratio} = \sqrt{\left(\frac{\pi^2 E}{\sigma_c}\right)} \tag{12}$$

with the limiting, minimum, slenderness ratio for which buckling occurs being when $\sigma_c = \sigma_y$, the yield stress. For a mild steel strut with $E = 200\,\text{GPa}$ and $\sigma_c = 240\,\text{MPa}$, this slenderness ratio is about 91. For slenderness ratios less than 91 crushing is the failure mode, for values above it buckling is the mode.

The above discussion can be refined by a more careful consideration of the elastic and inelastic behaviour of the material. Figure 5.8 shows the resulting stress at failure graph. The term *long* is used for a column or strut for which the slenderness ratio is such that Euler's equation is valid. The term *short* is used when the material fails by yielding or crushing. The term *intermediate* is used for columns which have too small a slenderness ratio for Euler's equation to hold and buckling is the governing mode of failure and too large for strength considerations alone to govern.

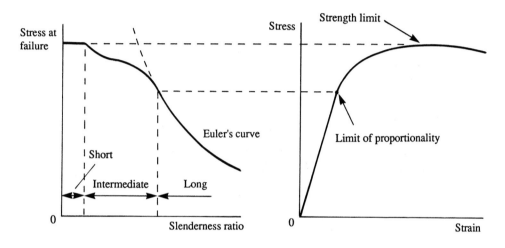

Fig. 5.8 Stress at failure.

Example

A structural steel universal beam section 254 mm by 254 mm has a sectional area of $2.12 \times 10^4 \, \text{mm}^2$ and a minimum second moment of area of $97.96 \times 10^6 \, \text{mm}^4$. If the steel has a yield stress of 250 MPa, what is the minimum slenderness ratio and the minimum length such a beam can have if failure is to be by buckling rather than crushing? The steel has a modulus of elasticity of 200 GPa.

Using equation [12]

$$\text{slenderness ratio} = \sqrt{\left(\frac{\pi^2 E}{\sigma_c}\right)} = \sqrt{\left(\frac{\pi^2 \times 200 \times 10^9}{250 \times 10^6}\right)} = 88.9$$

Since the second moment of area $= Ak^2$, the minimum length is given by

$$\text{slenderness ratio} = 88.9 = \frac{L_e}{k^2} = \frac{L_e}{97.96 \times 10^{-6}/2.12 \times 10^{-2}}$$

Hence $L_e = 0.411 \, \text{m}$.

5.3.1   Rankine–Gordon equation

The stress at failure graph in figure 5.7 shows that up to a certain slenderness ratio failure will occur by crushing and at higher slenderness ratio values by buckling. A general equation which reasonably describes the failure graph is

$$\frac{1}{\sigma} = \frac{1}{\sigma_y} + \frac{1}{\sigma_c} \qquad [13]$$

where σ is the actual stress to cause failure, σ_y the yield stress, i.e. the stress to cause failure by crushing, and σ_c is the stress to cause failure by buckling and which has been determined by the use of the Euler equation on the assumption that its validity applies to the strut concerned. Equation [13] is known as the *Rankine–Gordon* equation.

This equation can be rearranged to give

$$\sigma = \frac{\sigma_y \sigma_c}{\sigma_c + \sigma_y} = \frac{\sigma_y}{1 + \sigma_y/\sigma_c}$$

But equation [11] gives

$$\sigma_c = \frac{\pi^2 E}{(L_e/k)^2}$$

Hence

$$\sigma = \frac{\sigma_y}{1 + (\sigma_y/\pi^2 E)\,(L_e/k)^2} \qquad [14]$$

The term $(\sigma_y/\pi^2 E)$ depends on the material used and is replaced by a constant a, called the *Rankine constant*. The experimentally determined value for a may differ from the theoretical value to take account of imperfections.

$$\sigma = \frac{\sigma_y}{1 + a(L_e/k)^2} \qquad [15]$$

A typical value of a for mild steel is $1/7500$. In some cases a is quoted for a material, such as steel, with different values for different forms of end-connection. In such cases, the slenderness ratio used in equation [15] is the actual length divided by the radius of gyration. The value of $1/7500$ quoted above thus becomes the value for a pinned strut. For a fixed-end strut, where $L_e = \frac{1}{2}L$, then the value is $4 \times 1/7500$. Figure 5.9 shows how the graph given by the Rankine equation for a steel compares with that given by Euler and the yield stress criteria.

Frequently the requirement is not for the stress which would cause failure but for the maximum permissible working stress for a strut or column. Equation [15] then uses, instead of the yield stress, the maximum allowable stress for the material. This is generally the yield stress divided by some factor of safety.

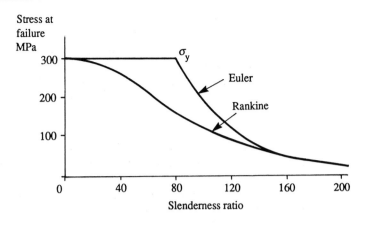

Fig. 5.9 Rankine equation.

Other equations have been developed to relate the stress at failure to the slenderness ratio. In some cases the equations incorporate a safety factor so that the allowable stress for a column can be predicted. Tables are also available which give the allowable stress as a function of the slenderness ratio.

Example

A pinned timber column has a rectangular cross-section 100 mm by 50 mm and a length of 2.0 m. Using the Rankine equation, with $a = 1.3 \times 10^{-3}$, what is the compressive maximum load which can be used if the allowable compressive stress for the timber is 30 MPa?

The second moment of area of the section is

$$I = \frac{bd^3}{12} = Ak^2 = bdk^2$$

Hence the smallest radius of gyration for the section is $\sqrt{(d^2/12)} = \sqrt{(0.050^2/12)}$ m. Using equation [15]

$$\sigma = \frac{\sigma_y}{1 + a(L_e/k)^2} = \frac{30 \times 10^6}{1 + 1.3 \times 10^{-3}\,(2.0^2 \times 12/0.050^2)}$$
$$= 1.16\,\text{MPa}$$

The maximum allowable force is thus $1.16 \times 10^6 \times 0.050 \times 0.010 = 580\,\text{N}$.

Problems

(1)   Assuming Euler's equation holds, what is the buckling load for a column of length 3.0 m and 20 mm in diameter when both ends are pin-jointed? The modulus of elasticity is 200 GPa.

(2) Assuming Euler's equation holds, what is the maximum central deflection that will occur for a strut of length 2.0 m and diameter 20 mm with both ends pinned if the material has a yield stress of 300 MPa? The modulus of elasticity is 200 GPa.

(3) A hollow rectangular section column of length 4.0 m has external dimensions of 100 mm by 50 mm and a wall thickness of 10 mm. What will be the critical axial load if the material has a modulus of elasticity of 200 GPa? Assume Euler's equation to apply and the column to be pin-jointed at both ends.

(4) What is the critical load for an I-section pin-ended strut of length 3.0 m if it has second moments of area of $I_x = 95 \times 10^6 \, \text{mm}^4$ and $I_y = 31 \times 10^6 \, \text{mm}^4$? Assume that Euler's equation can be used. The modulus of elasticity is 200 GPa.

(5) What is the critical load for a rectangular section bar 40 mm by 50 mm and length 3.0 m when it is pinned at both ends and subject to axial compressive forces? The modulus of elasticity is 200 GPa.

(6) What is the critical load for a universal section of length 5.0 m when used as a pin-ended strut if the section has second moments of area of $I_x = 299 \times 10^6 \, \text{mm}^4$ and $I_y = 98 \times 10^6 \, \text{mm}^4$? The modulus of elasticity is 200 GPa.

(7) Derive an equation for the increase in temperature needed to produce buckling for a pin-ended strut of length L, cross-sectional area A, coefficient of linear expansion α, and second moment of area I.

(8) What is the critical load for a strut of length 2.0 m and 20 mm in diameter when both ends are fixed? The material has a modulus of elasticity of 200 GPa.

(9) What is the critical load for a strut of length 2.0 m, and square cross-section of side 10 mm, if it is built-in at both ends? The material has a modulus of elasticity of 200 GPa.

(10) What is the critical stress for a pinned strut of length 2.0 m and square cross-section of side 50 mm? The material has a modulus of elasticity of 200 GPa.

(11) A steel column is an I-section with a length of 6.0 m and second moments of area of $387 \times 10^6 \, \text{mm}^4$ and $125 \times 10^6 \, \text{mm}^4$. The steel has a modulus of elasticity of 200 GPa. What is the critical load when (a) both ends are pinned, (b) both ends are fixed, (c) one end is fixed and the other free, (d) one end is fixed and the other pinned?

(12) Determine the maximum uniformly distributed load that can be applied to the horizontal member in figure 5.10 if the vertical pinned strut is not to buckle. The vertical strut has a length of 2.0 m and a rectangular cross-section of 20 mm by 30 mm and a modulus

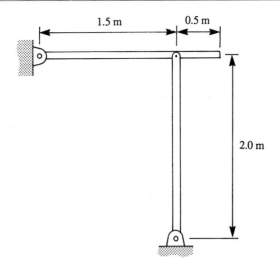

Fig. 5.10 Problem 12.

of elasticity of 200 GPa. What would the result be if the vertical strut had been fixed at the lower end and pinned at the upper end?

(13) What is the minimum slenderness ratio for a column 6.0 m long and with a rectangular cross-section of 200 mm by 250 mm?

(14) A pinned steel bar has a rectangular cross-section 40 mm by 50 mm. The steel has a yield stress of 240 MPa and a modulus of 200 GPa. What is the minimum length for which Euler's equation can be used to determine the failing load?

(15) For a steel strut of circular cross-section with one end fixed and the other pinned, show that Euler's equation cannot be applied if the length is less than thirty two times the strut diameter. The steel has a yield stress of 250 MPa and a modulus of 200 GPa.

Note that in the following problems the Rankine constant values have been quoted for the effective length.

(16) A tubular cast iron column with an external diameter of 250 mm and a length of 5.0 m has fixed ends. What should be the tube thickness if it is to support a maximum compressive load of 1 MN? The cast iron has a maximum working stress of 80 MPa. Use the Rankine equation with $a = 1/1600$.

(17) A tubular cast iron column has an external diameter of 200 mm and an internal diameter of 180 mm and a length of 5.0 m. The column is rigidly fixed at both ends. What will be the maximum permissible compressive load if the cast iron has a maximum working stress of 80 MPa? Use the Rankine equation with $a = 1/1600$.

(18) A steel strut with built-in ends has a rectangular cross-section with a

width twice the thickness and a length of 2.0 m. The maximum allowable stress for the material is 120 MPa. What will be the dimensions of the section if the Rankine constant for the steel is 1/7500 and the strut is to withstand a load of 80 kN?

(19)   A built-up section is made by rigidly joining two 200 mm × 75 mm channel sections back-to-back 100 mm apart with two 250 mm by 12.5 mm thick plates riveted to the flanges. The channel sections have second moments of area about the axis at right angles to the web of 19×10^6 mm^4 and about the parallel axis of 1.5×10^6 mm^4, and a centroid which is 21 mm from the base of the web. A 6.0 m length of the section is to be used as a column with fixed ends. What is the maximum working load that can be carried by the column if the material has a yield stress of 300 MPa, a Rankine constant a of 1/7500 and a safety factor of 2.5 is to be used?

Chapter 6
Torsion

6.1 Torsion of circular shafts

The twisting of a shaft about its longitudinal axis, as a result of the application of a torque, is called *torsion*. With a circular shaft the twisting is said to be *pure torsion* because the cross-section of the shaft retains its shape, all circular sections remain circular during the twisting with diameters unchanged. A torque applied to one end of a shaft twists it and this twisting action is communicated along the shaft. Thus if one end of a shaft is rotated as a result of a torque the twisting action transmits power along the shaft.

During the twisting of a circular shaft there will be rotation about the longitudinal axis of one end of the shaft with respect to the other, as illustrated in figure 6.1. Because of this rotation a longitudinal line on the surface of the shaft will rotate through a small angle ϕ. Because the angle is small, to a reasonable approximation we can take arc AB to be equal to $L\phi$ (arc length of a circle = radius × angle subtended by the arc). But we also have arc AB = $r\theta$, where r is the radius of the shaft. Hence

$$L\phi = r\theta$$

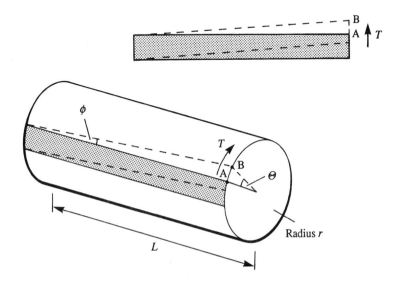

Fig. 6.1 Torsion of a circular shaft.

If we consider a strip on the surface of the shaft, as in figure 6.1, then it can be seen that ϕ is the shear strain experienced by the shaft. Thus

$$\text{shear strain } \phi = \frac{r\theta}{L} \qquad [1]$$

Since the modulus of rigidity G = shear stress τ/shear strain ϕ, then

$$\text{shear stress } \tau = G\phi = G\frac{r\theta}{L} \qquad [2]$$

We could have carried out the above derivations of the shear strain and shear stress at the surface of the shaft for a radial distance r which is internal in the shaft. The same relationships are obtained. Thus the shear strain and shear stress are proportional to the radius. The maximum values of the shear strain and shear stress for a circular shaft will thus occur at the surface of the shaft.

The relationship between the torque and the angle of twist can be determined from the fact that the sum of the moments of the tangential stresses acting over the cross-section must equal the applied torque. Consider an element in the cross-section of radius r and area δA, as in figure 6.2. If at this radius the shear stress is τ, then

$$\text{shearing force} = \tau \delta A$$

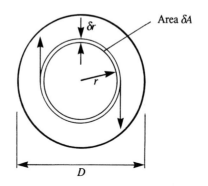

Fig. 6.2 Moment of force due to shearing stress.

The moment of this force about the shaft axis is

$$\text{moment} = \tau \delta A \times r$$

Thus substituting for τ using equation [2] gives

$$\text{moment} = \frac{G\theta}{L}r^2 \, \delta A$$

The total moment for the shaft, i.e. the torque T, will be given by the sum of the moments of the shearing forces on all areas of the ring segments that go to make up the shaft section. Hence,

$$T = \frac{G\theta}{L} \int_0^{D/2} r^2 \, dA \qquad [3]$$

The integral term is called the *polar second moment of area J* about the shaft axis. Hence

$$T = \frac{G\theta J}{L} \qquad [4]$$

But equation [2] gives $\tau/r = G\theta/L$, thus equation [4] can be written as

$$\frac{T}{J} = \frac{G\theta}{L} = \frac{\tau}{r} \qquad [5]$$

The term *torsional stiffness* is sometimes used for the torque per radian twist, i.e. T/θ.

6.1.1 Polar second moment of area

The polar second moment of area J of a shaft of diameter D about its axis is given by

$$J = \int_0^{D/2} r^2 \, dA \qquad [6]$$

Thus for the solid shaft in figure 6.2, if the segment of area δA has a thickness δr then $\delta A = 2\pi r \delta r$ and

$$J = \int_0^{D/2} r^2 \times 2\pi r \, dr = \left[\frac{2\pi r^4}{4} \right]_0^{D/2}$$

Thus for a solid shaft of diameter D

$$J = \frac{\pi D^4}{32} \qquad [7]$$

The units of J are m^4 if D is in m.
   For a hollow shaft with external diameter D and internal diameter d

$$J = \int_{d/2}^{D/2} r^2 \times 2\pi r \, dr = \left[\frac{2\pi r^4}{4} \right]_{d/2}^{D/2}$$

Hence

$$J = \frac{\pi (D^4 - d^4)}{32} \qquad [8]$$

Example

What is the diameter of a solid circular shaft which can be used to transmit a torque of 40 kN m if the shear stress is not to exceed 80 MPa? Also, what is the angle of twist in a 1 m length of shaft? The shaft material has a modulus of rigidity of 80 GPa.

For a solid shaft $J = \pi D^4/32$, hence equation [5] gives

$$\frac{T}{J} = \frac{\tau}{r}$$

$$\frac{40 \times 10^3}{\pi D^4/32} = \frac{80 \times 10^6}{D/2}$$

Hence $D = 0.137$ m. The angle of twist in a 1 m length of shaft is given by equation [4], $T = G\theta J/L$ as

$$\theta = \frac{TL}{GJ} = \frac{40 \times 10^3 \times 1.0}{80 \times 10^9 \times \pi \times 0.137^4/32} = 0.0145 \text{ rad}$$

Example

What is the external diameter of a hollow shaft needed to transmit a torque of 40 kN m if the shear stress is not to exceed 80 MPa and the shaft has an external diameter 1.5 times the internal diameter?

For the hollow shaft $D = 1.5d$ and so $d = 0.667D$. Thus

$$J = \frac{\pi(D^4 - d^4)}{32} = \frac{\pi(D^4 - 0.667^4 D^4)}{32} = 0.0787D^4$$

Equation [5] thus gives

$$\frac{T}{J} = \frac{\tau}{r}$$

$$\frac{40 \times 10^3}{0.0787D^4} = \frac{80 \times 10^6}{D/2}$$

Hence $D = 0.147$ m.

Example

What is the maximum shear stress in a 10 mm diameter bolt when it is tightened by a spanner which applies a torque of 8 N m?

For a solid shaft, the bolt, $J = \pi D^4/32$, thus equation [5] gives

$$\tau = \frac{Tr}{J} = \frac{8 \times 0.005}{\pi \times 0.010^4/32} = 40.7 \text{ MPa}$$

6.1.2 Polar modulus of section

For a shaft of radius r, polar second moment of area J, the torque T and shear stress τ are given by equation [5] as

$$\frac{T}{J} = \frac{\tau}{r}$$

Both J and r only involve the radius of a shaft, or the internal and external radius of a hollow shaft. Thus they are sometimes replaced by a quantity Z_p called the *polar modulus of section*, with

$$Z_p = \frac{J}{r} \qquad [9]$$

For a solid shaft where $J = \pi D^4/32$, with $D = 2r$, then $Z_p = \pi D^3/16$. The unit of Z_p is m^3 if D is in m. Equation [5] can thus be written as

$$T = Z_p \tau \qquad [10]$$

6.2 Transmission of power

For a rotating shaft of radius r the distance travelled by a point on its surface in one revolution is $2\pi r$ and in n revolutions $2\pi rn$. Thus if n revolutions are made per second, then the distance travelled in 1 s is $2\pi rn$. The torque T to which the shaft is subject is Fr, where F is the tangential force at radius r. Thus, since the work done is the product of the force and the distance moved in the direction of the force by its point of application

$$\text{work done per second} = F \times 2\pi rn = (T/r) \times 2\pi rn = 2\pi nT$$

But the work done per second is the power. Hence the power P transmitted is

$$P = 2\pi nT$$

Since the angular velocity ω of the shaft is $2\pi n$, then

$$P = T\omega \qquad [11]$$

If the unit of ω is rad/s and the torque is in N m, then the power is in watts.

Example

What is the power that can be transmitted by a solid shaft of diameter 100 mm rotating at 5 rev/s if the shear stress must not exceed 60 MPa?

For a solid shaft $J = \pi D^4/32$, hence using equation [5]

$$T = \frac{J\tau}{r} = \frac{\pi D^4 \tau}{32D/2} = \frac{\pi \times 0.100^3 \times 60 \times 10^6}{16} = 1.18 \times 10^4 \, \text{N m}$$

Hence the power transmitted is, using equation [11],

$$P = T\omega = 1.18 \times 10^4 \times 2\pi \times 5 = 370 \, \text{kW}$$

Example

A hollow drive shaft has an external diameter of 50 mm and an internal diameter of 40 mm. What is the maximum shear stress developed when the shaft is transmitting a power of 50 kW and rotating at 180 rev/min?

Since the power is $2\pi nT$, with $T = J\tau/r$ and $J = \pi(D^4 - d^4)/32$, then

$$P = 2\pi n \times \frac{\pi(D^4 - d^4)\tau}{32D/2}$$

and so

$$\tau = \frac{8DP}{\pi^2 n(D^4 - d^4)} = \frac{8 \times 0.050 \times 50 \times 10^3}{\pi^2 \times 3 \times (0.050^4 - 0.040^4)} = 183\,\text{MPa}$$

6.3 Compound shafts

Consider a shaft loaded by means of a number of couples. If the shaft is in equilibrium then the sum of all the torques acting on it must be zero. Thus for the shaft shown in figure 6.3(a), which is a shaft loaded by two couples and fixed at each end,

$$T_A - T_1 + T_2 - T_D = 0 \qquad [12]$$

where T_A is the reactive torque at the end A of the shaft and T_D the reactive torque at the end D. The shaft can be considered to have three elements AB, BC and CD. Each element is in equilibrium and so the three free-body diagrams are as shown in figure 6.3(b). For element AB to be in equilibrium there must be a torque of T_A at each end. This means that for point A, where the torque is T_1, the torque acting on element BC must be $(T_A - T_1)$. For element CD the torque acting on each end of the element is $(T_A - T_1 + T_2) = T_D$. The torque that is transmitted at each section is the algebraic sum of all the torques to one side of that section. The total angle of twist for the shaft is the algebraic sum of the angles of twist of each section. For a shaft fixed at both ends, this means that the total angle of twist must be zero.

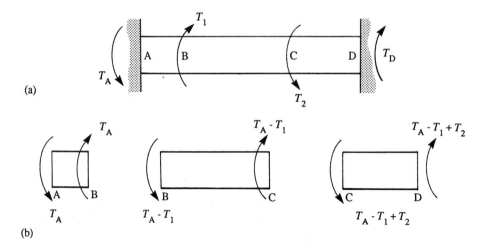

Fig. 6.3 Shaft loaded by two couples.

$$\phi = \phi_{AB} + \phi_{BC} + \phi_{CD} \tag{13}$$

Consider a stepped solid shaft with a length L_1 of radius r_1 in series with a length L_2 and radius r_2 (figure 6.4). If the shaft is acted on by a torque T then the total torque is transmitted by each section of the shaft. Thus $T_1 = T_2 = T$. But for section 1 we have $T = \tau_1 J_1 / r_1$ and for section 2 we have $T = \tau_2 J_2 / r_2$. But $J_1 = \pi d_1^4 / 32$ and $J_2 = \pi d_2^4 / 32$. Hence

$$\frac{\tau_1 \pi (2r_1)^4}{32 r_1} = \frac{\tau_2 \pi (2r_2)^4}{32 r_2}$$

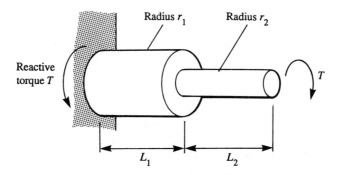

Fig. 6.4 A stepped shaft.

Hence the ratio of the maximum shear stresses in the two sections is

$$\frac{\tau_1}{\tau_2} = \left(\frac{r_2}{r_1}\right)^3 \tag{14}$$

The total twist θ of the shaft is the sum of the twists produced by each shaft. Thus

$$\theta = \theta_1 + \theta_2$$

But $\theta_1 = \tau_1 L_1 / r_1 G$ and $\theta_2 = \tau_2 L_2 / r_2 G$. Thus

$$\theta = \frac{1}{G}\left(\frac{\tau_1 L_1}{r_1} + \frac{\tau_2 L_2}{r_2}\right) \tag{15}$$

Alternatively we could have used $\theta_1 = TL_1 / GJ_1$ and $\theta_2 = TL_2 / GJ_2$ to have obtained

$$\theta = \frac{T}{G}\left(\frac{L_1}{J_1} + \frac{L_2}{J_2}\right) \tag{16}$$

A compound shaft may consist of two coaxial shafts, one placed inside the other, and which are firmly bonded together to act as a single member (figure 6.5). Because the shaft is in static equilibrium, the total torque T acting on it will give rise to resisting torques of T_A and T_B in the two elements with

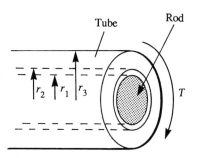

Fig. 6.5 Shafts in parallel.

$$T = T_A + T_B \tag{17}$$

The polar second moment of area for the inner shaft, radius r_1 is $\pi(2r_1)^4/32$ and for the tubular shaft, with inner radius r_2 and external radius r_3, $\pi[(2r_3)^4 - (2r_2)^4]/32$. Thus equation [17] can be written as

$$T = \frac{\tau_A \pi (r_3^4 - r_2^4)}{2r_3} + \frac{\tau_B \pi r_1^4}{2r_1} \tag{18}$$

where τ_A is the maximum shear stress for the tubular shaft and τ_B is the maximum shear stress for the inner rod shaft.

Because the two elements are bonded together the angles of twist for each element must be the same.

$$\phi_A = \phi_B \tag{19}$$

Thus, if both shafts have the same length L,

$$\frac{T_A L}{G_A J_A} = \frac{T_B L}{G_B J_B} \tag{20}$$

Example

A solid shaft AD of diameter 100 mm and length 3.0 m is built-in at both ends. What are the reactive torques at the ends if the shaft is loaded by torques of -5 kN m at B which is 1.0 m from end A and $+8$ kN m at C which is 2.0 m from A? What is the maximum shear stress in the shaft?

The shaft is loaded as in figure 6.3(a) and thus the free-body diagrams for each element of the shaft are as in figure 6.3(b). Since the shaft is in equilibrium then the algebraic sum of the torques must be zero. Thus

$$T_A - 5 + 8 + T_B = 0$$

Because the shaft is fixed at both ends, the total angle of twist must be zero. For element AB the torque causing an angle of twist is T_A. For element BC the torque is $T_A - 5$ and for element CD the torque is $T_A - 5 + 8$. Thus, since the angle of twist for an element is given by equation [5] as TL/GJ, then

$$\frac{T_A \times 1}{GJ} + \frac{(T_A - 5) \times 1}{GJ} + \frac{(T_A - 5 + 8) \times 1}{GJ} = 0$$

$$T_A + T_A - 5 + T_A - 5 + 8 = 0$$

Hence $T_A = 0.67\,\text{kN m}$. Substituting this in the earlier equation gives $T_B = -3.67\,\text{kN}$.

As the shaft is of constant diameter, the maximum shear stress will occur in that element where the torque is a maximum. The torque of element AB is $0.67\,\text{kN m}$, on BC it is $-4.33\,\text{kN m}$, and on CD it is $-3.67\,\text{kN m}$. The maximum torque is thus on element BC. Since the polar second moment of area is $\pi D^4/32$, then the shear stress is given for element BC, by equation [5], as

$$\tau = \frac{Tr}{J} = \frac{TD/2}{\pi D^4/32} = \frac{16T}{\pi D^3} = \frac{16 \times 4.33 \times 10^3}{\pi \times 0.100^3} = 22.1\,\text{MPa}$$

Example

A stepped shaft ABC, built-in at end A, consists of two solid circular parts, AB of length 500 mm and radius 30 mm and BC of length 300 mm and radius 25 mm. A torque of 500 N m is applied at end C. If both materials have the shear modulus of rigidity of 80 GPa, what will be the maximum shear stresses in each part?

The torque acting on each part of the shaft must be the same, i.e. $T_1 = T_2 = 500\,\text{N m}$. For part 1 we have

$$T = \frac{\tau_1 J_1}{r_1} = \frac{\tau_1 \pi (r_1/2)^4}{32 r_1} = \frac{\tau_1 \pi r_1^3}{2}$$

Hence

$$\tau_1 = \frac{2T}{\pi r_1^3} = \frac{2 \times 500}{\pi \times 0.030^3} = 11.8\,\text{MPa}$$

Similarly for the other part,

$$\tau_2 = \frac{2T}{\pi r_2^3} = \frac{2 \times 500}{\pi \times 0.025^3} = 20.4\,\text{MPa}$$

Example

A solid aluminium alloy rod of diameter 60 mm fits inside a hollow steel shaft of internal diameter 60 mm and external diameter 70 mm. The two are connected to rigid end plates so that they twist together. What is the angle of twist per metre length of shaft when a torque of 1 kN m is applied to the composite shaft? For the aluminium the modulus of rigidity is 30 GPa and for the steel 80 GPa.

The arrangement is as shown in figure 6.4. Since the shaft is in static equilibrium, the sum of the resisting torques on the two elements must equal the applied torque. Thus

$$T_A + T_B = 1\,kN\,m$$

Since the two elements are joined, their angles of twist must be the same. The polar second moment of area of the rod is $\pi d^4/32$ and that of the tube is $\pi(D^4 - d^4)/32$. Thus, since equation [5] gives $\theta = TL/GJ$,

$$\frac{T_A L}{30 \times 10^9 \times \pi \times 0.060^4/32} = \frac{T_B L}{80 \times 10^9 \times \pi(0.070^4 - 0.060^4)/32}$$

Thus $T_A = 0.44 T_B$. Hence, using the equilibrium equation, $T_A = 0.69\,kN\,m$ and $T_B = 0.31\,kN\,m$.

The angle of twist for the composite shaft can be found by determining the angle of twist for either element. Thus, using equation [5],

$$\theta = \frac{0.69 \times 1}{30 \times 10^9 \times \pi \times 0.060^4/32} = 1.8 \times 10^{-5}\,rad$$

Problems

(1)   A solid steel shaft has a diameter of 50 mm and is acted on by a torque of 1 kN m. What is the maximum shear stress produced and the angle of twist in a 2 m length of the shaft? The steel has a modulus of rigidity of 80 GPa.

(2)   A solid steel shaft has a diameter of 60 mm and a length of 2 m. What is the maximum torque that can be applied to the shaft if the maximum allowable shear stress is to be 40 MPa and what is the resulting angle of twist? The shaft material has a modulus of rigidity of 80 GPa.

(3)   A solid steel shaft has a diameter of 60 mm. What is the maximum torque that can be applied to the shaft if the maximum allowable angle of twist is 1° per metre? The material has a modulus of rigidity of 80 GPa.

(4)   What is the torque which can be transmitted by a solid circular shaft of diameter 120 mm if the maximum shear stress is to be 70 MPa?

(5)   What is the external diameter of a hollow shaft which is needed to transmit a torque of 20 kN m with shear stresses not exceeding 80 MPa if the ratio of the external to internal diameters is to be 1.4?

(6)   How do the masses of equal length hollow and solid shaft compare if they are required to transmit the same torque with the same maximum shear stress? The hollow shaft has an inner diameter which is 2/3 of the outer diameter.

(7) A valve handle transmits torque to the valve through a solid shaft of diameter 10 mm. What is the maximum torque that can be applied to the handle if the shaft has a yield shear stress of 80 MPa?

(8) What is the maximum shear stress produced in a bolt of diameter 20 mm when it is tightened by a spanner which exerts a force of 50 N with a radius of action of 150 mm?

(9) A torque wrench is used to apply a torque of 40 N m to a hollow shaft of external diameter 100 mm and wall thickness 20 mm. What are the shear stresses developed at the inner and outer walls of the shaft?

(10) What is the maximum shear stress in a solid circular shaft of diameter 120 mm when it is transmitting a power of 20 kW at 120 rev/min?

(11) A solid shaft is required to transmit 1 MW at 4 rev/s and the shear stress must not exceed 60 MPa. What is the minimum diameter shaft required?

(12) A hollow shaft with an external diameter of 200 mm and an internal diameter of 125 mm is used to transmit 1 MW at 3 rev/s. What are the shear stresses at the inner and outer surfaces of the shaft?

(13) What will be the power transmitted by a hollow shaft with an external diameter of 250 mm and an internal diameter of 150 mm at 120 rev/min if the maximum shearing stress is limited to 70 MPa?

(14) A hollow shaft with an internal diameter of 30 mm and a wall thickness of 12 mm is to be used to transmit a power of 50 kW. What will be the maximum frequency of rotation of the shaft if the shear stress must not exceed 50 MPa?

(15) A solid shaft of diameter 150 mm transmits a power of 500 kW at 2 rev/s. What is the maximum shear stress produced in the shaft and the angle of twist per metre if the modulus of rigidity of the material is 80 GPa?

(16) A shaft AB of constant diameter and length 4.0 m is built-in at both ends. What are the reactive torques at the fixed ends if torques of −1 kN m is applied at 1.0 m from A, −1 kN m at 2.0 m from A and +1.0 at 3.0 m from A?

(17) A pipe has an external diameter of 40 mm and a wall thickness of 3 mm. If it is built-in at one end, what will be the maximum shear stress developed in the pipe when at 0.2 m from the fixed end there is a torque of +20 N m, at 0.6 m a torque −10 N m and at 1.0 m a torque of +40 N m?

(18) A solid shaft AB has a diameter of 20 mm and a length of 2.0 m and is built-in at both ends. What will be the reactive torques at the ends when it is subject to a torque of +600 N m a distance of 0.2 m from A and −400 N m at 1.7 m from A?

(19) A stepped shaft of length $2L$ and built-in at both ends, is made of two elements, a length L with a diameter of d and a length L of diameter $2d$. At what position along the shaft should a torque T be applied so that the reactive torques at each end of the shaft are the same?

(20) A stepped shaft ABC, built-in at both ends A and C, has a length of 500 mm and is made of two parts. Part AB has a length of 300 mm and a diameter of 50 mm, being made of material with a modulus of rigidity of 20 GPa. Part BC has a length of 200 mm and a diameter of 25 mm, being made of material with a modulus of rigidity of 80 GPa. A torque of 1 kN m is applied to the joint at B. What are the maximum shear stresses in the two parts of the shaft?

(21) A solid stepped shaft ABC, built-in at end A, has a length of 1.0 m. Part AB has a length of 0.6 m and a diameter of 60 mm and part BC a length of 0.4 m and a diameter of 50 mm. Both parts have the same material, with a modulus of rigidity of 80 GPa. A torque of 1 kN m is applied at C. What are the maximum shear stresses in each part of the shaft and the angle of twist at C?

(22) A solid stepped shaft ABC, built-in at both ends A and C, has a length L. Part AB has a diameter d_a, part BC a diameter d_b. A torque T is applied at C. What fraction of the length L should be AB if the maximum shear stress is to be the same in both parts of the shaft?

(23) A stepped shaft ABC of length 1500 mm is built-in at A. Part AB is 500 mm long and has a diameter of 50 mm. Part BC is 1000 mm long and has a diameter of 35 mm. The part AB has an axial 30 mm diameter hole drilled into it from end A to a depth of 200 mm. What will be the maximum torque that can be applied at the free end C if the maximum shear stress is to be 80 MPa and where will the maximum shear stress occur?

(24) An aluminium alloy solid shaft with a diameter of 60 mm is connected in series with a hollow steel shaft of the same external diameter. What is the internal diameter of the steel shaft if it is to have the same angle of twist per unit length as the aluminium alloy shaft? The modulus of rigidity for the aluminium alloy is 30 GPa and for the steel 80 GPa.

(25) A compound shaft has a steel tube closely fitting over a brass core. The diameter of the brass core is 25 mm and the external diameter of the steel tube is 40 mm. The modulus of rigidity for the brass is 40 GPa and that for the steel 80 GPa. What are the maximum shear stresses in the brass and the steel when the shaft is subject to a torque of 1 kN m?

(26) A steel shaft has an aluminium alloy tube closely fitting over it. If the maximum shear stress in the steel is to be 100 MPa and in the

aluminium alloy 75 MPa when the torque is 1 kN m, what will be the required diameters of the steel shaft and the aluminium alloy tube? The modulus of rigidity for the steel is 80 GPa and for the aluminium alloy 30 GPa.

(27)   A compound shaft consists of a solid aluminium alloy rod of diameter 50 mm, surrounded by a hollow steel shaft of internal diameter 50 mm and external diameter 65 mm. If the compound shaft twists as a single unit, what are the maximum shear stresses in the two materials when there is a torque of 2 kN m? The modulus of rigidity of the aluminium alloy is 30 GPa and of the steel 80 GPa.

Chapter 7

Strain energy

7.1 Strain energy

Consider the situation where a length of material is stretched as the result of a gradually applied load. At this point the discussion is restricted to a gradually applied load, so that the concern is only the transfer of energy as the work done by the loading and no kinetic energy is transferred (see section 10.8 for a discussion of energy). Work has to be done to increase the length of the material. Energy is thus transferred from the person doing the pulling to the object being stretched. The internal energy of the object increases. The internal energy increase produced by some loading process is called *strain energy*, symbol U. The strain energy is equal to the work done by the loading process.

If the loading is slowly removed from the stretched material, some or all of the strain energy may be recovered and work done. Thus if none of the stresses in the stretched material exceed the elastic limit, the material will return to its original length when the load is removed and all the strain energy is recovered. If the elastic limit is exceeded then some of the energy is used to produce deformation of the material and this energy is not recovered when the loading is removed.

It is not only stretching that can result in a material gaining strain energy as a result of work being done. Compression, twisting, shearing and bending also involve work being done and strain energy being gained by the material.

7.2 Strain energy in axially loaded bars

Consider a length of material which is being stretched in the direction of its axis by means of a gradually applied load and which does not exceed the elastic limit as a result of the loading. To cause a material to extend and have an extension x a force is applied which gradually increases from zero to F (figure 7.1), the extension being proportional to the force because the elastic limit is not being exceeded. The average force is $\frac{1}{2}F$ and thus the work done in stretching the material is

$$\text{work done} = \text{force} \times \text{distance} = \tfrac{1}{2}Fx$$

This work results in an increase in the internal energy of the material and so

$$\text{strain energy } U = \tfrac{1}{2}Fx \qquad\qquad [1]$$

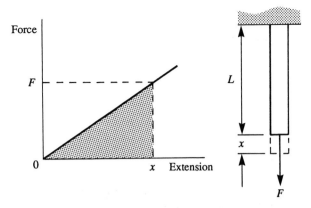

Fig. 7.1 Axially loaded bar.

This is the area under the force–extension graph between zero extension and the extension x.

The volume of the material being extended is AL, where A is the cross-sectional area and L the length. Hence equation [1] can be written as

$$\frac{\text{strain energy}}{\text{volume}} = \frac{Fx}{2AL}$$

The strain energy per unit volume is often referred to as the *strain energy density* or the *resilience*, symbol u. But $F/A = $ direct stress σ and $x/L = $ direct strain ϵ, hence

$$\text{strain energy per unit volume } u = \tfrac{1}{2}\sigma\epsilon \qquad [2]$$

This $\tfrac{1}{2}\sigma\epsilon$ is the area of a triangle of height σ and base ϵ and thus the strain energy per unit volume is the area under the stress–strain graph, assuming this to be a straight line relationship.

Since $\sigma = E\epsilon$, where E is the tensile modulus, we can also write equation [2] as

$$\text{strain energy per unit volume } u = \frac{\sigma^2}{2E} = \frac{E\epsilon^2}{2} \qquad [3]$$

For a uniform bar of insignificant weight being stretched by an axial load F, as in figure 7.2, the strain energy U due to the load is

$$U = \text{strain energy per unit volume} \times \text{volume}$$

$$= \frac{\sigma^2}{2E} AL = \frac{(F/A)^2}{2E} AL = \frac{F^2 L}{2AE} \qquad [4]$$

The greatest amount of strain energy that can be stored in a material without permanent deformation occurring is when the stress σ is equal to the elastic limit, or proof stress. The value of the strain energy with this stress is called the *proof resilience* and the strain energy per unit volume with this stress is called the *modulus of resilience*. Since the strain energy per unit volume is the area under the stress–strain graph then the modulus of resilience must be the area under the stress–strain graph up to the elastic limit.

Example

A crane has a steel cable of length 10 m and cross-sectional area 1200 mm^2. Neglecting the weight of the cable, what will be the strain energy stored in the cable when an object of mass 1000 kg is lifted by it? The tensile modulus of the steel is 210 GPa and the elastic limit is not exceeded.

The strain energy is the strain energy per unit volume multiplied by the volume of the cable. Thus, using equation [4]

$$\text{strain energy} = \frac{F^2 L}{2AE} = \frac{(1000 \times 9.8)^2 \times 10}{2 \times 1200 \times 10^{-6} \times 210 \times 10^9}$$
$$= 1.91 \, \text{J}$$

Example

An aluminium alloy has a proof stress of 450 MPa at a strain of 0.005. What is the modulus of resilience?

The modulus of resilience is the greatest amount of strain energy per unit volume that can be stored without permanent deformation occurring and is thus

$$\text{modulus of resilience} = \tfrac{1}{2} \times 450 \times 10^6 \times 0.005$$
$$= 1.125 \, \text{MJ/m}^3$$

7.2.1 Strain energy with composite bar

If a length of material does not have a uniform cross-sectional area then the strip of material has to be considered as being made up of a number of uniform parts and the strain energy computed for each part. The sum of these strain energies then gives the total strain energy of the strip.

To illustrate this, consider the strain energy of the bar shown in figure 7.2 if the weight of the bar can be neglected. The bar is composed of two parts. For part 1 the stress is F/A_1 and the length L_1, hence the strain energy for that part is, using equation [4],

$$U_1 = \frac{F^2 L_1}{2EA_1}$$

For part 2 the stress is F/A_2 and the length L_2. Hence the strain energy for that part is

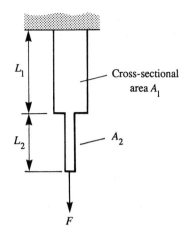

Fig. 7.2 Non-uniform bar.

$$U_2 = \frac{F^2 L_2}{2EA_2}$$

The total strain energy U of the bar is thus $U_1 + U_2$, i.e.

$$U = \frac{F^2 L_1}{2EA_1} + \frac{F^2 L_2}{2EA_2} \qquad [5]$$

Example

A tensile force of 30 kN acts on a bar having two different cross-sections, one of length 5.0 m and cross-sectional area 1000 mm^2 and the other of length 4.0 m and cross-sectional area 5000 mm^2 (figure 7.3). If the material has a tensile modulus of 205 GPa, what is the strain energy stored in the material as a result of the force?

Applying equation [4] to the first part of the bar

$$U_1 = \frac{F^2 L_1}{2EA_1}$$

For the second part

$$U_2 = \frac{F^2 L_2}{2EA_2}$$

The total strain energy U of the bar is thus $U_1 + U_2$, i.e.

$$U = \frac{F^2 L_1}{2EA_1} + \frac{F^2 L_2}{2EA_2}$$

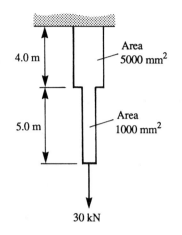

Fig. 7.3 Example.

$$U = \frac{(30 \times 10^3)^2 \times 4.0}{2 \times 205 \times 10^9 \times 5000 \times 10^{-6}}$$

$$+ \frac{(30 \times 10^3) \times 5.0}{2 \times 205 \times 10^9 \times 1000 \times 10^{-6}}$$

$$= 12.7 \, J$$

Example

What is the strain energy stored in the truss shown in figure 7.4 when the force F acts on it? Both members have the same cross-sectional area A and tensile modulus E.

The component of the force acting on each bar is the same and is $F/2\cos\theta$. Since $\cos\theta = 3/5$ then the strain energy in each bar is, using equation [4],

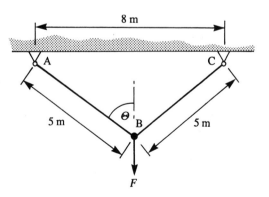

Fig. 7.4 Example.

$$\text{strain energy} = \frac{(5F/6)^2 \times 5}{2AE} = \frac{125F^2}{72AE}$$

Since there are two members, the total strain energy is thus $125F^2/36AE$.

7.2.2   *Strain energy with bar having weight*

Consider a uniform bar where the weight of the bar is significant and cannot be neglected. Consider an element of length δx of the bar at a distance x from the support (figure 7.5). The force acting on this element F_x is the weight of the bar below the element. This is $(L - x)A\rho g$, where A is the cross-sectional area, ρ the density and g the acceleration due to gravity. The strain energy U_x due to this element is thus

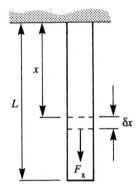

Fig. 7.5 Bar hanging under its own weight.

$$U_x = \text{strain density} \times \text{volume} = \frac{\sigma^2}{2E}A\delta x = \frac{(F_x/A)^2}{2E}A\delta x$$

$$= \frac{F_x^2\delta x}{2AE} = \frac{[(L - x)\rho g]^2 A\delta x}{2E}$$

The total strain energy due to the weight of the bar will be the sum of the strain energies due to each of these elements.

$$U = \int_0^L \frac{[(L - x)\rho g]^2 A}{2E}\, dx = \frac{\rho^2 g^2 A}{2E}[L^2 x - Lx^2 + x^3/3]_0^L$$

$$= \frac{\rho^2 g^2 AL^3}{6E} \tag{6}$$

If a bar is subject to its own weight and also a load F then the above analysis of a bar under its own weight requires modification. For the element at x the force F_x becomes

$$F_x = (L - x)A\rho g + F$$

Hence the strain energy for the element is

$$U_x = \frac{[(L - x)A\rho g + F]^2 \delta x}{2AE}$$

On integrating this gives for the total strain energy

$$U = \frac{\rho^2 g^2 A L^3}{6E} + \frac{\rho g L^2}{2E} + \frac{F^2 L}{2EA} \qquad [7]$$

The first term is the strain energy due to the weight, the third term is the strain energy due to the force alone. The inclusion of the second term means however that the total strain energy for a beam subject to more than one load cannot be obtained by just adding the strain energies due to the individual loads.

Example

A crane has a steel cable of length 10 m and cross-sectional area 1200 mm². What will be the strain energy in the cable due to its own weight if the steel has a density of 7.9×10^3 kg/m³? The tensile modulus of the steel is 210 GPa and the elastic limit is not exceeded.

Using equation [6]

$$U = \frac{\rho^2 g^2 A L^3}{6E} = \frac{7.9^2 \times 10^6 \times 9.8^2 \times 1200 \times 10^{-6} \times 10}{6 \times 210 \times 10^9}$$

$$= 5.7 \times 10^{-5} \, \text{J}$$

7.3 Dynamic loading

In the situations considered so far in this chapter it has been assumed that loads are applied very gradually, increasing from zero to the maximum value and thereafter remaining constant. However, quite often in practice. loads are applied suddenly and result in vibrations of the structure or they may change with time. Examples of these are a falling object striking a structure, cyclical loads produced by rotating machinery and variable loads produced by traffic or wind gusts. The term *dynamic loading* is used to describe loads which vary with time.

Consider an impact load produced by a collar of mass *M* falling from a height *h* on to a flange at the lower end of a bar, as in figure 7.6. When the collar hits the flange it causes the bar to elongate to its position of maximum displacement δ before it shortens and the bar length then oscillates before eventually attaining some equilibrium elongation. The bar behaves just like a spring would when a weight is suddenly attached to its end. To determine the strain energy acquired by the bar we need to consider all the energy that is transferred to it. There is the energy from the collar falling down to an

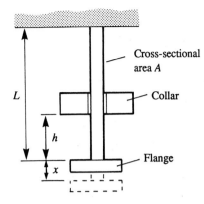

Fig. 7.6 Impact loading.

impact with the flange and then the extra potential energy from the collar falling the distance by which the bar stretches.

The potential energy of the collar, with respect to the flange, before it starts to fall is Mgh. As the collar falls this potential energy is converted into kinetic energy and at impact the kinetic energy is equal to the potential energy lost. To simplify the calculation, now assume that on impact the collar does not bounce but 'sticks' to the flange and that all the kinetic energy of the falling collar is converted into strain energy of the bar. Thus Mgh energy is converted to strain energy. There is also, however, strain energy produced by the collar falling through the distance by which the bar stretches. If it stretches by x then this extra potential energy is Mgx. Thus at its maximum elongation x the energy that has been transferred to the bar is $Mg(h + x)$. This is the strain energy U, and thus using equation [2],

$$U = \tfrac{1}{2}\epsilon^2 EAL = \tfrac{1}{2}(x^2/L^2)EAL$$

Thus

$$Mg(h + x) = \frac{x^2 EA}{2L}$$

$$EAx^2 - 2LMgx - 2LMgh = 0$$

This equation is a quadratic in x of the form $ax^2 + bx + c = 0$ and hence, since the roots of this equation are

$$x = \frac{-b \pm \sqrt{(b^2 - 4ac)}}{2a}$$

the roots of the equation are

$$x = \frac{+2LMg \pm \sqrt{(4L^2 M^2 g^2 + 8EALMgh)}}{2EA}$$

The positive maximum displacement is thus

$$x = \frac{MgL}{EA} + \sqrt{\left[\left(\frac{MgL}{EA}\right)^2 + \frac{2MghL}{EA}\right]} \qquad [8]$$

The weight Mg on the end of the bar would cause a static deflection x_{st} of

$$x_{st} = \frac{MgL}{EA}$$

Thus the maximum displacement, given by equation [8], can be expressed as

$$x = x_{st} + \sqrt{(x_{st}^2 + 2hx_{st})} \qquad [9]$$

Since the static elongation is generally very small the x_{st}^2 and x_{st} terms can usually be neglected and so

$$x \approx \sqrt{(2hx_{st})} = \sqrt{\left(\frac{2MghL}{EA}\right)} \qquad [10]$$

The maximum stress produced in the bar, if the stress is assumed to be uniform throughout the bar, can be assumed to occur when the elongation is a maximum. Thus, since maximum stress σ_{max} is E times the maximum strain, which is x/L, then

$$\sigma_{max} = \frac{Ex}{L} = \frac{Mg}{A} + \sqrt{\left[\left(\frac{Mg}{A}\right)^2 + \frac{2MghE}{AL}\right]} \qquad [11]$$

Since the static stress σ_{st} is Mg/A, then equation [11] can be written as

$$\sigma_{max} = \sigma_{st} + \sqrt{\left(\sigma_{st}^2 + \frac{2hE\sigma_{st}}{L}\right)} \qquad [12]$$

Since the terms in the above equation are all generally small compared with the term including h, this equation can be approximated to

$$\sigma_{max} = \sqrt{\left(\frac{2hE\sigma_{st}}{L}\right)} = \sqrt{\left(\frac{2MghE}{AL}\right)} \qquad [13]$$

Example

A vertical steel rod, tensile modulus 200 GPa, has a length of 3.0 m and a uniform cross-sectional area of 0.50×10^{-3} m². It is rigidly fixed at the top and the bottom end, which has a flange, is free to move. Calculate the maximum stress produced in the rod when a collar of mass 20 kg slides down the rod through a height of 50 mm on to the flange without rebounding.

Since the static stress σ_{st} is $Mg/A = 20 \times 9.8/0.50 \times 10^{-3} = 3.92 \times 10^5$ Pa, then equation [11] can be written as

$$\sigma_{max} = \sigma_{st} + \sqrt{\left(\sigma_{st}^2 + \frac{2hE\sigma_{st}}{L}\right)}$$

$$= 3.92 \times 10^5 + \sqrt{\left[(3.92 \times 10^5)^2 + \frac{2 \times 0.050 \times 200 \times 10^9 \times 3.92 \times 10^5}{3.0} \right]}$$

$$= 51.1 \, \text{MPa}$$

In fact the approximate form of the equation could have been used.

7.3.1   Suddenly applied loads

In the above discussion a load is applied to a rod by, what can be termed, an impact load. The load when it is applied to the rod has a velocity and hence kinetic energy. There are however other ways by which loads can be suddenly applied.

Consider the situation, involving a collar load being applied to the flanged end of a rod (as in figure 7.6), where the collar is placed gently on the flange and then suddenly released. In figure 7.6 we have $h = 0$. The potential energy lost by the load in stretching the rod through a distance x is Mgx. This is converted to strain energy U, and thus using equation [2],

$$U = \tfrac{1}{2}\epsilon^2 EAL = \tfrac{1}{2}(x^2/L^2)EAL$$

Thus

$$Mgx = \frac{x^2 EA}{2L}$$

$$x = \frac{2LMg}{EA} \qquad\qquad [14]$$

Hence, since the static deflection $x_{st} = MgL/AE$ then

$$x = 2x_{st} \qquad\qquad [15]$$

The maximum extension produced by this loading is thus twice the static extension.

Another possible loading situation is where a load at the end of a cable (figure 7.7) is being lowered with a constant velocity and the motion is suddenly stopped. Such a situation might occur with a hoist. When that occurs, the kinetic energy of the load is converted into strain energy, it generally being assumed that the cable is of insignificant mass, and hence insignificant kinetic energy, and that the kinetic energy of the drum can be ignored.

When the load is moving with constant velocity the cable will already have strain energy as a result of the static extension x_{st} of the cable by the load. This strain energy is, using equation [2],

$$U = \tfrac{1}{2}\epsilon^2 EAL = \tfrac{1}{2}(x_{st}^2/L^2)EAL$$

The total energy of the system when moving with a constant velocity v is thus

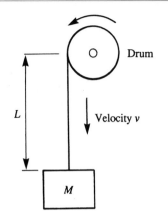

Fig. 7.7 Load being lowered.

$$\text{strain energy} = \tfrac{1}{2}EAx_{st}^2/L + \tfrac{1}{2}Mv^2$$

When the load stops moving, potential energy is lost in further extending the cable by x, this being Mgx. Thus the total energy converted to strain energy is

$$\text{total energy} = \tfrac{1}{2}EAx_{st}^2/L + \tfrac{1}{2}Mv^2 + Mgx \qquad [16]$$

The maximum extension x_{max} of the cable is $(x + x_{st})$. A cable with this extension must have a strain energy given by equation [2] of

$$U = \tfrac{1}{2}\sigma\epsilon AL = \tfrac{1}{2}(E\epsilon)\epsilon AL = \tfrac{1}{2}\epsilon^2 EAL = \tfrac{1}{2}(x_{max}^2/L^2)EAL$$

Thus

$$\tfrac{1}{2}EAx_{max}^2/L = \tfrac{1}{2}EAx_{st}^2/L + \tfrac{1}{2}Mv^2 + Mg(x_{max} - x_{st})$$

Since we have for the static situation $MgL/Ax_{st} = E$, then this equation can be rearranged as

$$\begin{aligned}
\tfrac{1}{2}Mv^2 &= (EA/2L)\,[x_{max}^2 - x_{st}^2 - (2MgL/EA)\,(x_{max} - x_{st})] \\
&= (EA/2L)\,[x_{max}^2 - x_{st}^2 - 2x_{st}x_{max} + 2x_{st}^2] \\
&= (EA/2L)\,(x_{max} - x_{st})^2
\end{aligned}$$

Hence

$$x_{max} - x_{st} = \sqrt{\left(\frac{Mv^2L}{EA}\right)} \qquad [17]$$

The maximum stress σ_{max} in the cable is given by $\sigma_{max}/\epsilon_{max} = E$ and so

$$\sigma_{max} = \frac{Ex_{max}}{L} = \frac{Ex_{st}}{L} + \frac{E}{L}\sqrt{\left(\frac{Mv^2L}{EA}\right)}$$

Since $MgL/Ax_{st} = E$, then this equation can be written as

$$\sigma_{max} = \frac{Mg}{A} + \sqrt{\left(\frac{Mv^2E}{LA}\right)} \qquad [18]$$

Example

A load of mass 20 kg is being lowered on the end of a cable at a speed of 0.4 m/s when the cable suddenly jams and the motion ceases. At this point the vertical length of the cable is 10 m. What is the maximum stress produced in the cable if it has a cross-sectional area of 400 mm² and a tensile modulus of 200 GPa?

This is the situation described by equation [18]. Thus

$$\sigma_{max} = \frac{Mg}{A} + \sqrt{\left(\frac{Mv^2E}{LA}\right)}$$

$$= \frac{20 \times 9.8}{400 \times 10^{-6}} + \sqrt{\left(\frac{20 \times 0.4^2 \times 200 \times 10^9}{10 \times 400 \times 10^{-6}}\right)}$$

$$= 13.1\,\text{MPa}$$

Example

A boat of mass 1000 kg drifts at 0.2 m/s and is stopped when the 10 m long tow-rope becomes taut. What is the maximum tension in the rope if it extends by 0.2 m in bringing the boat to a stop?

The kinetic energy of the boat is $\frac{1}{2}mv^2 = \frac{1}{2} \times 1000 \times 0.2^2 = 20\,\text{J}$. If this is all converted to strain energy in the rope then the strain energy stored in the rope is 20 J. But the strain energy stored in the rope when it is extended is given by equation [1] as

$$U = 20 = \tfrac{1}{2}Fx = \tfrac{1}{2}F \times 0.2$$

Hence $F = 200\,\text{N}$.

7.4   Strain energy due to shear

Consider a shear force V acting over an area A of a block of material and resulting in shear occurring, as in figure 7.8. The material deforms and the point of application of the shear force is displaced through a distance x relative to the bottom face of the block. If the material is not stretched beyond its elastic limit and obeys Hooke's law, then a graph of the force plotted against the displacement is a straight line (figure 7.9). The strain energy U stored in the element is thus equal to the area below this load–displacement graph, i.e.

$$U = \tfrac{1}{2}Vx$$

But the shear strain ϕ is x/h and the shear stress τ is V/A. Hence

$$U = \tfrac{1}{2}(\tau A)\,(\phi h) = \tfrac{1}{2}\tau\phi Ah$$

Since Ah is the volume of the material then the strain energy per unit

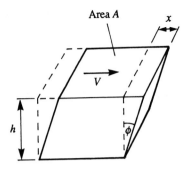

Fig. 7.8 Material subject to shear.

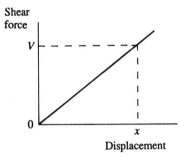

Fig. 7.9 Force−displacement graph.

volume u is

$$u = \tfrac{1}{2}\tau\phi \qquad\qquad\qquad [19]$$

The shear modulus $G = \tau/\phi$ and so equation [19] can also be written as

$$u = \frac{\tau\phi}{2} = \frac{G\phi^2}{2} = \frac{\tau^2}{2G} \qquad\qquad\qquad [20]$$

Generally the shear stress will not be constant for all parts of a body and thus the total strain energy has to be obtained by considering the body to be made up of small segments of volume, for each of which a constant shear stress can be assumed, and then summing the strain energies due to each of these segments.

7.4.1 *Strain energy due to torsion*

The strain energy stored in a cylindrical bar due to pure torsion can be obtained by considering the bar to be made up of elements, each subject to shear stress, and obtaining the sum of the shear strain energies due to each. An alternative method, which gives the same result and is given below, is to consider the torque−rotation relationship for the bar.

When a force is used to rotate an object then work is done. Thus if the point of application of a force F moves from A to B in figure 7.10 and so causes a rotation through an angle θ then

$$W = F \times \text{arc AB} = Fr\theta = T\theta \qquad [21]$$

since arc length = radius $r \times$ angle subtended and the torque T is the force multiplied by the radius arm, i.e. $T = Fr$.

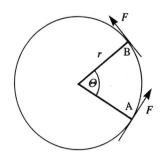

Fig. 7.10 Work done by a torque.

Consider a uniform circular bar with one end rigidly fixed and the other end twisted through some angle θ by a torque T. The angle of twist is proportional to the torque, provided the elastic limit is not exceeded, and thus gives a torque–twist graph of the form shown in figure 7.11. The work done in twisting the bar from zero twist to θ is thus

$$W = \tfrac{1}{2}T\theta$$

i.e. the area under the torque–twist graph. This energy is then stored in the shaft as strain energy U. Thus

$$U = \tfrac{1}{2}T\theta \qquad [22]$$

Since $T = G\theta J/L$ (see chapter 6, equation [4]), where G is the torsional modulus, J the polar second moment of area and L the length of the twisted bar, then equation [22] can be written as

$$U = \frac{T\theta}{2} = \frac{GJ\theta^2}{2L} = \frac{LT^2}{2GJ} \qquad [23]$$

If the torque varies along the length of the shaft, then since the energy δU stored in a short length δx is

$$\delta U = \frac{T^2\delta x}{2GJ}$$

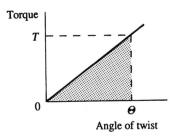

Fig. 7.11 Strain energy = area under graph.

the energy U stored in the entire shaft, length L, is

$$U = \int_0^L \frac{T^2}{2GJ} \, dx \qquad [24]$$

Example

What is the strain energy stored in a uniform circular solid shaft of length 2.0 m and diameter 100 mm when subject to pure torsion which gives rise to a maximum shear stress of 50 MPa? $G = 80$ GPa.

For a solid shaft of diameter D the polar second moment of area is $\pi D^4/32$. The maximum shear stress occurs at the surface, i.e. at $D/2$. Thus, since $T = J\tau/r$ (chapter 6, equation [5]), equation [23] gives

$$U = \frac{LT^2}{2GJ} = \frac{LJ^2\tau_{max}^2}{2GJ(D/2)^2} = \frac{L(\pi D^4/32)\tau_{max}^2}{2GD^2/4} = \frac{\tau_{max}^2 L\pi D^2/4}{4G}$$

Note that since $L\pi D^2/4$ is the volume of the shaft, the internal energy per unit volume for a solid circular shaft is $\tau_{max}^2/4G$.

$$U = \frac{(50 \times 10^6)^2 \times 2.0 \times \pi \times 0.100^2}{16 \times 80 \times 10^9} = 123 \text{ J}$$

7.5 Impact torsional loading

Sudden torsional loading can arise in situations such as when a rotating shaft is suddenly stopped. The loss in kinetic energy of the shaft then becomes strain energy. The kinetic energy of a rotating object is $\frac{1}{2}I\omega^2$, where I is its moment of inertia and ω its angular velocity. Thus, for example, if a rotating shaft is suddenly stopped then the kinetic energy converted to strain energy is $\frac{1}{2}I\omega^2$ and so, using equation [23] for the strain energy, the maximum torque T generated in the shaft is given by

$$\frac{I\omega^2}{2} = \frac{LT^2}{2GJ}$$

7.6   Strain energy due to bending

Consider a beam bent into the arc of a circle (figure 7.12(a)). The angle θ subtended by the bent beam is arc length/radius, i.e. L/R, where L is the length of the beam and R the radius of curvature. But $M/I = E/R$ (see chapter 3), where M is the bending moment, I the second moment of area and E the tensile modulus. Thus

$$\theta = \frac{L}{R} = \frac{ML}{EI} \qquad\qquad [25]$$

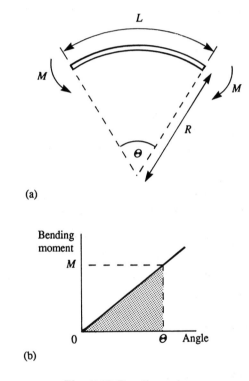

(a)

(b)

Fig. 7.12 Bending a beam.

For such a situation a graph of the angle plotted against the bending moment is a straight line (figure 7.12(b)). The work done in bending the beam to an angle θ is the area under the graph, i.e.

$$W = \tfrac{1}{2}M\theta$$

This is the energy transferred to the beam and stored in it as strain energy U. Thus

$$U = \tfrac{1}{2}M\theta$$

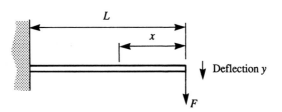

Fig. 7.13 Example.

Using equation [25] this equation can be written in the following forms

$$U = \frac{M^2L}{2EI} = \frac{EI\theta^2}{2L}$$ [26]

If the bending moment varies along the length of the beam then it is necessary to consider the beam as being effectively made up of small segments, each of which has a constant bending moment. We can then write equation [24] for each segment as

$$\delta\theta = \frac{M\delta x}{EI}$$

where $\delta\theta$ is the angle subtended by an element of the beam of length δx. The strain energy for that element δU is thus

$$\delta U = \frac{M^2\delta x}{2EI}$$

Hence the total strain energy U for the beam is the sum of all the strain energies due to the elements, i.e.

$$U = \int_0^L \frac{M^2dx}{2EI}$$ [27]

The above integration can only be applied over a length of a beam for which a continuous expression for M can be obtained. This will thus mean a separate integration for each section of a beam between two concentrated loads or reactions. The total strain energy for the beam is then the sum of these separate integrations.

The above gives the strain energy produced by a load, or loads, bending a beam. However such loads will also produce shear in the beam and thus there will be an additional amount of strain energy due to shear. If the beam is long and slender the strain energy due to shear can generally be neglected in comparison with the strain energy due to the bending.

The deflection of the beam can be obtained by equating the work done by the load in bending the beam with the strain energy. It should be noted that this method of determining the deflection is limited to those cases where only one concentrated load is applied and then will only give the deflection under the load.

Example

What is the strain energy stored in a slender cantilever of length L when carrying a concentrated load of F at the free end? What is the deflection of the free end (figure 7.13)?

The bending moment at a distance x from the free end is given by $M = -Fx$ and thus substituting for M in equation [27] gives

$$U = \int_0^L \frac{M^2 dx}{2EI} = \int_0^L \frac{(-Fx)^2 dx}{2EI} = \int_0^L \frac{F^2 x^2 dx}{2EI} = \frac{F^2 L^3}{6EI}$$

The deflection of the free end y can be obtained by equating the work done by the load with the strain energy. The work done is $\frac{1}{2}Fy$ and thus

$$U = \frac{Fy}{2} = \frac{F^2 L^3}{6EI}$$

$$y = \frac{FL^3}{3EI}$$

Example

A slender uniform beam of length L is supported at its ends and subject to a concentrated load F at its midpoint (figure 7.14). Determine the strain energy stored in the beam and the deflection of the midpoint.

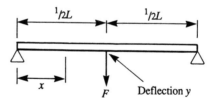

Fig. 7.14 Example.

Consider first just the left half of the beam. The reactions at each support are $\frac{1}{2}F$. Thus the bending moment a distance x from the left end is $-\frac{1}{2}Fx$. Thus substituting for M in equation [27] gives

$$U = \int_0^{L/2} \frac{M^2 dx}{2EI} = \int_0^{L/2} \frac{(-\frac{1}{2}Fx)^2 dx}{2EI} = \int_0^{L/2} \frac{F^2 x^2 dx}{8EI} = \frac{F^2 L^3}{192EI}$$

The right half of the beam will have the same strain energy and so the total strain energy is $F^2 L^3 / 96EI$.

The work done by the load in deflecting the beam is $\frac{1}{2}Fy$ and thus equating this to the strain energy gives

$$\frac{Fy}{2} = \frac{F^2 L^3}{96EI}$$

$$y = \frac{FL^3}{48EI}$$

Example

A slender uniform beam of length L is supported at its ends and subject to a concentrated load F at a distance a from one end and b from the other. Determine the strain energy stored in the beam and the deflection under the load.

Figure 7.15 shows the free-body diagram of the beam and its bending moment diagram. The reactions at the supports are determined from the equilibrium conditions as Fb/L and Fa/L. Consider first that part of the beam between $x = 0$ and $x = a$. The bending moment a distance x from the left end is $-Fbx/L$. Thus substituting for M in equation [27] gives

$$U = \int_0^a \frac{M^2 dx}{2EI} = \int_0^a \frac{(-Fbx/L)^2 dx}{2EI} = \int_0^a \frac{F^2 b^2 x^2 dx}{2EIL^2} = \frac{F^2 b^2 a^3}{6EIL^2}$$

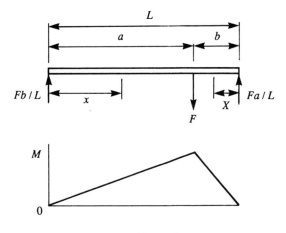

Fig. 7.15 Example.

Now consider the part of the beam between $x = a$ and $x = L$. For convenience it is worth taking the variable in this case to be a distance X from the right end and so between $X = 0$ and $X = b$. The bending moment for these values of X is $-FaX/L$. Thus substituting for M in equation [27] gives

$$U = \int_0^b \frac{M^2 dX}{2EI} = \int_0^b \frac{(-FaX/L)^2 dX}{2EI} = \int_0^b \frac{F^2 a^2 X^2 dX}{2EIL^2} = \frac{F^2 a^2 b^3}{6EIL^2}$$

The total strain energy is thus

$$U = \frac{F^2 b^2 a^3}{6EIL^2} + \frac{F^2 a^2 b^3}{6EIL^2} = \frac{F^2 a^2 b^2 (a + b)}{6EIL^2} = \frac{F^2 a^2 b^2}{6EIL}$$

If y is the deflection under the load then the work done by the load is $\frac{1}{2}Fy$ and so

$$\frac{Fy}{2} = \frac{F^2 a^2 b^2}{6EIL}$$

$$y = \frac{Fa^2 b^2}{3EIL}$$

Example

What is the strain energy stored in a slender cantilever as a result of it having a weight per unit length of w?

Consider the bending moment a distance x from the free end (figure 7.16). The weight of the beam of length x is wx and this can be considered to be located at its midpoint. Thus the bending moment is $-(wx)\frac{1}{2}x = -\frac{1}{2}wx^2$. Substituting for M in equation [27] gives

$$U = \int_0^L \frac{M^2 dx}{2EI} = \int_0^L \frac{(-\frac{1}{2}wx^2)^2 dx}{2EI} = \int_0^L \frac{w^2 x^4 dx}{8EI^2} = \frac{w^2 L^5}{40EI}$$

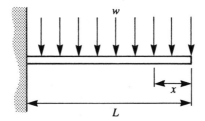

Fig. 7.16 Example.

7.7 Impact loading of beams

The maximum deflection of a beam when subject to an impact load can be determined by equating the work done by the load with the consequential strain energy stored in the beam. In this discussion the problem is simplified by assuming, as early in the chapter for axial impact loads, that a falling weight sticks to the beam and moves with it and also that the beam does not exceed the elastic limit and so Hooke's law is obeyed. Additionally it is also assumed that the deflected shape of the beam under an impact load is the same shape as could be obtained under a suitable static load.

Consider a beam of length L which is supported at its ends and has a mass m dropped from a height h on to its midpoint (figure 7.17). The potential energy lost by the mass in falling through the distance h is mgh and the further potential energy lost by it in deflecting the beam a distance y is mgy. Thus the total energy converted to strain energy is $(mgh + mgy)$.

$$U = mgh + mgy$$

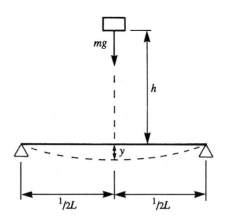

Fig. 7.17 Impact load on beam.

But the deflection y of such a simply supported beam is (see example above)

$$y = \frac{FL^3}{48EI}$$

Hence since the work done in bending the beam is $\frac{1}{2}Fy$ and this work done is the internal energy

$$U = \frac{Fy}{2} = \frac{(48EIy/L^3)y}{2} = \frac{24EIy^2}{L^3}$$

Thus equating this with the total energy converted to strain energy by the falling mass

$$mgh + mgy = \frac{24EIy^2}{L^3}$$

$$24EIy^2 - mGL^3y - mghL^3 = 0$$

This is a quadratic of the form $ay^2 + by + c = 0$ and so has the roots

$$y = \frac{mgL^3 \pm \sqrt{(m^2g^2L^6 + 4 \times 24EImghL^3)}}{2 \times 24EI}$$

$$= \frac{mgL^3}{48EI} \pm \sqrt{\left[\left(\frac{mgL^3}{48EI}\right)^2 + 2h\left(\frac{mgL^3}{48EI}\right)\right]} \tag{28}$$

The static deflection y_{st} of the beam due to the weight mg is $mgL^3/48EI$ and so equation [28] can be written, for its positive root, as

$$y = y_{st} + \sqrt{(y_{st}^2 + 2hy_{st})} \tag{29}$$

As h is generally large compared with y the equation can be approximated to

$$y \approx \sqrt{(2hy_{st})} \tag{30}$$

Example

What is the maximum bending stress produced in a beam of length L supported at its ends when a mass m is dropped from a height h on to its midpoint? The beam has a rectangular cross-section of area A and the deflection may be considered to be small compared with h.

The maximum deflection y of the midpoint is given by equation [28] as

$$y = \frac{mgL^3}{48EI} + \sqrt{\left[\left(\frac{mgL^3}{48EI}\right)^2 + 2h\left(\frac{mgL^3}{48EI}\right)\right]} \approx \sqrt{\left[\frac{2hmgL^3}{48EI}\right]}$$

For a rectangular cross-section of breadth b and depth d the second moment of area I is $bd^3/12$. Hence

$$y = \sqrt{\left(\frac{hmgL^3}{2Ebd^3}\right)}$$

The stress σ a distance a from the neutral axis is given by (see chapter 3, equation [13]) $\sigma/a = E/R$, where R is the radius of curvature of the bent beam. Thus the stress on the surface is $E(\frac{1}{2}d)/R$. Assuming that the beam is bent into the arc of a circle, the radius R is related to the deflection y by (using the theorem of intersecting chords, see figure 7.18)

$$(2R - y)y = \tfrac{1}{2}L \times \tfrac{1}{2}L$$
$$2Ry \approx \tfrac{1}{4}L^2$$

Hence the stress is

$$\sigma = \frac{Ed}{2R} = \frac{8Edy}{2L^2} = \frac{4Ed}{L^2}\sqrt{\left(\frac{hmgL^3}{2Ebd^3}\right)} = \sqrt{\left(\frac{8hmgE}{LA}\right)}$$

7.8 Strain energy due to combined loading

In the previous discussion the situations analysed have been ones where it

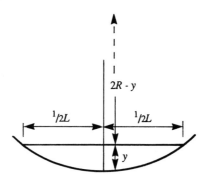

Fig. 7.18 Example.

could be assumed that the strain energy stored in a member was due to just one form of loading. In many situations, however, strain energy is produced in a member as a result of more than one type of action. Thus, for example, a horizontal cylindrical cantilever in the form of a quarter circle acted on by a vertical force at its free end, as in figure 7.19, will suffer bending, torsion and shear. Thus the strain energy in the material will be the sum of the energies due to bending, torsion and shear. For slender beams the strain energy due to shear can generally be neglected. Hence, using equations [24] and [27],

$$U = \int_0^L \frac{T^2}{2JG} \, dx + \int_0^L \frac{M^2}{2EI} \, dx$$

The torque T is $Fb = FR - FR \cos \theta$ and the moment M is $Fa = FR \sin \theta$. Since δx is approximately $R\delta\theta$ then

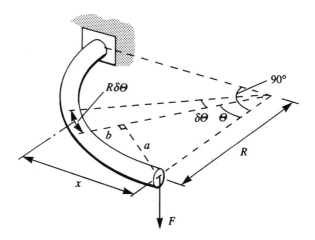

Fig. 7.19 Bending and torsion.

$$U = \int_0^{\pi/2} \frac{F^2 R^2 (1 - \cos \theta)^2}{2JG} R\, d\theta + \int_0^{\pi/2} \frac{F^2 R^2 \sin^2 \theta}{2EI} R\, d\theta$$

$$= \frac{F^2 R^3}{2JG} \int_0^{\pi/2} (1 - 2\cos \theta + \tfrac{1}{2} + \tfrac{1}{2}\cos 2\theta)\, d\theta$$

$$+ \frac{F^2 R^3}{2EI} \int_0^{\pi/2} (\tfrac{1}{2} - \tfrac{1}{2}\cos 2\theta)\, d\theta$$

$$= \frac{F^2 R^3}{2JG} \left(\frac{3\pi}{4} - 2 \right) + \frac{F^2 R^3 \pi}{8EI}$$

7.9   Castigliano's theorem

Consider an object or structure subject to a number of forces F_1, F_2, F_3, ... F_n with the displacements produced in the direction of each force being y_1, y_2, y_3, ... y_n. The strain energy U resulting from these displacements of the object is thus

$$U = \tfrac{1}{2}F_1 y_1 + \tfrac{1}{2}F_2 y_2 + \tfrac{1}{2}F_3 y_3 \dots \tfrac{1}{2}F_n y_n \qquad [31]$$

Now let F_1 increase to $F_1 + \delta F_1$ and consequently y_1 increase to $y_1 + \delta y_1$, y_2 to $y_2 + \delta y_2$, y_3 to $y_3 + \delta y_3$, ... y_n to $y_n + \delta y_n$. The strain energy increases to $U + \delta U$, with

$$U + \delta U = \tfrac{1}{2}(F_1 + \delta F_1)(y_1 + \delta y_1) + \tfrac{1}{2}F_2(y_2 + \delta y_2)$$
$$+ \tfrac{1}{2}F_3(y_3 + \delta y_3) + \dots \tfrac{1}{2}F_n(y_n + \delta y_n)$$

This can be expanded to

$$U + \delta U = \tfrac{1}{2}F_1 y_1 + \tfrac{1}{2}F_1 \delta y_1 + \tfrac{1}{2}\delta F_1 y_1 + \tfrac{1}{2}\delta F_1 \delta y_1 + \tfrac{1}{2}F_2 y_2$$
$$+ \tfrac{1}{2}F_2 \delta y_2 + \tfrac{1}{2}F_3 y_3 + \tfrac{1}{2}F_3 \delta y_3 + \dots \tfrac{1}{2}F_n y_n + \tfrac{1}{2}F_n \delta y_n$$

Subtracting equation [31] and neglecting the term $\tfrac{1}{2}\delta F_1 \delta y_1$ as being insignificant, then

$$\delta U = \tfrac{1}{2}F_1 \delta y_1 + \tfrac{1}{2}\delta F_1 y_1 + \tfrac{1}{2}F_2 \delta y_2 + \tfrac{1}{2}F_3 \delta y_3 + \dots + \tfrac{1}{2}F_n \delta y_n \qquad [32]$$

This increase in strain energy results from the forces doing further work when F_1 is increased to $F_1 + \delta F_1$. This extra work δW is

$$\delta W = F_1 \delta y_1 + F_2 \delta y_2 + F_3 \delta y_3 + \dots F_n \delta y_n$$

Thus, substituting into equation [32],

$$\delta U = \tfrac{1}{2}\delta W + \tfrac{1}{2}\delta F_1 y_1$$

But the internal energy change results from the work done, i.e. $\delta U = \delta W$, thus

$$\delta U = \delta F_1 y_1$$

$$y_1 = \frac{\delta U}{\delta F_1}$$

In the limit as δF_1 tends to 0, this becomes

$$y_1 = \frac{\partial U}{\partial F_1} \qquad [33]$$

Similarly if the other forces are changed we have

$$y_2 = \frac{\partial U}{\partial F_2}, \quad \ldots \quad y_n = \frac{\partial U}{\partial F_n} \qquad [34]$$

A ∂ is used to indicate that the differentiation is with respect to one variable in a situation where there is more than one variable. The differential coefficient obtained is said to be the partial derivative of the function. In the situation described above the deflection in the direction of a particular force due to all the applied forces is the partial derivative of the total strain energy of the system with respect to the force concerned. These equations [33] and [34] are known as *Castigliano's theorem*.

Consider the application of the theorem to the determination of the deflection of a slender beam. The strain energy U of the beam is given by equation [27] as

$$U = \int_0^L \frac{M^2 dx}{2EI}$$

Hence, using Castigliano's theorem, the deflection y in the direction of an applied force F is given by

$$y = \frac{\partial U}{\partial F} = \frac{\partial}{\partial F} \int_0^L \frac{M^2 dx}{2EI} = \frac{1}{2EI} \int_0^L \frac{\partial}{\partial F}(M^2) dx = \frac{1}{2EI} \int_0^L 2M \frac{\partial M}{\partial F} dx$$

$$y = \frac{1}{EI} \int_0^L M \frac{\partial M}{\partial F} dx \qquad [35]$$

The above equation refers only to the displacement in the direction of an applied force. There can however be situations where the deflection is required in some direction for which there is no applied force. In such a situation an imaginary force is considered to be applied in that direction and the analysis carried out using the theorem. Then the imaginary force is equated to zero to give the required deflection.

Example

Use Castigliano's theorem to determine the deflection of the free end of a cantilever when subject to a concentrated force F at that point.

For such a cantilever (as in figure 7.13), the bending moment M is related to the distance x from the free end by $M = -Fx$, hence $\partial M/\partial F = -x$. Thus equation [35] becomes

$$y = \frac{1}{EI} \int_0^L M \frac{\partial M}{\partial F} dx = \frac{1}{EI} \int_0^L (-Fx)(-x) dx = \frac{FL^3}{3EI}$$

Example

Use Castigliano's theorem to determine the deflection of the free end of a cantilever when subject to a uniform loading of w per unit length.

For such a cantilever (as in figure 7.16), since we require the displacement at a point where there is no concentrated load, a vertical concentrated force F is introduced at that point, i.e. the free end. The bending moment M is then, when x is measured from the free end, $-Fx - wx(\frac{1}{2}x)$. Hence $\partial M/\partial F = -x$. Now putting $F = 0$ gives $M = -\frac{1}{2}wx^2$ and $\partial M/\partial F = -x$. Thus equation [35] becomes

$$y = \frac{1}{EI} \int_0^L M\frac{\partial M}{\partial F}\,dx = \frac{1}{EI} \int_0^L (-\tfrac{1}{2}wx^2)\,(-x)\,dx = \frac{wL^4}{8EI}$$

Example

Use Castigliano's theorem to determine the deflection of the free end of the curved cantilever shown in figure 7.20 due to bending when subject to a concentrated force F at the free end.

The bending moment M is related to the distance x from the free end by $M = -Fx$. But $x = R\sin\theta$. Hence $M = -FR\sin\theta$ and $dM/dF = -R\sin\theta$. Also, to a reasonable approximation, $\delta x = R\delta\theta$. Thus equation [35] becomes

$$y = \frac{1}{EI} \int_0^L M\frac{\partial M}{\partial F}\,dx = \frac{1}{EI} \int_0^{\pi/2} (-FR\sin\theta)\,(-R\sin\theta)R\,d\theta$$

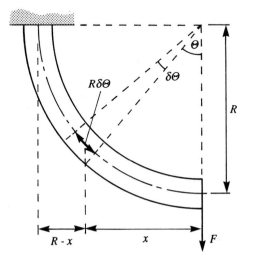

Fig. 7.20 Example.

$$= \frac{1}{EI} \int_0^{\pi/2} FR^3 \sin^2\theta \, d\theta = \frac{1}{EI} \int_0^{\pi/2} FR^3 \tfrac{1}{2} (1 - \cos 2\theta) \, d\theta$$

$$= \frac{FR^3}{2EI} [\theta - \tfrac{1}{2}\sin 2\theta]_0^{\pi/2} = \frac{\pi FR^3}{4EI}$$

Example

Use Castigliano's theorem to determine the horizontal and vertical deflections of the free end of the constant cross-section bar shown in figure 7.21 when the free end is subject to a concentrated horizontal force F_H.

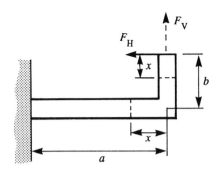

Fig. 7.21 Example.

There is no vertical force acting on the bar and so to determine the vertical deflection we need to suppose that there is a vertical force F_V which we can later equate to zero. Then the bending moment for the horizontal part of the bar due to both the vertical and horizontal forces is $M = F_H b + F_V x$ and for the vertical part $M = F_H x$. Then for the horizontal part of the bar $dM/dF_H = b$ and $dM/dF_V = x$, for the vertical part $dM/dF_H = x$ and $dM/dF_V = 0$.

For the horizontal deflection, equation [35] gives

$$y_H = \frac{1}{EI} \int_0^L M \frac{\partial M}{\partial F} dx = \frac{1}{EI} \int_0^b (F_H x) \, x \, dx + \frac{1}{EI} \int_0^a (F_H b + F_V x) \, b \, dx$$

$$= \frac{F_H b^3}{3EI} + \frac{F_H b^2 a}{EI} + \frac{F_V b a^2}{2EI}$$

Since $F_V = 0$ then

$$y_H = \frac{F_H b^2}{EI} \left(\frac{b}{3} + a \right)$$

For the vertical deflection, equation [35] gives

$$y_V = \frac{1}{EI} \int_0^b (F_H x)\, 0\, dx + \frac{1}{EI} \int_0^a (F_H b + F_V x)\, x\, dx$$

$$= 0 + \frac{F_H b a^2}{2EI} + \frac{F_V a^3}{3EI}$$

Since $F_V = 0$ then

$$y_V = \frac{F_H b a^2}{2EI}$$

Example

Determine for the horizontal curved bar shown in figure 7.19 the vertical deflection in the direction of the force F.

The strain energy due to torsion and bending is given by

$$U = \int_0^L \frac{T^2}{2JG}\, dx + \int_0^L \frac{M^2}{2EI}\, dx$$

Thus Castigliano's theorem gives for the vertical deflection

$$y = \frac{\partial U}{\partial F} = \int_0^L \frac{T}{JG} \frac{\partial T}{\partial F}\, dx + \int_0^L \frac{M}{EI} \frac{\partial M}{\partial F}\, dx$$

The torque T is $Fb = FR - FR \cos \theta$ and so $\partial T / \partial F = R - R \cos \theta$. The moment M is $Fa = FR \sin \theta$ and so $\partial M / \partial F = R \sin \theta$. Since δx is approximately $R \delta \theta$ then

$$y = \frac{1}{JG} \int_0^{\pi/2} (FR - FR \cos \theta)\,(R - R \cos \theta) R d\theta$$

$$+ \frac{1}{EI} \int_0^{\pi/2} FR^2 \sin^2 \theta\, R\, d\theta$$

$$= \frac{FR^3}{JG} \int_0^{\pi/2} (1 - \cos \theta)^2\, d\theta + \frac{FR^3}{EI} \int_0^{\pi/2} \sin^2 \theta\, d\theta$$

$$= \frac{FR^3}{JG} \int_0^{\pi/2} (1 - 2\cos \theta + \cos^2 \theta)\, d\theta + \frac{FR^3}{EI} \int_0^{\pi/2} \sin^2 \theta\, d\theta$$

$$= \frac{FR^3}{JG} \int_0^{\pi/2} (1 - 2\cos \theta + \tfrac{1}{2} + \tfrac{1}{2}\cos 2\theta)\, d\theta$$

$$+ \frac{FR^3}{EI} \int_0^{\pi/2} (\tfrac{1}{2} - \tfrac{1}{2}\cos 2\theta)\, d\theta$$

$$= \frac{FR^3}{JG} \left(\frac{3\pi}{4} - 2 \right) + \frac{FR^3 \pi}{4EI}$$

Problems

(1)   What is the strain energy stored in a bar of cross-section 40 mm by

50 mm and length 1.2 m as a result of it being subject to a tensile force of 100 kN? The tensile modulus of the material is 205 GPa.

(2)   What is the proof resilience of steel with a tensile modulus of 205 GPa and an elastic limit of 230 MPa?

(3)   What is the strain energy stored in a steel bar of length 1.0 m and 40 mm diameter for 0.4 m of its length and 60 mm diameter for the remainder when it is subject to a tensile force of 100 kN? The tensile modulus of elasticity is 205 GPa.

(4)   Derive an equation for the strain energy of a bar of length L when subject to a tensile load of F and having half its length of cross-sectional area A and the other half $4A$, the tensile modulus of elasticity being E.

(5)   What is the effect on the strain energy that can be stored by a bar with a constant circular cross-sectional area when subject to a tensile force F if the bar has a notch made which reduces the diameter to a third over a quarter of its length?

(6)   A cone of base diameter d and vertical height h hangs vertically under its own weight (figure 7.22). What is the strain energy stored in the cone if the cone material has a density of ρ and a tensile modulus E?

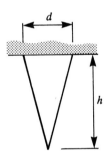

Fig. 7.22 Problem 6.

(7)   What is the strain energy stored in the truss shown in figure 7.23 when the force F is acting? The two members have the same cross-sectional area A and tensile modulus E.

(8)   What is the strain energy stored in the truss shown in figure 7.24 when the force F is acting? The two members have the same cross-sectional area A and tensile modulus E.

(9)   A vertical steel rod of constant diameter 25 mm and length 5.0 m is rigidly fixed at its upper end and the bottom, flanged, end is free to move. What will be the maximum deflection and the maximum

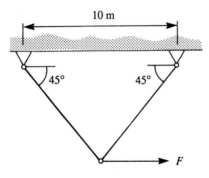

Fig. 7.23 Problem 7.

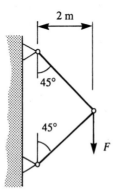

Fig. 7.24 Problem 8.

stress produced by a collar of mass 2.0 kg sliding down the rod through a height of 50 mm on to the flange without rebounding?

(10)   A mass on one end of a length of elastic cord of length 0.50 m is perched on the edge of a shelf, the other end of the cord being fixed to the edge. What will be the maximum distance the mass will be below the edge of the shelf when it falls off the shelf if the static length of the elastic cord in this situation is 0.55 m?

(11)   What is the maximum height from which a mass of 30 kg can fall on to a flange on the lower end of a vertical rod of length 1.0 m and diameter 20 mm, the upper end being rigidly fixed, if the stress in the rod is not to exceed 80 MPa? The tensile modulus of the rod is 200 GPa.

(12)   A load of 10 kN is to be suddenly applied to the free end of a vertical rod of length 500 mm, the upper end being rigidly fixed.

What should the diameter of the rod be if the maximum stress in the rod is not to exceed 100 MPa? The tensile modulus of the rod is 205 GPa.

(13)   A vertical steel rod of length 2.0 m and diameter 25 mm is rigidly fixed at its upper end and has a flange at its lower, free, end. What will be the maximum stress produced in the rod when a load of 20 kN is applied, (a) gradually to the lower end, (b) suddenly to the lower end, (c) by a collar falling from a height of 50 mm on to the flange?

(14)   What is the strain energy stored in a solid uniformly circular shaft of length 3.0 m and diameter 120 mm when subject to pure torsion if the maximum shear stress is 40 MPa? $G = 80$ GPa.

(15)   Determine the strain energy stored in a simply supported slender uniform beam of length L and its central deflection, when it is subject to the two identical concentrated loads F at $L/3$ and $2L/3$ from one end.

(16)   A simply supported beam ABC of length L has AB of length $\frac{1}{2}L$ and second moment of area I_1 and BC of length $\frac{1}{2}L$ and second moment of area I_2. Determine the strain energy stored in the beam when it is subject to a vertical load F at its midpoint and the consequential central deflection.

(17)   Determine the strain energy stored in a slender uniform beam of length L when supported at its ends and carrying a uniformly distributed load of w per unit length.

(18)   Determine, as a result of strain energy considerations, the deflection of a cantilever under its own weight if it is a slender I-section with a length of 2.0 m, a mass of 25 kg/m and a second moment of area of $2400 \times 10^{-8}\,\text{m}^4$.

(19)   Determine the relationship between the maximum deflection of a cantilever when a weight falls through a height h on to its end and the static deflection.

(20)   Determine the relationship between the maximum bending stress of a simple beam supported at its ends and the static bending stress and static deflection when a weight is dropped through a distance h on to the midpoint of the beam.

(21)   Use Castigliano's theorem to determine the vertical deflection of the slender cantilever shown in figure 7.25 due to bending as a consequence of the concentrated vertical force F applied at its free end.

(22)   Use Castigliano's theorem to determine how the separation of the ends of a slender semicircular frame, radius R, depends on the forces F acting at the ends (figure 7.26).

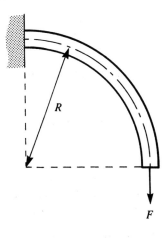

Fig. 7.25 Problem 21.

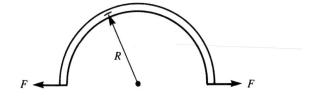

Fig. 7.26 Problem 22.

(23)   Use Castigliano's theorem to determine the vertical deflection of the end of the slender framework shown in figure 7.27 where a concentrated force F acts. The framework is of constant cross-section and tensile modulus throughout.

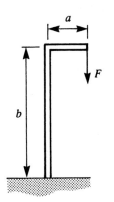

Fig. 7.27 Problem 23.

(24) Use Castigliano's theorem to determine the vertical deflection at the free end of a slender cantilever of uniform weight per unit length w. (Hint: consider there to be a concentrated force at the free end and then later equate this to zero.)

(25) Use Castigliano's theorem to determine the horizontal and vertical deflections of the free end of the slender quarter circle bar of radius R when acted on by a concentrated horizontal force at its free end (figure 7.28).

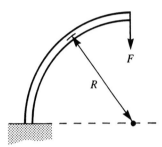

Fig. 7.28 Problem 25.

(26) Use Castigliano's theorem to determine the vertical deflection of the free end of a slender overhanging beam in figure 7.29 when subject to a concentrated load at the free end.

(27) Use Castigliano's theorem to determine the vertical deflection of the free end of the slender horizontal bent rod (figure 7.30) when subject to a vertical force at its free end. The rod is of constant cross-sectional area.

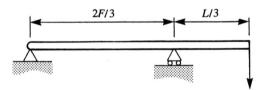

Fig. 7.29 Problem 26.

(28) A circular cross-section rod with a diameter of 10 mm is bent into the form of three-quarters of a circle, radius 150 mm. One end is fixed to a horizontal surface and the free end is constrained to move vertically, as illustrated in figure 7.31. Use Castigliano's theorem to determine the vertical deflection of the free end when it is subject to a vertical force of 100 N. The material has a modulus of elasticity of 200 GPa.

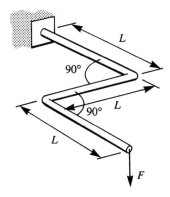

Fig. 7.30 Problem 27.

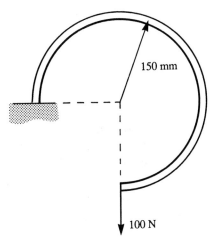

Fig. 7.31 Problem 28.

(29) A circular cross-section rod with a diameter of 10 mm is bent into the form of a quarter-circle, radius 150 mm. One end is fixed and the free end is acted on by a horizontal force of 100 N, as illustrated in figure 7.32. Use Castigliano's theorem to determine the horizon-

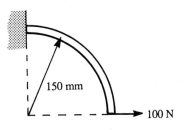

Fig. 7.32 Problem 29.

tal and vertical deflections of the free end. The material has a modulus of elasticity of 200 GPa.

(30)   A circular cross-section rod with a diameter of 10 mm and a length of 1250 mm is bent into an L-shape and used as a cantilever with the horizontal length of the L being 1000 mm and the vertical, free end, being 250 mm and pointing downwards. Use Castigliano's theorem to determine the horizontal deflection of the free end of the cantilever when it is subject to a horizontal force of 100 N. The material has a modulus of elasticity of 200 GPa.

Chapter 8

Stress and strain analysis

8.1 Complex stress situations

The loading and shape of most engineering components are generally such that at any point in the material the stress is more complex than just a tensile or compressive stress in one direction. A section might be subject to direct stresses in two perpendicular directions and a shear stress.

For example, in a drive shaft used to transmit power by means of a belt passing over a pulley, the shaft can be subject to bending stresses due to the load exerted by the belt plus pulley and a shear stress from the torque exerted by the belt. In other situations the shaft might be subject to axial stresses and a shear stress from a torque.

The stresses encountered in axially loaded bars, beams subject to a bending moment and shafts in torsion are examples of a state of stress known as *plane stress*. With such a state of stress we have a two-dimensional state of stress rather than three-dimensional. If we consider an element of the material in the form of a cube, then there are only stresses on the faces at right angles to the x- and y-axes, those on the faces at right angles to the z-axis being negligible.

8.2 Simple tension

Consider a long uniform bar, cross-sectional area A, subject to an axial tensile force F, as in figure 8.1. For any transverse plane AB through the bar there will be a pure normal stress of σ_x acting on it, with $\sigma_x = F/A$. Now consider an inclined plane BC which is at an angle θ to the y-direction. The force F will have a component of $F\cos\theta$ at right angles to this plane and

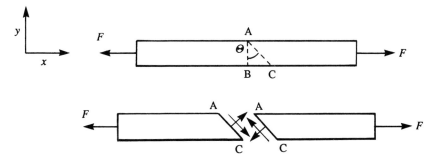

Fig. 8.1 Stresses on an inclined plane.

$F\sin\theta$ parallel to the plane BC. The area of the inclined plane is $A/\cos\theta = A\sec\theta$. Thus the normal stress σ_θ acting on the inclined plane is

$$\sigma_\theta = \frac{F\cos\theta}{A/\cos\theta} = \sigma_x\cos^2\theta \tag{1}$$

The stress acting parallel to the plane, i.e. the shear stress τ_θ, is

$$\tau_\theta = \frac{F\sin\theta}{A/\cos\theta} = \sigma_x\sin\theta\cos\theta \tag{2}$$

Since $\sin 2\theta = 2\sin\theta\cos\theta$, then equation [2] can be written as

$$\tau_\theta = \tfrac{1}{2}\sigma_x\sin 2\theta \tag{3}$$

With $\theta = 0°$, when the section concerned is at right angles to the bar, the shear stress $\tau_\theta = 0$ and the stress $\sigma_\theta = \sigma_x$. With $\theta = 90°$, when the section concerned is along the longitudinal axis of the bar, the shear stress $\tau_\theta = 0$ and the stress $\sigma_\theta = 0$. The maximum value of the shear stress is with $\sin 2\theta = 1$, i.e. when $\theta = 45°$, the shear stress then being $\tfrac{1}{2}\sigma_x$. On all inclined sections between $0°$ and $90°$ there are shear stresses, with the maximum value being on planes at $45°$ to the longitudinal axis of the bar. For materials whose shear strength is less than half the tensile strength, direct tensile loading will result in failure along the planes of maximum shear stress, i.e. at $45°$ to the longitudinal axis. Figure 8.2 shows how the normal and shear stresses vary with the angle of the plane.

Though the above discussion has been for tensile forces, similar equations can be developed for axial compressive forces. This is provided the bar does not behave as a column and become subject to buckling (see chapter 5).

Example

A bar of rectangular cross-section 20 mm by 20 mm is subject to an axial tensile load of 10 kN. What will be the normal stress and the shear stress on

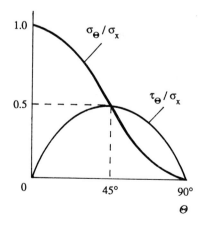

Fig. 8.2 Variation of normal and shear stresses with angle of plane.

a plane which makes an angle of 30° to a transverse plane through the bar?

Using equation [1]

$$\sigma_\theta = \sigma_x \cos^2 \theta = \frac{10 \times 10^3}{0.020 \times 0.020} \cos^2 30° = 18.75 \, \text{MPa}$$

Using equation [3]

$$\tau_\theta = \tfrac{1}{2}\sigma_x \sin 2\theta = \frac{10 \times 10^3}{2 \times 0.020 \times 0.020} \sin 60° = 10.8 \, \text{MPa}$$

8.3   Biaxial stress system

Consider a plane in a body which is subject to biaxial stresses, i.e. a two-dimensional state of stress exists with there being, for example, stresses acting in the x- and y-directions. Figure 8.3 illustrates such a situation.

Consider the equilibrium of the triangular element of that plane. Since the element is in static equilibrium then the algebraic sum of the forces acting in any direction must be zero. If the area of the hypotenuse face of the triangular element is A, then the x-direction face will have an area $A \sin \theta$ and the y-direction face an area $A \cos \theta$. The force acting at right angles to the hypotenuse face is $\sigma_\theta A$, the force at right angles to the x-direction face is $\sigma_y A \sin \theta$ and the force at right angles to the y-direction face is $\sigma_x A \cos \theta$. For equilibrium in the direction at right angles to the hypotenuse face we have to consider the resolved components of all the forces acting in that direction. The resolved component of the force at right angles to the x-direction face is $(\sigma_y A \sin \theta) \sin \theta$ and that of the force at right angles to the y-direction face is $(\sigma_x \cos \theta) \cos \theta$. Thus

$$\sigma_\theta A - \sigma_x A \cos^2 \theta - \sigma_y A \sin^2 \theta = 0 \tag{4}$$

Now consider the algebraic sum of the forces parallel to the inclined plane.

$$\tau_\theta A - \sigma_x A \sin \theta \cos \theta + \sigma_y A \sin \theta \cos \theta = 0 \tag{5}$$

Equations [4] and [5] can be written as

$$\sigma_\theta = \sigma_x \cos^2 \theta + \sigma_y \sin^2 \theta$$
$$\tau_\theta = (\sigma_x - \sigma_y) \sin \theta \cos \theta$$

But

$$\cos^2 \theta = \tfrac{1}{2}(\cos 2\theta + 1)$$
$$\sin^2 \theta = \tfrac{1}{2}(1 - \cos 2\theta)$$
$$\sin \theta \cos \theta = \tfrac{1}{2}\sin 2\theta$$

Thus

$$\sigma_\theta = \tfrac{1}{2}(\sigma_x + \sigma_y) + \tfrac{1}{2}(\sigma_x - \sigma_y) \cos 2\theta \tag{6}$$
$$\tau_\theta = \tfrac{1}{2}(\sigma_x - \sigma_y) \sin 2\theta \tag{7}$$

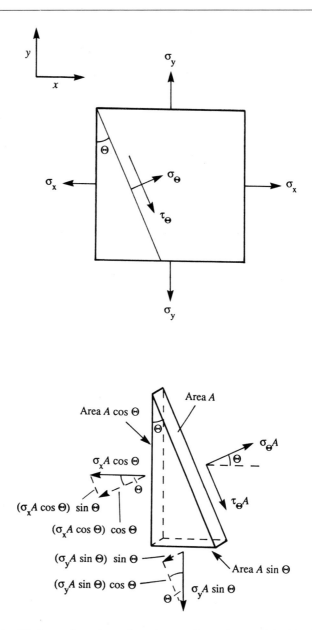

Fig. 8.3 Equilibrium of a triangular element in a plane subject to biaxial stress.

Example

If the stresses on two perpendicular planes through a point are 40 MPa tension and 20 MPa compression, what are the stresses on a plane at 60° to the tensile force plane?

Using equation [6]

$$\sigma_\theta = \tfrac{1}{2}(\sigma_x + \sigma_y) + \tfrac{1}{2}(\sigma_x - \sigma_y)\cos 2\theta$$
$$= \tfrac{1}{2}(40 - 20) + \tfrac{1}{2}(40 + 20)\cos 120° = -5\,\text{MPa}$$

Using equation [7]

$$\tau_\theta = \tfrac{1}{2}(\sigma_x - \sigma_y)\sin 2\theta = \tfrac{1}{2}(40 + 20)\sin 120° = 26\,\text{MPa}$$

Thus there is a shear stress of 26 MPa and a direct compressive stress of 5 MPa acting on the plane.

8.3.1 *Principal stresses*

Figure 8.4 shows how, for a biaxial stress situation, the normal stress σ_θ and the shear stress τ_θ given by equations [6] and [7], vary with the angle θ. There are two mutually perpendicular planes for which the shear stress is zero, these being when $\sin 2\theta = 0$, i.e. $\theta = 0°$ or $90°$. Such planes are called the *planes of principal stress* and the direct stresses acting on these planes the *principal stresses*.

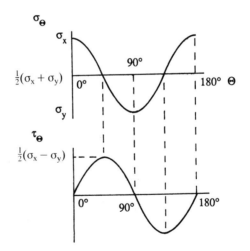

Fig. 8.4 Variation of normal and shear stresses with θ.

The planes of the principal stresses occur at $\theta = 0°$ and $\theta = 90°$. This gives, using equation [6], the principal stresses as

$$\sigma_\theta = \tfrac{1}{2}(\sigma_x + \sigma_y) + \tfrac{1}{2}(\sigma_x - \sigma_y)\cos 2\theta$$

with $\theta = 0°$

$$\sigma_\theta = \tfrac{1}{2}(\sigma_x + \sigma_y) + \tfrac{1}{2}(\sigma_x - \sigma_y) \times (+1) = \sigma_x$$

and with $\theta = 90°$

$$\sigma_\theta = \tfrac{1}{2}(\sigma_x + \sigma_y) + \tfrac{1}{2}(\sigma_x - \sigma_y) \times (-1) = \sigma_y$$

These are the maximum and minimum values of the direct stresses.

The maximum shear stress occurs on the plane for which equation [7] has a maximum value, i.e. when $\sin 2\theta = 1$ or $\theta = 45°$.

$$\text{max. } \tau_\theta = \tfrac{1}{2}(\sigma_x - \sigma_y) \sin 2\theta = \tfrac{1}{2}(\sigma_x - \sigma_y) \qquad [8]$$

Example

A cylindrical boiler has for its cylindrical plates a circumferential stress of 140 MPa and an axial stress of 70 MPa. What is the maximum shear stress?

The plates of the boiler are subject to biaxial stresses. Thus using equation [8]

$$\text{max. } \tau_\theta = \tfrac{1}{2}(\sigma_x - \sigma_y) = \tfrac{1}{2}(140 - 70) = 35 \text{ MPa}$$

8.4 General two-dimensional stress system

Consider a triangular element, as in figure 8.5, where there are direct stresses of σ_x and σ_y and shear stress acting. Over the two faces in the x-direction are shearing stresses τ_{xy} and over the faces in the y-direction τ_{yx}. Note that the convention has been adopted of using the first subscript of a shear stress to indicate the face on which the stress acts and the second subscript the direction on that face. If we image a set of Cartesian coordinates centred at the centre of the element then a shear stress is positive if it acts on the positive face of the element in the positive direction of an axis, positive if it acts on the negative face in the negative direction, negative if it acts on a positive face in the negative direction of an axis and negative if it acts on the negative face in the positive direction.

Because there is rotational equilibrium then we must have $\tau_{xy} = \tau_{yx}$. Because the element is in static equilibrium then the algebraic sum of the resolved components of the forces in any direction must be zero. Thus for the forces normal to the hypotenuse plane of the triangular element

$$\sigma_\theta A - (\sigma_x A \cos\theta)\cos\theta - (\sigma_y A \sin\theta)\sin\theta$$
$$- (\tau_{xy} A \cos\theta)\sin\theta - (\tau_{xy} A \sin\theta)\cos\theta = 0$$

which simplifies to

$$\sigma_\theta = \sigma_x \cos^2\theta + \sigma_y \sin^2\theta + 2\tau_{xy}\sin\theta\cos\theta \qquad [9]$$

For the forces parallel to the hypotenuse plane of the triangular element

$$\tau_\theta A + (\sigma_x A \cos\theta)\sin\theta - (\sigma_y A \sin\theta)\cos\theta$$
$$- (\tau_{xy} A \cos\theta)\cos\theta + (\tau_{xy} A \sin\theta)\sin\theta = 0$$

which simplifies to

$$\tau_\theta = -\sigma_x \cos\theta\sin\theta + \sigma_y \sin\theta\cos\theta + \tau_{xy}(\cos^2\theta - \sin^2\theta) \qquad [10]$$

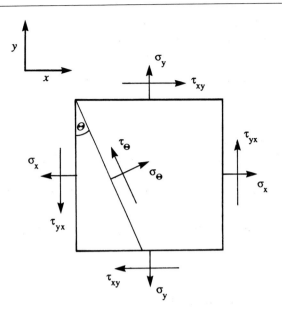

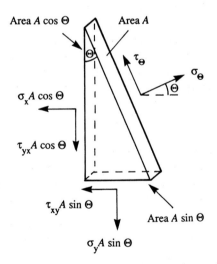

Fig. 8.5 Equilibrium of a triangular element in a general two-dimensional stress system.

Since

$$\cos^2 \theta = \tfrac{1}{2}(\cos 2\theta + 1)$$
$$\sin^2 \theta = \tfrac{1}{2}(1 - \cos 2\theta)$$
$$\sin \theta \cos \theta = \tfrac{1}{2}\sin 2\theta$$

then equations [9] and [10] can be written as

$$\sigma_\theta = \tfrac{1}{2}(\sigma_x + \sigma_y) + \tfrac{1}{2}(\sigma_x - \sigma_y)\cos 2\theta + \tau_{xy}\sin 2\theta \qquad [11]$$

$$\tau_\theta = -\tfrac{1}{2}(\sigma_x - \sigma_y)\sin 2\theta + \tau_{xy}\cos 2\theta \qquad [12]$$

Example

A plane element in a body is subject to stresses of $\sigma_x = 100\,\text{MPa}$, $\sigma_y = 40\,\text{MPa}$ and $\tau_{xy} = \tau_{yx} = 25\,\text{MPa}$. What are the stresses acting on an element at an angle of 45° to the y-direction?

Figure 8.6 shows the stress system. Using equation [11]

$$\begin{aligned}
\sigma_\theta &= \tfrac{1}{2}(\sigma_x + \sigma_y) + \tfrac{1}{2}(\sigma_x - \sigma_y)\cos 2\theta + \tau_{xy}\sin 2\theta \\
&= \tfrac{1}{2}(100 + 40) + \tfrac{1}{2}(100 - 40)\cos 90° + 25\sin 90° \\
&= 95\,\text{MPa}
\end{aligned}$$

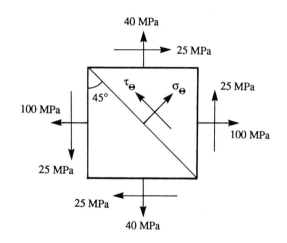

Fig. 8.6 Example.

Using equation [12]

$$\begin{aligned}
\tau_\theta &= -\tfrac{1}{2}(\sigma_x - \sigma_y)\sin 2\theta + \tau_{xy}\cos 2\theta \\
&= -\tfrac{1}{2}(100 - 40)\sin 90° + 25\cos 90° \\
&= -30\,\text{MPa}
\end{aligned}$$

Example

A cantilevered shaft of diameter 200 mm is subject to a torque of 100 kN m and a load of 300 kN at its free end (figure 8.7). Determine the normal and shear stresses a distance of 0.5 m from the free end on (a) a plane on its upper surface and which is at 30° to a transverse plane through the shaft and

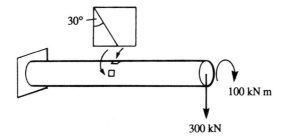

Fig. 8.7 Example.

(b) its neutral plane and which is at 30° to a transverse plane through the shaft.

(a) The bending stress σ_x on the upper surface is given by, equation [17] chapter 3,

$$\sigma_x = \frac{My}{I}$$

Since $M = 300 \times 0.5 = 150\,\text{kN}\,\text{m}$ and $I = \pi d^4/64$, then the stress when $y = $ radius r is

$$\sigma_x = \frac{150 \times 10^3 \times 0.100}{\pi \times 0.200^4/64} = 191\,\text{MPa}$$

The shear stress due to the torque is given by, equation [5] chapter 6,

$$\tau = \frac{Tr}{J}$$

Since $J = \pi d^4/32$, then

$$\tau = \frac{100 \times 10^3 \times 0.100}{\pi \times 0.200^4/32} = 64\,\text{MPa}$$

There is no shear stress on the upper surface due to the shear force resulting from the bending (see chapter 3). The stresses on the plane element are thus as shown in figure 8.8(a). Hence, using equations [11] and [12]

$$\sigma_\theta = \tfrac{1}{2}(\sigma_x + \sigma_y) + \tfrac{1}{2}(\sigma_x - \sigma_y)\cos 2\theta + \tau_{xy}\sin 2\theta$$
$$= \tfrac{1}{2}(191 + 0) + \tfrac{1}{2}(191 - 0)\cos 60° + 64\sin 60°$$
$$= 199\,\text{MPa}$$

$$\tau_\theta = -\tfrac{1}{2}(\sigma_x - \sigma_y)\sin 2\theta + \tau_{xy}\cos 2\theta$$
$$= -\tfrac{1}{2}(191 - 0)\sin 60° + 64\cos 60°$$
$$= -51\,\text{MPa}$$

(b) For the plane on the neutral plane there will be no bending stresses. Thus there will be only shear stresses acting on the element. These will be

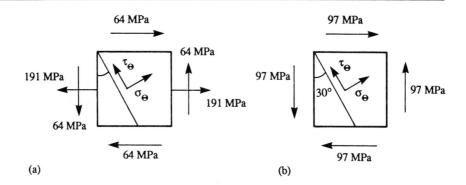

Fig. 8.8 Example.

due to the torque, having the same value as determined above of 64 MPa, and the shear force resulting from the bending. This is given by equation [29], chapter 3, as

$$\tau = \frac{VQ}{Ib}$$

For a solid circular section, the second moment of area I is $\frac{1}{4}\pi r^4$, $b = 2r$ and the first moment of area $Q = \frac{1}{2}\pi r^2(4r/3\pi)$. Thus

$$\tau = \frac{4V}{3\pi r^2}$$

Since $V = 300\,\text{kN}$, then

$$\tau = \frac{4 \times 300 \times 10^3}{3\pi \times 0.100^2} = 13\,\text{MPa}$$

The shear stresses are thus $13 + 64 = 97\,\text{MPa}$ and as shown in figure 8.8(b). Hence, using equations [11] and [12],

$$\sigma_\theta = \tfrac{1}{2}(\sigma_x + \theta y) + \tfrac{1}{2}(\sigma_x - \sigma_y)\cos 2\theta + \tau_{xy}\sin 2\theta$$
$$= 0 + 0 + 97\sin 60° = 84\,\text{MPa}$$
$$\tau_\theta = -\tfrac{1}{2}(\sigma_x - \sigma_y)\sin 2\theta + \tau_{xy}\cos 2\theta$$
$$= 0 + 97\cos 60° = 49\,\text{MPa}$$

8.4.1 *Principal stresses*

For the general two-dimensional stress system, the shear stress τ_θ is zero when, using equation [12],

$$0 = -\tfrac{1}{2}(\sigma_x - \sigma_y)\sin 2\theta + \tau_{xy}\cos 2\theta$$

that is, when

$$\tan 2\theta = \frac{2\tau_{xy}}{\sigma_x - \sigma_y} \qquad\qquad [13]$$

This gives two values of 2θ differing by $180°$ and hence two values of θ differing by $90°$. The two planes specified by these angles are called the *principal planes* and the values of the normal stresses for those planes the *principal stresses*.

The normal stress is a maximum when $d\sigma_\theta/d\theta = 0$. Thus, differentiating equation [11] gives

$$\frac{d\sigma_\theta}{d\theta} = \frac{d}{d\theta} [\tfrac{1}{2}(\sigma_x + \sigma_y) + \tfrac{1}{2}(\sigma_x - \sigma_y)\cos 2\theta + \tau_{xy}\sin 2\theta]$$

$$0 = -(\sigma_x - \sigma_y)\sin 2\theta + 2\tau_{xy}\cos 2\theta$$

Thus the maximum is when

$$\tan 2\theta = \frac{2\tau_{xy}}{\sigma_x - \sigma_y}$$

which is the same as equation [13]. Thus the principal stresses are also the maximum and minimum normal stresses.

Since $\tan 2\theta$ is the perpendicular divided by the hypotenuse for a right-angled triangle of angle 2θ, as illustrated in figure 8.9, then the hypotenuse has a length which is given by the Pythagoras theorem as $\sqrt{[(\sigma_x - \sigma_y)^2 + 4\tau_{xy}^2]}$. Thus

$$\sin 2\theta = \pm \frac{2\tau_{xy}}{\sqrt{[(\sigma_x - \sigma_y)^2 + 4\tau_{xy}]}}$$

$$\cos 2\theta = \pm \frac{\sigma_x - \sigma_y}{\sqrt{[(\sigma_x - \sigma_y)^2 + 4\tau_{xy}]}}$$

Hence the principal stresses are given by equation [11] as

$$\sigma_\theta = \tfrac{1}{2}(\sigma_x + \sigma_y) + \tfrac{1}{2}(\sigma_x - \sigma_y)\cos 2\theta + \tau_{xy}\sin 2\theta$$

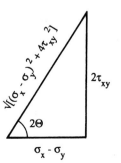

Fig. 8.9 Tan 2θ.

$$= \tfrac{1}{2}(\sigma_x + \sigma_y) \pm \frac{\tfrac{1}{2}(\sigma_x - \sigma_y)(\sigma_x - \sigma_y)}{\sqrt{[(\sigma_x - \sigma_y)^2 + 4\tau_{xy}]}}$$

$$\pm \frac{\tau_{xy} \times 2\tau_{xy}}{\sqrt{[(\sigma_x - \sigma_y)^2 + 4\tau_{xy}]}}$$

$$= \tfrac{1}{2}(\sigma_x + \sigma_y) \pm \tfrac{1}{2}\sqrt{[(\sigma_x - \sigma_y)^2 + 4\tau_{xy}^2]} \qquad [14]$$

The principal stresses are thus

$$\sigma_1 = \tfrac{1}{2}(\sigma_x + \sigma_y) + \tfrac{1}{2}\sqrt{[(\sigma_x - \sigma_y)^2 + 4\tau_{xy}^2]} \qquad [15]$$

$$\sigma_2 = \tfrac{1}{2}(\sigma_x + \sigma_y) - \tfrac{1}{2}\sqrt{[(\sigma_x - \sigma_y)^2 + 4\tau_{xy}^2]} \qquad [16]$$

These stresses occur on mutually perpendicular planes.

A useful check on calculations is that, as a result of adding equations [15] and [16] we have

$$\sigma_1 + \sigma_2 = \sigma_x + \sigma_y \qquad [17]$$

8.4.2 Maximum shear stress

The principal planes give the directions of zero shear stress. Thus if we consider the equilibrium of a triangular element in a plane subject to the biaxial principal stresses σ_1 and σ_2, then the situation is similar to that considered in section 8.3 and so equation [7] gives the shear stress

$$\tau_\theta = \tfrac{1}{2}(\sigma_1 - \sigma_2)\sin 2\theta \qquad [18]$$

The maximum shear stress is thus when $2\theta = 90°$, i.e. on planes at 45° to the principal planes, and has the value

$$\text{max. } \tau = \tfrac{1}{2}(\sigma_1 - \sigma_2) \qquad [19]$$

The maximum shear stress is half the algebraic difference between the principal stresses.

Equation [19] can be written in terms of the stresses σ_x and σ_y by substituting for σ_1 and σ_2 using equations [15] and [16]. Thus

$$\text{max. } \tau = \tfrac{1}{2}\{\tfrac{1}{2}(\sigma_x + \sigma_y) + \tfrac{1}{2}\sqrt{[(\sigma_x - \sigma_y)^2 + 4\tau_{xy}^2]}$$
$$- \tfrac{1}{2}(\sigma_x + \sigma_y) - \tfrac{1}{2}\sqrt{[(\sigma_x - \sigma_y)^2 + 4\tau_{xy}^2]}\}$$
$$= \tfrac{1}{2}\sqrt{[(\sigma_x - \sigma_y)^2 + 4\tau_{xy}^2]} \qquad [20]$$

Example

A plane element in a body is subject to stresses of $\sigma_x = 50\,\text{MPa}$, $\sigma_y = 30\,\text{MPa}$, and $\tau_{xy} = \tau_{yx} = 20\,\text{MPa}$. Determine the principal stresses and their direction and the maximum shear stress and its direction.

Using equations [15] and [16]

$$\sigma_1 = \tfrac{1}{2}(\sigma_x + \sigma_y) + \tfrac{1}{2}\sqrt{[(\sigma_x - \sigma_y)^2 + 4\tau_{xy}^2]}$$

$$= \tfrac{1}{2}(50 + 30) + \tfrac{1}{2}\sqrt{[(50 - 30)^2 + 4 \times 20^2]} = 62.4\,\text{MPa}$$
$$\sigma_2 = \tfrac{1}{2}(\sigma_x + \sigma_y) - \tfrac{1}{2}\sqrt{[(\sigma_x - \sigma_y)^2 + 4\tau_{xy}^2]}$$
$$= \tfrac{1}{2}(50 + 30) - \tfrac{1}{2}\sqrt{[(50 - 30)^2 + 4 \times 20^2]} = 17.6\,\text{MPa}$$

As a check, $\sigma_1 + \sigma_2 = 62.4 + 17.6 = 80\,\text{MPa}$, which is $\sigma_x + \sigma_y$.
Using equation [13]

$$\tan 2\theta = \frac{2\tau_{xy}}{\sigma_x - \sigma_y} = \frac{2 \times 20}{50 - 30} = 2$$

Hence $2\theta = 63.4°$ and $\theta = 31.7°$. The principal planes are thus at $31.7°$ from the y-axis and $(90° + 31.7°) = 121.7°$ and thus we have $62.4\,\text{MPa}$ at 31.7 and $17.6\,\text{MPa}$ at $121.7°$.

The maximum shear stress is given by equation [19] as

$$\text{max. } \tau = \tfrac{1}{2}(\sigma_1 - \sigma_2) = \tfrac{1}{2}(62.4 - 17.6) = 22.4\,\text{MPa}$$

Thus maximum shear stress is at $45°$ to the principal planes and thus at $\theta = 76.7°$ and $166.7°$.

8.4.3 Combined loadings

Structural members are often subject to more than one type of loading. For example, a shaft may be subject to both a torque and an axial load or perhaps a torque and a bending moment. The stress analysis of such members can be carried out by determining the stresses due to each load acting separately and then combining the stresses at any particular point in the structure by means of the equations derived earlier in this chapter.

Consider a shaft of diameter d subject to a torque T and an axial thrust F (figure 8.10). For a plane area on the surface of the shaft, the stress σ_x is the stress F/A due to the axial thrust and is thus

$$\sigma_x = \frac{F}{\pi d^2/4} = \frac{4F}{\pi d^2}$$

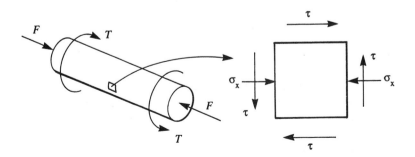

Fig. 8.10 Combined thrust and torque.

The shear stress acting on the plane area is given by, equation [5] chapter 6, as

$$\tau = \frac{Tr}{J}$$

Since $J = \pi d^4/32$, then

$$\tau = \frac{T(d/2)}{\pi d^4/32} = \frac{16T}{\pi d^3}$$

The principal stresses are given by equation [14] as

$$\sigma = \tfrac{1}{2}(\sigma_x + \sigma_y) \pm \tfrac{1}{2}\sqrt{[(\sigma_x - \sigma_y)^2 + 4\tau_{xy}^2]}$$

The maximum direct stress is thus

$$\begin{aligned}
\text{max. } \sigma &= \tfrac{1}{2}(\sigma_x + \sigma_y) + \tfrac{1}{2}\sqrt{[(\sigma_x - \sigma_y)^2 + 4\tau_{xy}^2]} \\
&= \frac{2F}{\pi d^2} + \tfrac{1}{2}\sqrt{\left(\frac{16F^2}{\pi^2 d^4} + \frac{4 \times 16^2 \times T^2}{\pi^2 d^6}\right)} \\
&= \frac{2}{\pi d^2}\left[F + \sqrt{\left(F^2 + \frac{64T^2}{d^2}\right)}\right]
\end{aligned}$$

[21]

The maximum shear stress is given by equation [20] as

$$\begin{aligned}
\text{max. } \tau &= \tfrac{1}{2}\sqrt{[(\sigma_x - \sigma_y)^2 + 4\tau_{xy}^2]} \\
&= \tfrac{1}{2}\sqrt{\left(\frac{16F^2}{\pi^2 d^4} + 6\frac{4 \times 16^2 T^2}{\pi^2 d^6}\right)} \\
&= \frac{2}{\pi d^2}\sqrt{\left(F^2 + \frac{64T^2}{d^2}\right)}
\end{aligned}$$

[22]

Now consider a shaft of diameter d subject to a torque T and a bending moment M and the stresses acting on a plane element on the upper surface (figure 8.11). The maximum bending stresses will occur at the upper and lower surfaces of the shaft (on the bottom and top surfaces of the shaft there will be no shear stresses due to the shear force V). These bending stresses are given by equation [17] chapter 3,

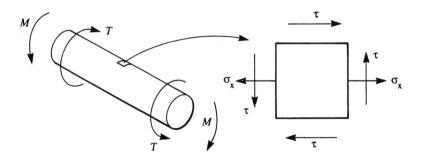

Fig. 8.11 Combined bending and torque.

$$\sigma_x = \frac{My}{I}$$

Since $I = \pi d^4/64$, then the stress when $y = $ radius r is

$$\sigma_x = \frac{M(d/2)}{\pi d^4/64} = \frac{32M}{\pi d^3}$$

The shear stress is given by, equation [5] chapter 6,

$$\tau = \frac{Tr}{J}$$

Since $J = \pi d^4/32$, then

$$\tau = \frac{T(d/2)}{\pi d^4/32} = \frac{16T}{\pi d^3}$$

The stresses acting on a plane element when the bending results in a direct tensile stress is thus as shown in figure 8.11. The maximum stress is a principal stress and thus given by equation [15] as

$$\begin{aligned}
\text{max. } \sigma &= \tfrac{1}{2}(\sigma_x + \sigma_y) + \tfrac{1}{2}\sqrt{[(\sigma_x - \sigma_y)^2 + 4\tau_{xy}^2]} \\
&= \frac{16M}{\pi d^3} + \tfrac{1}{2}\sqrt{\left(\frac{32^2 M^2}{\pi^2 d^6} + \frac{4 \times 16^2 T^2}{\pi^2 d^6}\right)} \\
&= \frac{16}{\pi d^3}\left[M + \sqrt{(M^2 + T^2)}\right]
\end{aligned}$$

[23]

The maximum shear stress is given by equation [20] as

$$\begin{aligned}
\text{max. } \tau &= \tfrac{1}{2}\sqrt{[(\sigma_x - \sigma_y)^2 + 4\tau_{xy}^2]} \\
&= \tfrac{1}{2}\sqrt{\left[\frac{32^2 M^2}{\pi^2 d^6} + \frac{4 \times 16^2 T^2}{\pi^2 d^6}\right]} \\
&= \frac{16}{\pi d^3}\sqrt{(M^2 + T^2)}
\end{aligned}$$

[24]

If instead of considering the stresses acting on a plane area a distance r from the neutral axis we had considered a plane area on the side of the shaft at the neutral plane, then there would be no bending stresses, i.e. $\tau_x = 0$, and thus the element would be just in pure shear. The shear stresses would be the sum of those due to the torque T and those due to the shear force V.

Example

A solid circular shaft of diameter 200 mm is subject to a torque of 10 kN m and a thrust of 50 kN. What is the maximum shear stress?

The compressive stress σ_x is

$$\sigma_x = -\frac{F}{A} = -\frac{50 \times 10^3}{\tfrac{1}{4}\pi \times 0.200^2} = -1.59 \text{ MPa}$$

The shear stress due to the torque is

$$\tau = \frac{Tr}{J} = \frac{T(d/2)}{\pi d^4/32} = \frac{16T}{\pi d^3} = \frac{16 \times 10 \times 10^3}{\pi \times 0.200^3} = 6.37\,\text{MPa}$$

The maximum shear stress is given by equation [20] as

$$\text{max. } \tau = \tfrac{1}{2}\sqrt{[(\sigma_x - \sigma_y)^2 + 4\tau_{xy}^2]}$$
$$= \tfrac{1}{2}\sqrt{[(-1.59)^2 + 4 \times 6.37^2]} = 6.42\,\text{MPa}$$

Alternatively we could have used equation [22].

Example

Figure 8.12 shows a shaft ABC supported by bearings at A and B and cantilevered to C where it carries a pulley and belt drive. The shaft is solid and has a diameter of 75 mm. The distance BC is 300 mm and the sum of the belt tensions and the pulley weight acting vertically downwards on the pulley is 5 kN. The pulley transmits a torque of 4 kN m. What are the maximum tensile stress and the maximum shear stress?

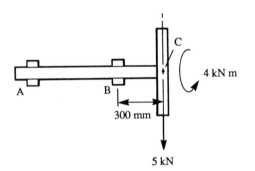

Fig. 8.12 Example.

The bending moment at B is $5 \times 0.300 = 1.5\,\text{kN m}$. The maximum bending stresses are on the upper and lower surfaces of the shaft just to the right of B where 1.5 kN m can be considered to be the bending moment. These bending stresses are given by equation [17] chapter 3,

$$\sigma_x = \frac{My}{I}$$

Since $I = \pi d^4/64$, then the stress when $y = $ radius r is

$$\sigma_x = \frac{M(d/2)}{\pi d^4/64} = \frac{32M}{\pi d^3} = \frac{32 \times 1.5 \times 10^3}{\pi \times 0.075^3} = 36.2\,\text{MPa}$$

The shear stress is given by, equation [5] chapter 6,

$$\tau = \frac{Tr}{J}$$

Since $J = \pi d^4/32$, then

$$\tau = \frac{T(d/2)}{\pi d^4/32} = \frac{16T}{\pi d^3} = \frac{16 \times 4 \times 10^3}{\pi \times 0.075^3} = 48.3\,\text{MPa}$$

The principal stresses are thus given by equation [4] as

$$\sigma = \tfrac{1}{2}(\sigma_x + \sigma_y) \pm \tfrac{1}{2}\sqrt{[(\sigma_x - \sigma_y)^2 + 4\tau_{xy}^2]}$$

When σ_x is tensile, then the maximum tensile stress is

$$\text{max. } \sigma = \tfrac{1}{2} \times 36.2 + \tfrac{1}{2}\sqrt{[36.2^2 + 4 \times 48.3^2]} = 69.7\,\text{MPa}$$

and the maximum shear stress is given by equation [20] as

$$\text{max. } \tau = \tfrac{1}{2}\sqrt{[(\sigma_x - \sigma_y)^2 + 4\tau_{xy}^2]}$$
$$= \tfrac{1}{2}\sqrt{(36.2^2 + 4 \times 48.3^2)} = 51.6\,\text{MPa}$$

8.5 Mohr's stress circle

A geometrical interpretation of the equations for the direct stress and shear stress on a plane at angle θ and those for the principal stresses and the shear stress is provided by *Mohr's stress circle*. The procedure for drawing such a circle is as follows. Note that often the signs of τ are reversed.

(1) A point P is chosen as the origin.
(2) Horizontal distances PB corresponding to σ_x and PA to σ_y are marked out, with tensile values being to the right and compressive values to the left of P.
(3) Vertical distances corresponding to τ are marked out with $-\tau$ being vertically upwards from A, to give AD, and $+\tau$ vertically downwards from B, to give BE. This conforms with section 8.4.
(4) DE is then taken to represent the diameter of a circle with a centre C on the direct stress line.
(5) The direct stress σ_θ and shear stress τ_θ at some angle θ are then found by setting off a radius CF of the circle at an angle 2θ to DE. The horizontal coordinate of F is σ_θ and the vertical coordinate τ_θ.
(6) The principal stresses are the distances from the origin P to where the circle cuts the horizontal axis.
(7) The maximum shear stress is the vertical distance from the centre of the circle to the circle circumference, i.e. the circle radius.

Confirmation that the distances represent the quantities indicated can be obtained by considering the geometry of the figure. Thus, the horizontal projection of CF is $R\cos\phi$ and, since the centre of the circle C is a distance $\tfrac{1}{2}(\sigma_x + \sigma_y)$ from P,

$$\sigma_\theta = \tfrac{1}{2}(\sigma_x + \sigma_y) + R\cos\phi \qquad [25]$$

where R is the radius of the circle. The vertical projection of CF is $R\sin\phi$.

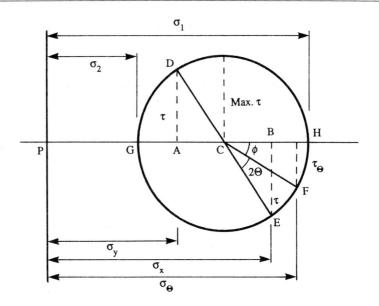

Fig. 8.13 Mohr's stress circle.

Thus

$$\tau_\theta = R \sin \phi \qquad [26]$$

But $CB = R \cos(2\theta + \phi)$ and since $CB = \sigma_x - \frac{1}{2}(\sigma_x + \sigma_y) = \frac{1}{2}(\sigma_x - \sigma_y)$, then

$$\frac{1}{2}(\sigma_x - \sigma_y) = R \cos(2\theta + \phi)$$
$$= R(\cos 2\theta \cos \phi - \sin 2\theta \sin\phi) \qquad [27]$$

Multiplying this equation by $\cos 2\theta$ gives

$$\frac{1}{2}(\sigma_x - \sigma_y)\cos 2\theta = R(\cos^2 2\theta \cos \phi - \cos 2\theta \sin 2\theta \sin \phi) \qquad [28]$$

We can also write $EB = R \sin(2\theta + \phi)$ and since $EB = \tau$ then

$$\tau = R \sin(2\theta + \phi)$$
$$= R(\sin 2\theta \cos \phi + \cos 2\theta \sin \phi) \qquad [29]$$

Multiplying this equation by $\sin 2\theta$ gives

$$\tau \cos 2\theta = R(\sin^2 2\theta \cos \phi + \sin 2\theta \cos 2\theta \sin \phi) \qquad [30]$$

Adding equations [28] and [30] gives

$$\frac{1}{2}(\sigma_x - \sigma_y)\cos 2\theta + \tau \cos 2\theta = R \cos \phi[\cos^2 2\theta + \sin^2 2\theta] = R \cos \phi$$

Thus equation [25] can be written as

$$\sigma_\theta = \frac{1}{2}(\sigma_x + \sigma_y) + \frac{1}{2}(\sigma_x - \sigma_y)\cos 2\theta + \tau \cos 2\theta$$

This is equation [11]. If equation [27] had been multiplied by $\sin 2\theta$ and equation [29] by $\cos 2\theta$, then adding them would have given equation [12], namely (τ_θ would be negative if the signs of τ were reversed)

$$\tau_\theta = -\tfrac{1}{2}(\sigma_x - \sigma_y)\sin 2\theta + \tau \cos 2\theta$$

The shear stress is zero when CF lies along the axis through PAB. This is the condition for the principal stresses σ_1 and σ_2. Thus $\sigma_1 = PH$ and $\sigma_2 = PG$. The angles between the principal planes and the direct stresses σ_x and σ_y are given by the angles between the line CE and the direct stress axis.

The maximum value of the shear stress occurs on a plane which is 90° displaced from the principal planes and is thus given by the vertical line through the circle centre C and so equal to the radius of the circle.

Example

Draw Mohr's circle of stress for a plane for which $\sigma_x = 50\,\text{MPa}$, $\sigma_y = 30\,\text{MPa}$ and $\tau = 20\,\text{MPa}$. Hence determine the principal stresses and the maximum in-plane shear stress.

Direct stress and shear stress axes are drawn. Then the coordinates for $(\sigma_x, -\tau)$ and (σ_y, τ) marked. The line joining these coordinates gives the diameter of the circle which can then be drawn. Figure 8.14 shows the circle. The principal stresses are thus where the circle cuts the direct stress axis, this being at about 62 MPa and 18 MPa. The angle 2θ between σ_x and the principal plane for σ_1 is 63° and thus the angle θ is about 32°. The maximum shear stress is the radius of the circle and is thus about 22 MPa.

Example

Draw Mohr's circle of stress for a plane for which $\sigma_x = 30\,\text{MPa}$,

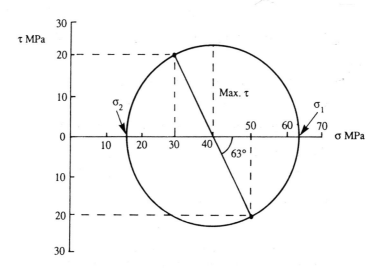

Fig. 8.14 Example.

$\sigma_y = -10\,\text{MPa}$ and $\tau = 20\,\text{MPa}$. Hence determine the principal stresses and the maximum in-plane shear stress.

Direct stress and shear stress axes are drawn. Then the coordinates for $(\sigma_x, -\tau)$ and (σ_y, τ) marked. In this case the coordinates for the σ_y stress are to the left of the origin. The line joining these coordinates gives the diameter of the circle which can then be drawn. Figure 8.15 shows the circle. The principal stresses are thus where the circle cuts the direct stress axis, this being at about 38 MPa and $-18\,\text{MPa}$. The maximum shear stress is the radius of the circle and is thus about 28 MPa.

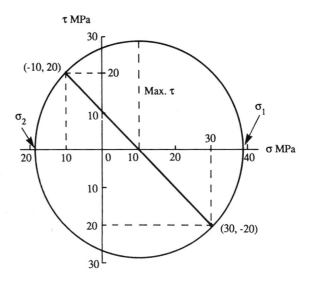

Fig. 8.15 Example.

8.6 Principal strains and stresses

Consider a plane element subject to tensile principal stresses σ_1 and σ_2 (figure 8.16). The strain produced in the direction of σ_1 by σ_1 is σ_1/E. There will also be a strain in that direction due to σ_2 of $-v\sigma_1/E$, where v is Poisson's ratio (see section 2.4). Thus the total strain ϵ_1 in the σ_1 direction is

$$\epsilon_1 = \frac{\sigma_1}{E} - \frac{v\sigma_2}{E} \tag{31}$$

Similarly, the strain ϵ_2 in the σ_2 direction is

$$\epsilon_2 = \frac{\sigma_2}{E} - \frac{v\sigma_1}{E} \tag{32}$$

Equations [31] and [32] can be solved simultaneously to give

$$\sigma_1 = \frac{E}{1 - v^2}(\epsilon_1 + v\epsilon_2) \qquad\qquad [33]$$

$$\sigma_2 = \frac{E}{1 - v^2}(\epsilon_2 + v\epsilon_1) \qquad\qquad [34]$$

Example

The principal strains at a point on a loaded plate are found to be 320×10^{-6} and 200×10^{-6}. What are the principal stresses if the modulus of elasticity is 200 GPa and Poisson's ratio 0.3?

Using equations [33] and [34]

$$\sigma_1 = \frac{E}{1 - v^2}(\epsilon_1 + v\epsilon_2) = \frac{200 \times 10^9}{1 - 0.3^2}(320 + 0.3 \times 200)10^{-6}$$

$$= 83.5 \, \text{MPa}$$

$$\sigma_2 = \frac{E}{1 - v^2}(\epsilon_2 + v\epsilon_1) = \frac{200 \times 10^9}{1 - 0.3^2}(200 + 0.3 \times 320)10^{-6}$$

$$= 65.1 \, \text{MPa}$$

8.6.1 Strain energy

Consider an element subject to principal stresses σ_1 and σ_2, as in figure 8.16. The strain ϵ_1 in the direction of principal stress σ_1 is given by equation [31] as

$$\epsilon_1 = \frac{\sigma_1}{E} - \frac{v\sigma_2}{E}$$

Hence since the strain energy per unit volume is $\frac{1}{2}$ stress $\times$ strain (see equation [2], chapter 7), then

$$u_1 = \frac{\sigma_1}{2E}(\sigma_1 - v\sigma_2)$$

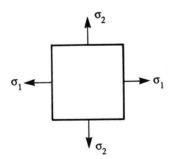

Fig. 8.16 Principal stresses.

The strain ϵ_2 in the direction of the principal stress σ_2 is given by equation [32] as

$$\epsilon_2 = \frac{\sigma_2}{E} - \frac{v\sigma_1}{E}$$

Hence the strain energy per unit volume due to this is

$$u_2 = \frac{\sigma_2}{2E}(\sigma_2 - v\sigma_1)$$

The total strain energy per unit volume due to both principal stresses is thus

$$u = u_1 + u_2 = \frac{1}{2E}(\sigma_1^2 + \sigma_2^2 - 2v\sigma_1\sigma_2) \qquad [35]$$

8.7 Plane strain

Consider a plane rectangular element which is subject to direct strains ϵ_x, ϵ_y and shear strain ϕ and for which the strain is required in the direction of the diagonal, it being at angle θ. The diagonal is considered to have an initial length d and thus the issue is as to what is the change in length of the diagonal as a result of the three strains. For the strain ϵ_y acting alone, as in figure 8.17(a), then the rectangle has a length $d\sin\theta$ in that direction and so the change in length in that direction is $\epsilon_y d\sin\theta$. To a reasonable approximation, considering the strains are actually very small, the change in length of the diagonal is $(\epsilon_y d\sin\theta)\sin\theta$ (see figure 8.17(a)). For the strain ϵ_x acting alone, as in figure 8.17(b), then the rectangle has a length $d\cos\theta$ in that direction and so the change in length in that direction is $\epsilon_x d\cos\theta$. To a reasonable approximation, the change in length of the diagonal is $(\epsilon_x d\cos\theta)\cos\theta$. For the shear strain ϕ acting alone, as in figure 8.17(c) then the change in length of the diagonal is, to a reasonable approximation, $(\phi d\sin\theta)\cos\theta$. Thus the total change in length of the diagonal when all three strains are acting is

$$\epsilon_y d\sin^2\theta + \epsilon_x d\cos^2\theta + \phi d\sin\theta\cos\theta$$

The strain is this change divided by d, thus

$$\epsilon_\theta = \epsilon_y \sin^2\theta + \epsilon_x \cos^2\theta + \phi\sin\theta\cos\theta$$

Substituting using

$$\cos^2\theta = \tfrac{1}{2}(\cos 2\theta + 1)$$
$$\sin^2\theta = \tfrac{1}{2}(1 - \cos 2\theta)$$
$$\sin\theta\cos\theta = \tfrac{1}{2}\sin 2\theta$$

gives

$$\epsilon_\theta = \tfrac{1}{2}\epsilon_y(1 - \cos 2\theta) + \tfrac{1}{2}\epsilon_x(\cos 2\theta + 1) + \tfrac{1}{2}\phi\sin 2\theta$$
$$\epsilon_\theta = \tfrac{1}{2}(\epsilon_x + \epsilon_y) + \tfrac{1}{2}(\epsilon_x - \epsilon_y)\cos 2\theta + \tfrac{1}{2}\phi\sin 2\theta \qquad [36]$$

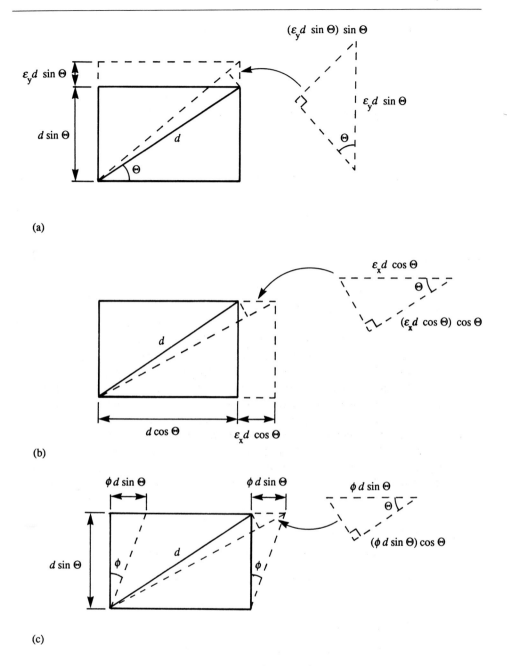

Fig. 8.17 Plane strain.

The shear strain τ_θ at angle θ can be determined by considering the angular rotations of the diagonal in figure 8.17 produced by ϵ_x, ϵ_y and τ. The result is

$$\tfrac{1}{2}\tau_\theta = -\tfrac{1}{2}(\epsilon_x - \epsilon_y)\sin 2\theta + \tfrac{1}{2}\phi\cos 2\theta \qquad\qquad [37]$$

The equations for plane strain, equations [36] and [37], are similar in form to those for plane stress, equations [11] and [12].

The principal strains are the maximum and minimum values of ϵ_θ. Thus differentiating equation [36] with respect to θ gives

$$\frac{d\epsilon_\theta}{d\theta} = -(\epsilon_x - \epsilon_y)\sin 2\theta + \phi \cos 2\theta$$

Equating $d\epsilon_\theta/d\theta$ to zero gives, for the angles at which the principal strains occur,

$$\tan 2\theta = \frac{\phi}{\epsilon_x - \epsilon_y} \qquad [38]$$

This can be represented by the triangle shown in figure 8.18 and so equation [36] can be written for the principal strains as

$$\epsilon_\theta = \frac{\epsilon_x + \epsilon_y}{2} \pm \frac{\epsilon_x - \epsilon_y}{2} \times \frac{\epsilon_x - \epsilon_y}{\sqrt{[(\epsilon_x - \epsilon_y)^2 + \phi^2]}}$$

$$\pm \frac{\phi^2}{\sqrt{[(\epsilon_x - \epsilon_y)^2 + \phi^2]}}$$

$$\epsilon_1 = \tfrac{1}{2}(\epsilon_x + \epsilon_y) + \tfrac{1}{2}\sqrt{[(\epsilon_x - \epsilon_y)^2 + \phi^2]} \qquad [39]$$

$$\epsilon_2 = \tfrac{1}{2}(\epsilon_x + \epsilon_y) - \tfrac{1}{2}\sqrt{[(\epsilon_x - \epsilon_y)^2 + \phi^2]} \qquad [40]$$

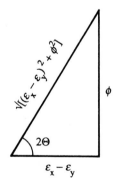

Fig. 8.18 Tan 2θ.

Example

An element of material is subject to plane strains of $\epsilon_x = 300 \times 10^{-6}$, $\epsilon_y = 100 \times 10^{-6}$ and $\phi = 150 \times 10^{-6}$. What are the principal in-plane strains?

Using equations [39] and [40]

$$\epsilon_1 = \tfrac{1}{2}(\epsilon_x + \epsilon_y) + \tfrac{1}{2}\sqrt{[(\epsilon_x - \epsilon_y)^2 + \phi^2]}$$
$$= \tfrac{1}{2}(300 + 100)10^{-6} + \tfrac{1}{2}\sqrt{[(300 - 100)^2 10^{-12} + 150^2 \times 10^{-12}]}$$

$$= 200 \times 10^{-6} + 125 \times 10^{-6}$$
$$= 325 \times 10^{-6}$$
$$\epsilon_2 = \tfrac{1}{2}(\epsilon_x + \epsilon_y) - \tfrac{1}{2}\sqrt{[(\epsilon_x - \epsilon_y)^2 + \phi^2]}$$
$$= 200 \times 10^{-6} - 125 \times 10^{-6}$$
$$= 75 \times 10^{-6}$$

The angles to the principal strain directions are given by equation [38] as

$$\tan 2\theta = \frac{\phi}{\epsilon_x - \epsilon_y} = \frac{150}{300 - 100}$$

Hence the two values of 2θ between $0°$ and $360°$ are $36.9°$ and $216.9°$. The principal strain ϵ_1 is thus at angle $18.5°$ and ϵ_2 at $108.5°$.

8.7.1 Strain gauge rosettes

The electrical resistance strain gauge consists basically of a length of fine wire attached to a backing paper (figure 8.19). This is then stuck to the surface being investigated, like a postage stamp being stuck on an envelope. When the surface suffers strain then the wire of the gauge suffers the same strain. The resistance of the wire changes as a consequence of the strain, the fractional change in resistance being proportional to the strain.

$$\text{fractional change in resistance} = G \times \text{strain} \qquad [41]$$

where G, the constant of proportionality, is called the *gauge factor*. A measurement of the change in resistance of a strain gauge can thus be used as a measure of the strain. The change in resistance is usually measured by means of a Wheatstone bridge.

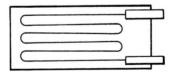

Fig. 8.19 Electrical resistance strain gauge.

To determine the principal strains, strain gauge rosettes are used. Figure 8.20 shows the basic forms. In figure 8.20(a) three strain gauges are glued to the surface with their axes inclined at $45°$. The strains ϵ_a, ϵ_b and ϵ_c are thus measured in directions inclined at θ, $\theta + 45°$, and $\theta + 90°$ to principal strain ϵ_1. Thus equation [36] can be written for each measured strain as

$$\epsilon_a = \tfrac{1}{2}(\epsilon_1 + \epsilon_2) + \tfrac{1}{2}(\epsilon_1 - \epsilon_2)\cos 2\theta \qquad [42]$$

$$\epsilon_b = \tfrac{1}{2}(\epsilon_1 + \epsilon_2) + \tfrac{1}{2}(\epsilon_1 - \epsilon_2)\cos 2(\theta + 45°)$$

$$= \tfrac{1}{2}(\epsilon_1 + \epsilon_2) - \tfrac{1}{2}(\epsilon_1 - \epsilon_2)\sin 2\theta \qquad [43]$$

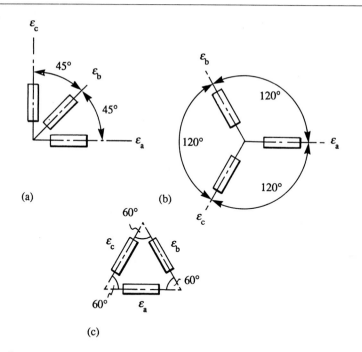

Fig. 8.20 Strain gauge rosettes, (a) 45°, (b) and (c) 120°.

$$\epsilon_c = \tfrac{1}{2}(\epsilon_1 + \epsilon_2) + \tfrac{1}{2}(\epsilon_1 - \epsilon_2)\cos 2(\theta + 90°)$$
$$= \tfrac{1}{2}(\epsilon_1 + \epsilon_2) - \tfrac{1}{2}(\epsilon_1 - \epsilon_2)\cos 2\theta \qquad [44]$$

The principal strains and θ can thus be found from measurements of ϵ_a, ϵ_b and ϵ_c.

Another form of rosette uses three strain gauges with their axes inclined at 120°, figure 8.20(b) and (c) showing two versions of this rosette. The strains ϵ_a, ϵ_b and ϵ_c are thus measured in directions inclined at θ, $\theta + 120°$ and $\theta + 240°$ to principal strain ϵ_1. Thus equation [36] can be written for each measured strain as

$$\epsilon_a = \tfrac{1}{2}(\epsilon_1 + \epsilon_2) + \tfrac{1}{2}(\epsilon_1 - \epsilon_2)\cos 2\theta \qquad [45]$$

$$\epsilon_b = \tfrac{1}{2}(\epsilon_1 + \epsilon_2) + \tfrac{1}{2}(\epsilon_1 - \epsilon_2)\cos 2(\theta + 120°)$$
$$= \tfrac{1}{2}(\epsilon_1 + \epsilon_2) - \tfrac{1}{2}(\epsilon_1 - \epsilon_2)(\tfrac{1}{2}\cos 2\theta - \tfrac{1}{2}\sqrt{3}\sin 2\theta) \qquad [46]$$

$$\epsilon_c = \tfrac{1}{2}(\epsilon_1 + \epsilon_2) + \tfrac{1}{2}(\epsilon_1 - \epsilon_2)\cos 2(\theta + 240°)$$
$$= \tfrac{1}{2}(\epsilon_1 + \epsilon_2) - \tfrac{1}{2}(\epsilon_1 - \epsilon_2)(\tfrac{1}{2}\cos 2\theta + \tfrac{1}{2}\sqrt{3}\sin 2\theta) \qquad [47]$$

Example

A 45° strain gauge rosette is stuck to the surface of a steel shaft. Measurements of the changes in resistance of each gauge when the shaft was loaded gave strains of $\epsilon_a = 500 \times 10^{-6}$, $\epsilon_b = 350 \times 10^{-6}$ and $\epsilon_c = -300 \times 10^{-6}$. What are the in-plane principal strains and their orientations? If the material has a modulus of elasticity of 200 GPa and Poisson's ratio of 0.3, what are the principal stresses?

Using equations [42], [43] and [44],

$$\epsilon_a = 500 \times 10^{-6} = \tfrac{1}{2}(\epsilon_1 + \epsilon_2) + \tfrac{1}{2}(\epsilon_1 - \epsilon_2)\cos 2\theta \qquad [48]$$

$$\epsilon_b = 350 \times 10^{-6} = \tfrac{1}{2}(\epsilon_1 + \epsilon_2) - \tfrac{1}{2}(\epsilon_1 - \epsilon_2)\sin 2\theta \qquad [49]$$

$$\epsilon_c = -300 \times 10^{-6} = \tfrac{1}{2}(\epsilon_1 + \epsilon_2) - \tfrac{1}{2}(\epsilon_1 - \epsilon_2)\cos 2\theta \qquad [50]$$

Adding equations [48] and [50] gives

$$\epsilon_a + \epsilon_c = 200 \times 10^{-6} = \epsilon_1 + \epsilon_2 \qquad [51]$$

Substituting this value in equations [48] and [49] gives

$$\epsilon_a = 500 \times 10^{-6} = 100 \times 10^{-6} + \tfrac{1}{2}(\epsilon_1 - \epsilon_2)\cos 2\theta \qquad [52]$$

$$\epsilon_b = 350 \times 10^{-6} = 100 \times 10^{-6} - \tfrac{1}{2}(\epsilon_1 - \epsilon_2)\sin 2\theta \qquad [53]$$

Hence

$$\tan 2\theta = -\frac{250 \times 10^{-6}}{400 \times 10^{-6}}$$

Hence $2\theta = -32.0°$ or $328.0°$ and $\theta = 164.0°$.
   Since $\cos^2 2\theta + \sin^2 2\theta = 1$, then equations [52] and [53] give

$$\tfrac{1}{2}(\epsilon_1 - \epsilon_2) = \sqrt{[(400 \times 10^{-6})^2 + (250 \times 10^{-6})^2]}$$

$$\epsilon_1 - \epsilon_2 = 943 \times 10^{-6}$$

Thus, with equation [51] this gives $\epsilon_1 = 572 \times 10^{-6}$ and $\epsilon_2 = -372 \times 10^{-6}$.
   The principal stresses are given by equations [33] and [34] as

$$\sigma_1 = \frac{E}{1 - \nu^2}(\epsilon_1 + \nu\epsilon_2) = \frac{200 \times 10^9}{1 - 0.3^2}(572 - 0.3 \times 372)10^{-6}$$

$$= 101\,\text{MPa}$$

$$\sigma_2 = \frac{E}{1 - \nu^2}(\epsilon_2 + \nu\epsilon_1) = \frac{200 \times 10^9}{1 - 0.3^2}(-372 + 0.3 \times 572)10^{-6}$$

$$= -44\,\text{Mpa}$$

8.7.2   Mohr's strain circle

The equations for plane strain, equations [36] and [37], are similar in form to those for plane stress, equations [11] and [12]. Thus a Mohr's strain circle can be constructed in a similar way to Mohr's stress circle (see section 8.5). The procedure for drawing such a circle is as follows.

(1) A point P is chosen as the origin (figure 8.21).
(2) Horizontal distances PB corresponding to ϵ_x and PA to ϵ_y are marked

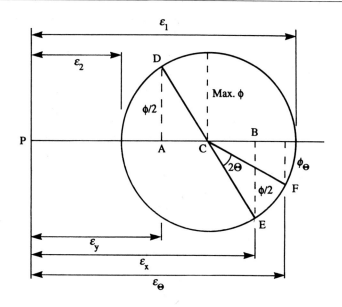

Fig. 8.21 Mohr's strain circle.

out, with tensile values being to the right and compressive values to the left of P.

(3) Vertical distances corresponding to $\frac{1}{2}\phi$ are marked out with $-\frac{1}{2}\phi$ being vertically upwards from A, to give AD, and $+\frac{1}{2}\phi$ vertically downwards from B, to give BE. Note that the signs of ϕ may be reversed.

(4) DE is then taken to represent the diameter of a circle with a centre C on the direct stress line.

(5) The direct strain ϵ_θ and shear strain ϕ_θ at some angle θ are then found by setting off a radius CF of the circle at an angle 2θ to DE. The horizontal coordinate of F is ϵ_θ and the vertical coordinate ϕ_θ.

(6) The principal strains are the distances from the origin P to where the circle cuts the horizontal axis.

(7) The maximum shear strain is the vertical distance from the centre of the circle to the circle circumference, i.e. the circle radius.

8.8 Theories of failure

If a member is subject to a uniaxial stress it is common to use as a guide to the prediction of failure the yield stress if the material is ductile and the strength if the material is brittle. There are a number of theories used for predicting failure when the stress is more complex, the main theories used being outlined below.

(1) *Maximum principal stress theory*

This theory, due to Rankine, considers that failure will occur when the maximum principal stress reaches the tensile strength stress value that occurs for the material when subject to simple tension. This theory is found to be only applicable to brittle materials.

(2) *Maximum shear stress theory*

This theory, due to Guest and Tresca, considers that failure will occur when the maximum shear stress in the complex stress situation reaches the value of the maximum shear stress that occurs for the material at the elastic limit in simple tension. This theory is used with ductile materials.

If we assume that the greatest shear stress will occur in the x-y plane, equation [20] gives

$$\text{max. } \tau = \tfrac{1}{2}\sqrt{[(\sigma_x - \sigma_y)^2 + 4\tau_{xy}^2]}$$

Thus, in the case of simple tension when $\sigma_y = 0$ and $\tau_{xy} = 0$ then max. $\tau = \tfrac{1}{2}\sigma_x$. Failure thus occurs when σ_x is the elastic limit σ_e, i.e. $\tau = \tfrac{1}{2}\sigma_e$. The maximum shear stress is given by equation [19], in terms of the principal stresses, as

$$\text{max. } \tau = \tfrac{1}{2}(\sigma_1 - \sigma_2)$$

Thus failure is when

$$\sigma_1 - \sigma_2 = \sigma_e \qquad\qquad [54]$$

(3) *Strain energy theory*

This theory, due to Haigh, considers that failure occurs when the strain energy per unit volume is equal to the strain energy at the elastic limit in simple tension. This theory is used with ductile materials.

Assuming that we have a state of plane stress, the strain energy per unit volume is given by equation [35] as

$$u = \frac{1}{2E}(\sigma_1^2 + \sigma_2^2 - 2\nu\sigma_1\sigma_2)$$

If σ_e is the elastic limit stress in simple tension, then for such a uniaxial stress the strain energy per unit volume is given by equation [3] chapter 7 as $\sigma_e^2/2E$. Thus failure occurs when

$$\sigma_e^2 = \sigma_1^2 + \sigma_2^2 - 2\nu\sigma_1\sigma_2 \qquad\qquad [55]$$

Example

A solid circular shaft has a diameter of 100 mm and is transmitting torque and subject also to a bending moment of 5 kN m. What is the maximum torque possible if failure is considered to occur according to (a) the maximum principal stress theory if the material is brittle and fails in simple tension of 150 MPa, (b) the maximum shear stress theory if the material is ductile and has an elastic limit in simple tension of 200 MPa, (c) the strain energy theory if the material is ductile and has an elastic limit of 200 MPa and Poisson's ratio of 0.3?

The maximum and minimum principal stresses are given by [23] as

$$\sigma = \frac{16}{\pi d^3} \left[M \pm \sqrt{(M^2 - T^2)} \right]$$

$$= \frac{16}{\pi \times 0.100^3} \left[5 \times 10^3 \pm \sqrt{(25 \times 10^6 + T^2)} \right]$$

$$= 5.09 \times 10^3 [5 \times 10^3 \pm \sqrt{(25 \times 10^6 + T^2)}]$$

(a) According to the maximum principal stress theory, failure will occur when

$$150 \times 10^6 = 5.09 \times 10^3 [5 \times 10^3 + \sqrt{(25 \times 10^6 + T^2)}]$$

This is thus when $T = 23.9 \, \text{kN m}$.

(b) According to the maximum shear stress theory and equation [54], failure occurs when

$$\sigma_1 - \sigma_2 = \sigma_e$$

This is thus when

$$5.09 \times 10^3 [5 \times 10^3 + \sqrt{(25 \times 10^6 + T^2)}]$$
$$- 5.09 \times 10^3 [5 \times 10^3 - \sqrt{(25 \times 10^6 + T^2)}] = 200 \times 10^6$$

This is thus when $T = 19.0 \, \text{kN m}$.

(c) According to the strain energy theory and equation [55], failure occurs when

$$\sigma_e^2 = \sigma_1^2 + \sigma_2^2 - 2v\sigma_1\sigma_2$$

$$(200 \times 10^6)^2 = \{5.09 \times 10^3 [5 \times 10^3 + \sqrt{(25 \times 10^6 + T^2)}]\}^2$$
$$+ \{5.09 \times 10^3 [5 \times 10^3 - \sqrt{(25 \times 10^6 + T^2)}]\}^2$$
$$- 2 \times 0.3 \times 5.09 \times 10^3 [5 \times 10^3 + \sqrt{(25 \times 10^6 + T^2)}]$$
$$\times 5.09 \times 10^3 [5 \times 10^3 - \sqrt{(25 \times 10^6 + T^2)}]$$

This is thus when $T = 23.6 \, \text{kN m}$

8.9 Elastic constants

The elastic constants which determine the deformations produced by a stress system acting on a material are the modulus of elasticity E, the modulus of rigidity G, the bulk modulus K and Poisson's ratio v (see chapter 2 for definitions). These constants are not independent of each other.

8.9.1 Relationship between K, E and v

Consider a rectangular element of the material, as in figure 8.22. The element is subject to compressive stresses of σ on each face. Consider first just the compressive stress in the x-direction. As a consequence there is a

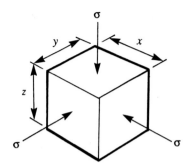

Fig. 8.22 Volume subject to stresses.

strain in the direction of the stress of $-\epsilon$ and in the transverse directions of $+\text{ve}$. Thus the length in the direction of the stress decreases from x to $x - \epsilon x$ and increases in the transverse directions to $y + \text{ve}y$ and $z + \text{ve}z$. The original volume of the element was xyz and is now

$$\text{new volume} = (x - \epsilon x)(y + \text{ve}y)(z + \text{ve}z)$$

Thus the change in volume divided by the original volume is

$$\frac{\Delta V}{V} = \frac{xyz(1 - \epsilon)\,(1 + \text{ve})\,(1 + \text{ve}) - xyz}{xyz}$$

Neglecting terms involving ϵ^2, gives

$$\frac{\Delta V}{V} = -\epsilon + 2\text{ve}$$

But $\epsilon = \sigma/E$. Thus

$$\frac{\Delta V}{V} = -\frac{\sigma}{E}\,(1 - 2v)$$

Now if the element is subject to compressive stresses of σ on all the faces, the change in volume per unit volume would be trebled. Thus the bulk strain, which is the change in volume per unit volume in such a situation, is

$$\text{bulk strain} = -\frac{3\sigma}{E}\,(1 - 2v)$$

The bulk modulus K is $-\sigma/\text{bulk strain}$ and thus

$$E = 3K(1 - 2v) \qquad\qquad [56]$$

The volume of a material diminishes under a hydrostatic pressure, i.e. compressive stresses on all faces, thus K is positive. Thus we must have

$$1 - 2v > 0$$

This means $v < \tfrac{1}{2}$. Poisson's ratio is thus always less than $\tfrac{1}{2}$.

8.9.2 *Relationship between* **E**, **G** *and* v

Consider a rectangular plane element which is subject to just shear stresses on its faces, as in figure 8.23. The direct strain ϵ_{45} at an angle of 45° is given by equation [36] as

$$\epsilon_\theta = \tfrac{1}{2}(\epsilon_x + \epsilon_y) + \tfrac{1}{2}(\epsilon_x - \epsilon_y)\cos 2\theta + \tfrac{1}{2}\phi\sin 2\theta$$

$$\epsilon_{45} = 0 + 0 + \tfrac{1}{2}\phi\sin 90° = \tfrac{1}{2}\phi \qquad [57]$$

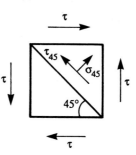

Fig. 8.23 Element subject to shear.

The direct stress σ_{45} in the 45° direction is given by equation [11] as

$$\sigma_\theta = \tfrac{1}{2}(\sigma_x + \sigma_y) + \tfrac{1}{2}(\sigma_x - \sigma_y)\cos 2\theta + \tau_{xy}\sin 2\theta$$

$$\sigma_{45} = 0 + 0 + \tau\sin 90° = \tau$$

The direct stress in a direction at right angles to σ_{45}, i.e. σ_{135}, is according to equation [11]

$$\sigma_{135} = \tau\sin 135° = -\tau$$

The strain in the 45° direction is thus

$$\epsilon_{45} = \frac{\sigma_{45}}{E} - \frac{v\sigma_{45}}{E} = \frac{\tau}{E}(1 + v)$$

Thus, since the shear strain is τ/G, equation [57] becomes

$$\frac{\tau}{E}(1 + v) = \frac{\tau}{2G}$$

Hence

$$G = \frac{E}{2(1 + v)} \qquad [58]$$

For most metals v is about 0.3 and thus $E \approx 2.6G$.

Example

Estimate the value of G for a steel which has $E = 205\,\text{GPa}$ and $v = 0.30$.

Using equation [58]

$$G = \frac{E}{2(1 + v)} = \frac{205}{2(1 + 0.30)} = 78.8\,\text{GPa}$$

8.9.3 Elastic properties

Materials which have the same properties in all directions are said to be *isotropic*. Such a material would thus have, for example, the same modulus of elasticity in longitudinal and transverse directions. If the properties differ in various directions then the material is said to be *anisotropic*. A special form of anisotropic materials is where the properties in a particular direction are the same throughout the material and the properties in a perpendicular direction, though the same throughout the material, are different from those in the other direction. Such materials are called *orthotropic*. An example of such a material is concrete which is reinforced with parallel steel bars.

Problems

(1) A bar of cross-section 25 mm by 25 mm is subject to an axial tensile load of 15 kN. What are the normal stress and shear stress acting on a plane at (a) 30°, (b) 45° to a transverse plane?

(2) A bar with a cross-sectional area of 800 mm^2 is subject to axial forces of 50 kN. What are the normal and shear stresses acting on a plane at 60° to a transverse plane and what is the maximum shear stress?

(3) What is the maximum shear stress acting in a rectangular section bar 50 mm by 75 mm when subject to an axial force of 300 kN?

(4) A short steel bar with a rectangular cross-section 20 mm by 20 mm is subject to an axial compressive load of 40 kN. What are the normal and shear stresses acting on a plane at 60° to a transverse plane?

(5) A plane element in a body is subject to the stresses indicated below, what are the stresses acting on the element in a direction at the angle to the *y*-direction indicated?

(a) $\sigma_x = 20\,\text{MPa}$, $\sigma_y = 0$, $\tau_{xy} = \tau_{yx} = 30\,\text{MPa}$, $\theta = 45°$

(b) $\sigma_x = 50\,\text{MPa}$, $\sigma_y = 30\,\text{MPa}$, $\tau_{xy} = \tau_{yx} = 20\,\text{MPa}$, $\theta = 45°$

(c) $\sigma_x = 40\,\text{MPa}$, $\sigma_y = 20\,\text{MPa}$, $\tau_{xy} = \tau_{yx} = 30\,\text{MPa}$, $\theta = 30°$

(d) $\sigma_x = 20\,\text{MPa}$, $\sigma_y = 10\,\text{MPa}$, $\tau_{xy} = \tau_{yx} = 30\,\text{MPa}$, $\theta = 30°$

(e) $\sigma_x = -80\,\text{MPa}$, $\sigma_y = 50\,\text{MPa}$, $\tau_{xy} = \tau_{yx} = 30\,\text{MPa}$, $\theta = 30°$

(6) A plane element is subject to the stresses indicated below. What

are the principal stresses and their directions and the maximum shear stresses and their directions?

(a) $\sigma_x = 100\,MPa$, $\sigma_y = 0$, $\tau_{xy} = \tau_{yx} = 25\,MPa$

(b) $\sigma_x = 80\,MPa$, $\sigma_y = 30\,MPa$, $\tau_{xy} = \tau_{yx} = 30\,MPa$

(c) $\sigma_x = 40\,MPa$, $\sigma_y = 60\,MPa$, $\tau_{xy} = \tau_{yx} = 30\,MPa$

(d) $\sigma_x = 80\,MPa$, $\sigma_y = -30\,MPa$, $\tau_{xy} = \tau_{yx} = -30\,MPa$

(e) $\sigma_x = -20\,MPa$, $\sigma_y = 90\,MPa$, $\tau_{xy} = \tau_{yx} = 60\,MPa$

(7) A solid shaft with a diameter of 75 mm is subject to an axial tensile load of 200 kN and a torque of 3.0 kN m. What will be the maximum tensile stress and the maximum in-plane shear stress?

(8) A solid circular cross-section shaft of diameter 200 mm is subject to an axial compressive load of 500 kN and a torque of 20 kN m. What will be the maximum compressive stress, the maximum tensile stress and the maximum in-plane shear stress?

(9) A solid circular shaft of diameter 50 mm is subject to a torque of 2 kN m and an axial compressive loading of 200 kN. What will be the maximum compressive stress, the maximum tensile stress and the maximum in-plane shear stress?

(10) A solid circular shaft of diameter 60 mm is subject to a torque of 2 kN m and an axial thrust. What is the maximum value possible for the thrust if the in-plane shear stress is not to exceed 50 MPa?

(11) A hollow circular shaft of internal diameter 22 mm and external diameter 30 mm is subject to an axial tensile force of 15 kN and a torque of 90 N m. What will be the maximum compressive stress and maximum tensile stress?

(12) A solid circular shaft of diameter 200 mm is subject to a bending moment of 40 kN m when transmitting 2 MW at 250 rev/min. What is the maximum in-plane shear stress?

(13) A solid circular shaft of diameter 50 mm is subject to a bending moment of 3 kN m and a torque of 4 kN m. What will be the maximum tensile stress, the maximum compressive stress and the maximum shear stress?

(14) A flywheel of mass 500 kg is mounted on a shaft of diameter 75 mm and midway between bearings 600 mm apart. What will be the maximum tensile stress, the maximum compressive stress and the maximum shear stress if the shaft transmits a power of 20 kW at 5 rev/s?

(15) Use Mohr's stress circle to solve problem 5.

(16) Use Mohr's stress circle to solve problem 6.

(17)   The principal strains at a point on a loaded plate are found to be 300×10^{-6} and 160×10^{-6}. What are the principal stresses if the modulus of elasticity is 200 GPa and Poisson's ratio 0.3?

(18)   What are the principal strains for a plane element subject to the following strains?

(a) $\epsilon_x = 120 \times 10^{-6}$, $\epsilon_y = 400 \times 10^{-6}$, $\phi = 300 \times 10^{-6}$

(b) $\epsilon_x = 120 \times 10^{-6}$, $\epsilon_y = -400 \times 10^{-6}$, $\phi = 300 \times 10^{-6}$

(c) $\epsilon_x = -100 \times 10^{-6}$, $\epsilon_y = 200 \times 10^{-6}$, $\phi = 200 \times 10^{-6}$

(19)   Measurement of the resistances of the three gauges in a 45° strain gauge rosette on a bracket gave, when the bracket was loaded, strains of $\epsilon_a = 400 \times 10^{-6}$, $\epsilon_b = 400 \times 10^{-6}$ and $\epsilon_c = 100 \times 10^{-6}$. What are the principal strains and principal stresses? The modulus of elasticity of the bracket material is 200 GPa and Poisson's ratio is 0.3.

(20)   Measurement of the resistances of the three gauges in a 45° strain gauge rosette on a steel shaft gave, when it was loaded, strains of $\epsilon_a = 800 \times 10^{-6}$, $\epsilon_b = 400 \times 10^{-6}$ and $\epsilon_c = -100 \times 10^{-6}$. What are the principal strains and principal stresses? The modulus of elasticity of the shaft material is 200 GPa and Poisson's ratio is 0.3.

(21)   Measurement of the resistances of the three gauges in a 120° strain gauge rosette on a steel shaft gave, when it was loaded, strains of $\epsilon_a = 250 \times 10^{-6}$, $\epsilon_b = 400 \times 10^{-6}$ and $\epsilon_c = 300 \times 10^{-6}$. What are the principal strains and principal stresses? The modulus of elasticity of the shaft material is 200 GPa and Poisson's ratio is 0.3.

(22)   A strain gauge rosette has three strain gauges which measure the strains in three directions inclined at 60° to each other. If the strains were for the 0° gauge 550×10^{-6}, for the 60° gauge -100×10^{-6} and for the 120° gauge 150×10^{-6}, what are the principal strains and principal stresses? The modulus of elasticity is 200 GPa and Poisson's ratio 0.3

(23)   Use Mohr's strain circle to solve problem 18.

(24)   A solid circular shaft is subject to a torque of 50 kN m. What is the smallest diameter the shaft can have according to the maximum principal stress theory? The shaft material when tested in tension has a tensile strength of 200 MPa.

(25)   A bolt is subject to an axial tensile force of 10 kN and a transverse shear force of 5 kN. What is the smallest diameter the bolt can have according to (a) the maximum shear stress theory, (b) the strain energy theory? A tensile test on the material gives an elastic limit of 200 MPa. Poisson's ratio $= 0.3$.

(26)   A solid circular shaft is simply supported at two bearings and

subject to belt pulls on pulleys which result in the shaft being subject to a bending moment of 0.5 kN m and a torque of 0.6 kN m. What is the smallest diameter, with a factor of safety of 3, the shaft can have according to (a) the maximum principal stress theory, (b) the maximum shear stress theory? Tensile tests on a sample of the material gave a tensile strength of 250 MPa and an elastic limit of 200 MPa.

(27)   The principal plane stresses acting on an element are 50 MPa and − 80 MPa. What is the factor of safety according to the maximum shear stress theory if the material has an elastic limit of 400 MPa?

(28)   The direct stresses acting on an element, in directions at right angles to each other, are 60 MPa and − 45 MPa and there is a shear stress of 37.5 MPa. What is the factor of safety according to the maximum principal stress theory if the material has a tensile strength of 240 MPa?

(29)   Estimate the rigidity modulus for an alloy for which the modulus of elasticity is 45 GPa and Poisson's ratio 0.31.

(30)   A steel is found to have a modulus of elasticity of 205 GPa and a modulus of rigidity of 80.6 GPa. Estimate the values of Poisson's ratio and the bulk modulus.

Chapter 9

Cylindrical shells

9.1 Thin cylindrical shell

When a closed cylindrical shell is subject to an internal gauge pressure p, perhaps due to it containing a liquid or a gas under pressure, then stresses are produced in the shell. Note that the term gauge pressure is used for the difference in pressure between that inside the shell and outside the shell when outside is at normal atmospheric pressure. If we think of this internal pressure endeavouring to cause the volume of the cylinder to expand then there will be a *circumferential or hoop stress* σ_c as a result of the circumference endeavouring to get bigger, a *longitudinal stress* σ_l as a result of the pressure on the ends of the cylinder causing the length of the cylinder to get bigger and a *radial stress* σ_r due to the pressure acting at right angles to the surface and endeavouring to squash the wall of the shell (figure 9.1). This radial stress has a maximum value equal to the gauge pressure p at the inner surface of the cylinder and decreases to zero at the exterior wall where the gauge pressure is zero.

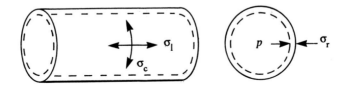

Fig. 9.1 Stresses with a cylindrical shell.

If the ratio of the thickness of the wall of the cylindrical shell to its internal diameter is less than about 1/20 (see section 9.4.2) then the radial stress is small and can be neglected. Also, the circumferential and longitudinal stresses will be reasonably uniform over the thickness of the cylinder wall. Such a shell is called a *thin shell*. When the ratio is greater than 1/20 the shell is said to be a *thick shell*.

9.1.1 Stresses with thin cylindrical shell

Consider the equilibrium of the upper half of a length L of the cylinder, remote from the closed ends (figure 9.2(a)). The pressure p in the cylinder

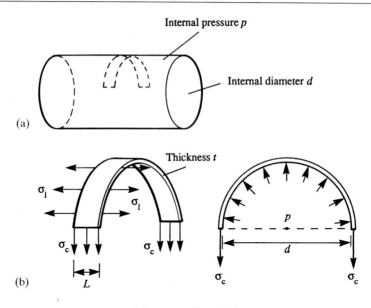

(a)

(b)

Fig. 9.2 Circumferential stresses.

results in forces acting at right angles to the inner surfaces of the cylinder (figure 9.2(b)). The total vertical force acting on this half-cylinder is thus the sum of the resolved components of these forces in the vertical direction. This is the same as the sum of the forces acting on the horizontal plane through the diameter, i.e. $p \times$ area of diametral plane $= pLd$. Because there is equilibrium, this upward directed force must be balanced by the forces provided by the circumferential stresses. These act over an area $2tL$. Thus

$$\sigma_c \times 2tL = pLd$$

$$\sigma_c = \frac{pd}{2t} \qquad [1]$$

Now consider the equilibrium of the cylinder when it is sectioned by a transverse plane, as in figure 9.3. Longitudinal forces are provided by the pressure p acting on the end faces of the cylinder. Since these have an area $\pi d^2/4$, then the force is $p\pi d^2/4$. Because there is equilibrium, this force is balanced by the longitudinal stress within the walls of the cylinder. The walls have an area of πdt. Thus

$$\sigma_l \times \pi dt = p\pi d^2/4$$

$$\sigma_l = \frac{pd}{4t} \qquad [2]$$

Note that, for the closed cylinder, the longitudinal stress is half the circumferential stress.

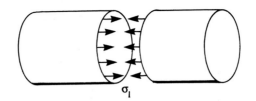

Fig. 9.3 Longitudinal stress.

Example

A thin cylindrical pipe has an inside diameter of 150 mm and a wall thickness of 4 mm. What will be the circumferential and longitudinal stresses when the pipe has an internal gauge pressure of 5 MPa?

Using equation [1]

$$\sigma_c = \frac{pd}{2t} = \frac{5 \times 10^6 \times 0.150}{2 \times 0.004} = 93.8\,\text{MPa}$$

The longitudinal stress is half the circumferential stress and is thus 46.9 MPa.

Example

An open cylindrical water storage tank has a diameter of 4.0 m and a height of 6.0 m and is filled with water, density 1000 kg/m³, to a height of 5.8 m. What will be the wall thickness required if the maximum permissible stress for the tank material is 80 MPa?

The maximum gauge pressure in the tank is $h\rho g = 5.8 \times 1000 \times 9.8 = 5.684 \times 10^4$ Pa and occurs in the tank wall at the base of the tank. The longitudinal stress will be zero because the tank is not closed at both ends. The maximum circumferential stress is thus given by equation [1] as

$$\sigma_c = \frac{pd}{2t} = \frac{5.684 \times 10^4 \times 4.0}{2t}$$

But this maximum stress must be 80 MPa. Hence $t = 1.42$ mm.

9.1.2   Jointed thin cylinders

If a thin, closed, cylindrical shell (figure 9.4) is made up of joined, perhaps riveted, plates then if the efficiency of a longitudinal joint is ς_l the circumferential stress at the joint is

$$\sigma_c' = \frac{pd}{2\varsigma_l t} \qquad\qquad [3]$$

The efficiency is the ratio of the strength of the joint to the strength of the

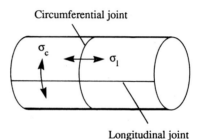

Fig. 9.4 Cylindrical shell made up of joined plates.

unjoined plate. If the efficiency of a circumferential joint is ζ_c then the longitudinal stress at the joint is

$$\sigma_1 = \frac{pd}{4\zeta_c t}$$ [4]

Example

A compressed air container is cylindrical with an internal diameter of 2.0 m and a wall thickness of 10 mm and is made up from four half cylindrical plates riveted at their joints. The longitudinal joint efficiency is 80% and the circumferential joint efficiency is 50%. What is the maximum permissible gauge pressure for the compressed air if the maximum permissible tensile stress in the container material is 100 MPa?

Using equation [3]

$$\sigma_c = \frac{pd}{2\zeta_l t} = \frac{2.0p}{2 \times 0.80 \times 0.010}$$

Thus, since the maximum stress is 100 MPa, the maximum pressure for the longitudinal joints is 800 kPa. Using equation [4]

$$\sigma_1 = \frac{pd}{4\zeta_c t} = \frac{2.0p}{4 \times 0.50 \times 0.010}$$

Thus, since the maximum stress is 100 MPa, the maximum pressure for the circumferential joints is 1000 kPa. Thus the maximum pressure is determined by the longitudinal joints and is 800 kPa.

Example

What is the joint efficiency of a single riveted lap joint formed by rivets of diameter 20 mm and pitch 50 mm joining plates of thickness 15 mm if the maximum working stress for the plate is 100 MPa and the shear strength of a rivet is 80 MPa? Assume that failure is by shearing of the rivets.

The load required to shear a rivet is $\tau \times$ cross-sectional area, i.e.

$$\text{load to shear rivet} = 80 \times 10^6 \times \tfrac{1}{4}\pi \times 0.020^2 = 25.1\,\text{kN}$$

The strength of the original plate per rivet pitch is the maximum working stress multiplied by the cross-sectional area of the pitch, i.e. pitch $\times$ plate thickness. Thus

$$\text{plate strength} = 100 \times 10^6 \times 0.050 \times 0.015 = 75.0\,\text{kN}$$

Thus the joint efficiency is $25.1/75.0 = 33.5\%$.

9.2   Thin spherical shell

For a thin spherical shell under internal pressure, consider the upper half of the sphere (figure 9.5). The pressure will result in forces acting at right angles to the internal surface of the hemisphere. The total force in the vertical direction resulting from the pressure is thus the sum of the vertical resolved components of these forces. This is the same as the forces acting over the area of the horizontal plane of the diameter. Thus the total force is $p \times$ area of diametral plane $= p \times \tfrac{1}{4}\pi d^2$. Thus for equilibrium, since the tangential stress σ acts over an area equal to the circumference multiplied by the shell thickness,

$$\sigma \times \pi dt = p \times \tfrac{1}{4}\pi d^2$$

$$\sigma = \frac{pd}{4t} \qquad\qquad [5]$$

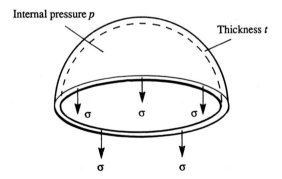

Fig. 9.5 Thin spherical shell under pressure.

Example

A spherical gas container has an inner diameter of 2.0 m and is subject to an internal gauge pressure of 400 kPa. What should the minimum thickness of the container material be if the maximum permissible stress for it is 30 MPa?

Using equation [5], $\sigma = pd/4t$, and thus

$$t = \frac{pd}{4\sigma} = \frac{400 \times 10^3 \times 2.0}{4 \times 30 \times 10^6} = 6.7\,\text{mm}$$

9.3 Volume changes of shells

The internal volume of a cylindrical shell of internal diameter d and length L is $\frac{1}{4}\pi d^2 L$. The effect of there being an internal gauge pressure is to increase both the diameter and the length. Thus if the diameter increases by δd and the length by δL, the change in volume will be

$$\begin{aligned}
\text{change in volume} &= \tfrac{1}{4}\pi[(d + \delta d)^2\,(L + \delta L) - d^2 L] \\
&= \tfrac{1}{4}\pi[d^2 L + d^2\delta L + 2\delta d d L + 2\delta d d \delta L + (\delta d)^2 L \\
&\quad + (\delta d)^2\delta L - d^2 L]
\end{aligned}$$

Neglecting products of small quantities gives

$$\text{change in volume} = \tfrac{1}{4}\pi[d^2\delta L + 2\delta d d L]$$

The fractional change in volume, or the so-called *volumetric strain*, is

$$\begin{aligned}
\frac{\text{change in volume}}{\text{original volume}} &= \frac{\tfrac{1}{4}\pi[d^2\delta L + 2\delta d d L]}{\tfrac{1}{4}\pi d^2 L} \\
&= \frac{\delta L}{L} + 2\frac{\delta d}{d}
\end{aligned}$$

But $\delta L/L$ is the longitudinal strain. The circumference of the cylinder is πd and thus the circumferential strain is $\delta d/d$. Thus

$$\text{volumetric strain} = \text{longitudinal strain} + 2 \times \text{circumferential strain}$$
$$[6]$$

For a sphere, the original volume is $4\pi r^3/3 = \pi d^3/6$. If the diameter changes by δd then

$$\frac{\text{change in volume}}{\text{original volume}} = \frac{(\pi/6)\,[(d + \delta d)^3 - d^3]}{d^3}$$

Neglecting products of small quantities,

$$\frac{\text{change in volume}}{\text{original volume}} = 3\frac{\delta d}{d}$$

Thus

$$\text{volumetric strain} = 3 \times \text{circumferential strain} \qquad [7]$$

Example

A cylindrical pressure vessel has a length of 2.0 m, an internal diameter of 500 mm and walls of thickness 10 mm. What is the increase in the diameter, the increase in length and the increase in volume when the vessel has an internal gauge pressure of 1 MPa? The modulus of elasticity of the wall material is 200 GPa and Poisson's ratio is 0.3.

The circumferential stress is given by equation [1] as

$$\sigma_c = \frac{pd}{2t} = \frac{1 \times 10^6 \times 0.500}{2 \times 0.010} = 25 \text{ MPa}$$

The longitudinal stress is given by equation [2] as

$$\sigma_l = \frac{pd}{4t} = \frac{1 \times 10^6 \times 0.500}{4 \times 0.010} = 12.5 \text{ MPa}$$

The circumferential strain due to the circumferential stress is σ_c/E. In addition there will be a circumferential strain of $-v\sigma_l/E$ due to the longitudinal stress, with v being Poisson's ratio (see section 2.4). Thus

$$\epsilon_c = \frac{1}{E}(\sigma_c - v\sigma_l)$$

$$= \frac{25 \times 10^6 - 0.3 \times 12.5 \times 10^6}{200 \times 10^9} = 1.06 \times 10^{-4} \qquad [8]$$

The longitudinal strain due to the longitudinal stress is σ_l/E and due to the circumferential stress is $-v\sigma_c/E$. Thus

$$\epsilon_l = \frac{1}{E}(\sigma_l - v\sigma_c)$$

$$= \frac{12.5 \times 10^6 - 0.3 \times 25 \times 10^6}{200 \times 10^9} = 2.5 \times 10^{-5} \qquad [9]$$

The circumferential strain is $\pi\delta d/\pi d$. Thus the change in internal diameter δd is $\epsilon_c d$.

$$\delta d = \epsilon_c d = 1.06 \times 10^{-4} \times 0.500 = 5.3 \times 10^{-5} \text{ m}$$

The longitudinal strain is $\delta L/L$. Thus the change in length δL is

$$\delta L = \epsilon_l L = 2.5 \times 10^{-5} \times 2.0 = 5.0 \times 10^{-5} \text{ m}$$

Using equation [6],

$$\text{volumetric strain} = \text{longitudinal strain} + 2 \times \text{circumferential strain}$$

$$= 2.5 \times 10^{-5} + 2 \times 1.06 \times 10^{-4}$$

$$= 2.37 \times 10^{-4}$$

Thus the change in volume is volumetric strain $\times$ original volume, i.e.

$$\delta V = 2.37 \times 10^{-4} \times 2.0 \times \pi \times 0.500 = 7.45 \times 10^{-4} \text{ m}^3$$

Example

A steel cylindrical boiler has an internal diameter of 1.0 m, a length of 2.0 m and walls of thickness 25 mm. In a hydraulic test the boiler is filled with water at atmospheric pressure. What further volume of water needs to be pumped into the boiler to test it at a gauge pressure of 10 MPa? The steel has a modulus of elasticity of 200 GPa and a Poisson's ratio of 0.3. The water has a bulk modulus of 2 GPa.

The circumferential stress is given by equation [1] as

$$\sigma_c = \frac{pd}{2t} = \frac{10 \times 10^6 \times 1.0}{2 \times 0.025} = 200 \, \text{MPa}$$

The longitudinal stress is given by equation [2] as

$$\sigma_l = \frac{pd}{4t} = \frac{10 \times 10^6 \times 1.0}{4 \times 0.025} = 100 \, \text{MPa}$$

The circumferential strain is given by equation [8] as

$$\epsilon_c = \frac{1}{E}(\sigma_c - v\sigma_l) = \frac{200 \times 10^6 - 0.3 \times 100 \times 10^6}{200 \times 10^9}$$

$$= 8.5 \times 10^{-4}$$

The longitudinal strain is given by equation [9] as

$$\epsilon_l = \frac{1}{E}(\sigma_l - v\sigma_c) = \frac{100 \times 10^6 - 0.3 \times 200 \times 10^6}{200 \times 10^9}$$

$$= 2.0 \times 10^{-4}$$

Using equation [6],

$$\text{volumetric strain} = \text{longitudinal strain} + 2 \times \text{circumferential strain}$$
$$= 2.0 \times 10^{-4} + 2 \times 8.5 \times 10^{-4}$$
$$= 1.9 \times 10^{-3}$$

Thus the change in volume is volumetric strain $\times$ original volume, i.e.

$$= 1.9 \times 10^{-3} \times 2.0 \times \pi \times 1.0 = 1.19 \times 10^{-2} \, \text{m}^3$$

The bulk modulus K is defined as $-p$/volumetric strain, the negative sign indicating that an increase in pressure p results in a decrease in volume (see section 2.5).

$$K = -\frac{p}{\delta V/V} \tag{10}$$

Thus the increase in the pressure to 10 MPa will result in a decrease in the volume of the water already in the boiler. This decrease is

$$\delta V = -\frac{Vp}{K} = -\frac{2.0 \times \pi \times 1.0 \times 10 \times 10^6}{2.0 \times 10^9}$$

$$= -3.14 \times 10^{-2}\,\text{m}^3$$

Thus the additional amount of water required at a pressure of 10 MPa is $1.19 \times 10^3 + 3.14 \times 10^{-2} = 4.33 \times 10^{-2}\,\text{m}^3$. At atmospheric pressure this water will have a volume of $(4.33 \times 10^{-2} + \delta V)$ where

$$\delta V = \frac{Vp}{K} = \frac{4.33 \times 10^{-2} \times 10 \times 10^6}{2.0 \times 10^9} = 2.2 \times 10^{-4}\,\text{m}^3$$

Thus the additional water required at atmospheric pressure is $4.33 \times 10^{-2} + 2.2 \times 10^{-4} = 4.35 \times 10^{-2}\,\text{m}^3$.

9.4 Thick cylindrical shell

With a thick-walled cylinder there will be circumferential, longitudinal and radial stresses (see section 7.1). Consider such a cylinder which is long in comparison with its diameter. With this condition we can assume that the longitudinal stress and longitudinal strain are uniform across the thickness of the cylinder wall.

Figure 9.6 shows a cross-section of a thick cylinder. The inner wall has a radius r_1 and the outer wall a radius r_2. Inside the cylinder the pressure is p_1 and outside the cylinder p_2. Consider the equilibrium of an element of the cross-section at radius r, the element subtending an angle $\delta\theta$ at the centre

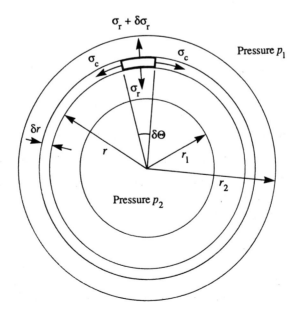

Fig. 9.6 Thick cylinder.

and having a thickness δr and a longitudinal length L. For the element the radial stress is taken to be σ_r on the side of it at radius r and $\sigma_r + \delta\sigma_r$ at radius $r + \delta r$. For equilibrium of the forces acting on the element in the radial direction we have, when determining arc length as radius $\times$ angle subtended at centre,

$$(\sigma_r + \delta\sigma_r)(r + \delta r)L\delta\theta - \sigma_r rL\delta\theta + \text{radial components of}$$
$$\text{forces due to } \sigma_c = 0$$

The radial components of the forces due to the circumferential stresses acting on each side of the element are each $\sigma_c L \delta r \sin\frac{1}{2}\theta$. Thus

$$(\sigma_r + \delta\sigma_r)(r + \delta r)L\delta\theta - \sigma_r rL\delta\theta - 2\sigma_c L\delta r \sin\tfrac{1}{2}\theta = 0$$

Neglecting the products of small quantities and taking the angle of the element to be small enough to make the approximation $\sin\frac{1}{2}\delta\theta \approx \frac{1}{2}\delta\theta$,

$$\sigma_r \delta r + r\delta\sigma_r - \sigma_c \delta r = 0$$

In the limit this becomes

$$\sigma_r + r\frac{d\sigma_r}{dr} = \sigma_c \qquad\qquad [11]$$

The longitudinal strain is σ_l/E due to the longitudinal stress, plus $-v\sigma_c/E$ due to the circumferential stress, plus $-v\sigma_r/E$ due to the radial stress. Thus

$$\epsilon_l = \frac{\sigma_l}{E} - v\left(\frac{\sigma_c}{E} + \frac{\sigma_r}{E}\right) \qquad\qquad [12]$$

But we have assumed that the longitudinal stress and strain are constant over the section. Thus, for equation [12], we must have $(\sigma_c + \sigma_r)$ constant. Let

$$\sigma_c + \sigma_r = 2a \qquad\qquad [13]$$

where a is a constant. Substituting for σ_c in equation [11]

$$\sigma_r + r\frac{d\sigma_r}{d_r} = 2a - \sigma_r$$

and then multiplying throughout by r gives

$$2\sigma_r r + r^2 \frac{d\sigma_r}{dr} - 2ar = 0$$

$$\frac{d}{dr}(\sigma_r r^2 - ar^2) = 0$$

Hence, on integrating,

$$\sigma_r r^2 - ar^2 = \text{a constant} = b$$

$$\sigma_r = a + \frac{b}{r^2} \qquad\qquad [14]$$

Substituting for σ_r in equation [13] gives

$$\sigma_c = a - \frac{b}{r^2} \qquad\qquad [15]$$

Equations [14] and [15] are known as *Lamé's equations*. The constants a and b can be determined from the dimensions of the cylinder and the loading conditions.

Example

A steel pipe of internal diameter 60 mm and external diameter 100 mm is subject to an internal pressure of 12 MPa and an external pressure of 5 MPa. What are the radial and circumferential stresses at the inner and outer surfaces?

When $r = 30$ mm then $\sigma_r = -12$ MPa. Thus equation [14] gives

$$\sigma_r = -12 \times 10^6 = a + \frac{b}{r^2} = a + \frac{b}{0.030^2}$$

When $r = 50$ mm then $\sigma_r = -5$ MPa. Thus

$$\sigma_r = -5 \times 10^6 = a + \frac{b}{r^2} = a + \frac{b}{0.050^2}$$

Subtracting the above equations gives

$$-7 \times 10^6 = \frac{b}{0.030^2} - \frac{b}{0.050^2}$$

Thus $b = -9844$. Substituting this in one of the equations gives $a = -1.06 \times 10^6$. The circumferential stress is given by equation [15] as

$$\sigma_c = a - \frac{b}{r^2}$$

Hence at the outer surface where $r = 50$ mm

$$\sigma_c = -1.06 \times 10^6 - \frac{9844}{0.050^2} = +2.9 \text{ MPa}$$

and at the inner surface where $r = 30$ mm

$$\sigma_c = -1.06 \times 10^6 - \frac{9844}{0.030^2} = +9.9 \text{ MPa}$$

9.4.1 Internal pressure only

Consider a thick cylinder subject to just an internal gauge pressure p. Then at the inner surface of the cylinder, radius r_1, the radial stress is $-p$ and at the outer surface, radius r_2, the radial stress is zero. Equation [14] thus becomes with these conditions

$$-p = a + \frac{b}{r_1^2}$$

and

$$0 = a + \frac{b}{r_2^2}$$

Subtracting these two equations gives

$$b = -\frac{r_1^2 r_2^2}{r_2^2 - r_1^2} p$$

and

$$a = -\frac{b}{r_2^2} = \frac{r_1^2}{r_2^2 - r_1^2} p$$

Thus equations [14] and [15] give

$$\sigma_r = \frac{pr_1^2}{r_2^2 - r_1^2}\left(1 - \frac{r_2^2}{r^2}\right) \tag{16}$$

$$\sigma_c = \frac{pr_1^2}{r_2^2 - r_1^2}\left(1 + \frac{r_2^2}{r^2}\right) \tag{17}$$

Figure 9.7 shows the distribution of these stresses across the tube section. The maximum value of the radial stress is when $r = r_1$ and is $-p$. The maximum value of the circumferential stress is when $r = r_1$ and is

$$\text{max. } \sigma_c = \frac{r_1^2 + r_2^2}{r_2^2 - r_1^2} p \tag{18}$$

The longitudinal stress for a closed thick cylinder can be obtained from a consideration of the equilibrium of a transverse section. The pressure p on

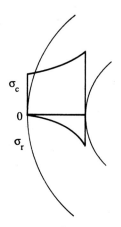

Fig. 9.7 Stress distribution with internal pressure.

the internal end face, radius r_1 provides a longitudinal force of $p \times \pi r_1^2$. This is balanced by the longitudinal stress acting over the wall area of $\pi(r_2^2 - r_1^2)$. Thus

$$\sigma_1 \times \pi(r_2^2 - r_1^2) = p \times \pi r_1^2$$

$$\sigma_1 = \frac{pr_1^2}{r_2^2 - r_1^2} \qquad [19]$$

Since σ_r, σ_c and σ_1 are the principal stresses, the shear stress is half the difference between principal stresses and thus the maximum shear stress is either $\frac{1}{2}(\sigma_c - \sigma_r)$ or $\frac{1}{2}(\sigma_1 - \sigma_r)$ according to whichever is the greater. Since σ_c is greater than σ_1 then the maximum shear stress is

$$\tau = \frac{\sigma_c - \sigma_r}{2} = \frac{(a - b/r^2) - (a + b/r^2)}{2} = -\frac{b^2}{r^2}$$

$$= \frac{r_1^2 r_2^2}{r^2(r_2^2 - r_1^2)}p$$

The maximum value is when $r = r_1$ and thus

$$\text{max. } \tau = \frac{pr_2^2}{r_2^2 - r_1^2} \qquad [20]$$

Example

The cylinder of a hydraulic ram has an internal diameter of 100 mm. What wall thickness will be required to withstand an internal gauge pressure of 20 MPa if the maximum permissible tensile stress is 60 MPa and the maximum permissible shear stress is 50 MPa? What will be the increase in diameter resulting from the pressure? The modulus of elasticity is 200 GPa and Poisson's ratio 0.3.

Since the circumferential stress is greater than the longitudinal stress, the maximum tensile stress will be given by equation [18] as

$$60 \times 10^6 = \frac{r_1^2 + r_2^2}{r_2^2 - r_1^2}p = \frac{0.050^2 + r_2^2}{r_2^2 - 0.050^2} \times 20 \times 10^6$$

$$3r_2^2 - 0.0075 = 0.0025 + r_2^2$$

Hence $r_2 = 0.0707$ m.
The maximum permissible shear stress is given by equation [20] as

$$50 \times 10^6 = \frac{pr_2^2}{r_2^2 - r_1^2} = \frac{20 \times 10^6 \times r_2^2}{r_2^2 - 0.050^2}$$

$$5r_2^2 - 0.0125 = 2r_2^2$$

Hence $r_2 = 0.0645$ m. Since this is less than the value required for the maximum tensile stress, the maximum tensile stress is the limiting factor. The required wall thickness is thus $0.0707 - 0.050 = 0.0207$ m $= 20.7$ mm.

The circumferential strain is $2\pi\delta r_1/2\pi r_1 = \delta r_1/r_1$. The circumferential strain resulting from the tensile circumferential stress is a tensile strain of σ_c/E, from the compressive radial stress is a tensile strain of $\nu\sigma_r/E$, and from the tensile longitudinal stress is a compressive strain of $\nu\sigma_l$. Thus

$$\epsilon_c = \frac{\delta r_1}{r_1} = \frac{\sigma_c}{E} + \frac{\nu\sigma_r}{E} - \frac{\nu\sigma_l}{E} \qquad [21]$$

The radial stress is given, at $r = r_1$ by equation [16] as

$$\sigma_r = \frac{pr_1^2}{r_2^2 - r_1^2}\left(1 - \frac{r_2^2}{r_1^2}\right) = \frac{20 \times 10^6 \times 0.050^2}{0.0707^2 - 0.050^2}\left(1 - \frac{0.0707^2}{0.050^2}\right)$$

Hence $\sigma_r = -20.0\,\text{MPa}$. The circumferential stress is given, at $r = r_1$, by equation [17] as

$$\sigma_c = \frac{pr_1^2}{r_2^2 - r_1^2}\left(1 + \frac{r_2^2}{r_1^2}\right) = \frac{20 \times 10^6 \times 0.050^2}{0.0707^2 - 0.050^2}\left(1 + \frac{0.0707^2}{0.050^2}\right)$$

Hence $\sigma_c = 60.0\,\text{MPa}$. The longitudinal stress is given, at $r = r_1$, by equation [19] as

$$\sigma_l = \frac{pr_1^2}{r_2^2 - r_1^2} = \frac{20 \times 10^6 \times 0.050^2}{0.0707^2 - 0.050^2} = 20.0\,\text{MPa}$$

Hence equation [21] becomes

$$\frac{\delta r_1}{0.050} = \frac{60.0 \times 10^6}{200 \times 10^9} + \frac{0.3 \times 20.0 \times 10^6}{200 \times 10^9} - \frac{0.3 \times 20.0 \times 10^6}{200 \times 10^9}$$

and $\delta r_1 = +1.5 \times 10^{-5}\,\text{m}$.

9.4.2 Error in thin cylinder equation

With a cylinder of wall thickness t, then $r_2 = r_1 + t$ and equation [19] becomes

$$\sigma_l = \frac{p(r_1 + t)^2}{(r_1 + t)^2 - r_1^2}$$

If the thickness to internal diameter ratio is 1/10, i.e. $t/r_1 = 1/5$, then the above equation gives

$$\sigma_l = \frac{p(5t + t)^2}{(5t + t)^2 - 25t^2} = \frac{36}{11}p = 3.27p$$

The thin cylinder equation [2] gives $\sigma_l = pd/4t = 10p/4 = 2.5p$. The thin cylinder equation thus underestimates the longitudinal stress in this situation by a factor of 1.3.

If the thickness to internal diameter ratio is 1/20, i.e. $t/r_1 = 1/10$, the above equation gives

$$\sigma_1 = \frac{p(10t + t)^2}{(10t + t)^2 - 100t^2} = \frac{121}{21}p = 5.8p$$

The thin cylinder equation [2] gives $\sigma_1 = pd/4t = 20p/4 = 5.0p$. The thin cylinder equation thus underestimates the longitudinal stress in this situation by a factor of 1.2.

9.5 Compound tube

As indicated by figure 9.7, the circumferential stress for a cylindrical tube decreases appreciably as the radius increases. With, for example, the external radius of a cylinder being twice the inner radius then the circumferential stress at the inner surface will be given by equation [17] as

$$\sigma_c = \frac{pr_1^2}{r_2^2 - r_1^2}\left(1 + \frac{r_2^2}{r^2}\right) = \frac{5p}{3}$$

and at the outer surface as

$$\sigma_c = \frac{pr_1^2}{r_2^2 - r_1^2}\left(1 + \frac{r_2^2}{r^2}\right) = \frac{2p}{3}$$

The circumferential stress at the inner surface is two and a half times that at the outer surface. The inner surface circumferential stress thus provides a limiting factor with regard to the internal pressures that can be used with a tube, the material near the outside surface of the tube not being stressed to anywhere near its limit.

One way by which the circumferential stresses can be made more even across the wall of a tube is to use a compound tube in which an outer cylindrical tube is shrink fitted over an inner cylindrical tube. In the state prior to assembly the inner tube is slightly too large to slide inside the outer tube, i.e. there is a slight radial interference. The outer tube is then heated so that it expands and can then fit over the inner tube. When it cools it provides a shrink fit. The shrinking subjects the inner tube to circumferential compression and the outer tube to circumferential tension. Figure 9.8 shows the resulting form of the shrinkage stresses. The compound cylindrical tube is thus pre-stressed. When such a compound cylindrical tube is subject to an internal pressure the pressure stress is superposed on the shrinkage stress with the resultant stress being a more even distribution across the compound tube walls. The high circumferential stress at the inner wall has been reduced. Such a compound cylindrical tube can thus be used at higher pressures than otherwise would be the case.

The radial pressure at the common interface between the two cylindrical tubes as a result of the shrinkage is related to the radial interference before the tubes are fitted together. The radial interference is the sum, at the common interface, of the displacement of the inner cylinder inwards and the outer cylinder outwards in order that they fit together. But the circumferential strain for the inner cylinder is

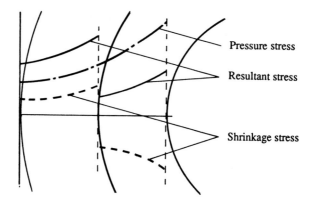

Fig. 9.8 Stress distribution with compound tube.

$$\epsilon_{ci} = \frac{\sigma_{ci}}{E_i} - v_i \frac{\sigma_{ri}}{E_i} \qquad [22]$$

where E_i is the modulus of elasticity, v_i Poisson's ratio, σ_{ci} the outer surface circumferential stress and σ_{ri} the outer surface radial stress of the inner cylinder. The circumferential strain for the outer cylinder is

$$\epsilon_{co} = \frac{\sigma_{co}}{E_o} - v_o \frac{\sigma_{ro}}{E_o} \qquad [23]$$

where E_o is the modulus of elasticity, v_o Poisson's ratio, σ_{co} the inner surface circumferential stress and σ_{ro} the inner surface radial stress of the outer cylinder. The change in radius of the inner surface of the outer cylinder is $\epsilon_{co}r$. The change in radius of the outer surface of the inner cylinder is $\epsilon_{ci}r$. The inner surface of the outer cylinder has the same radius r as that of the outer surface of the inner cylinder. Thus the total difference of radii before shrinking, i.e. the interference δ, is

$$\delta = r(\epsilon_{co} - \epsilon_{ci}) \qquad [24]$$

Example

A steel cylindrical tube of external radius 30 mm and internal radius 20 mm has a steel tube of external radius 40 mm shrunk on to it. The interference between the two tubes is to be such that the final maximum stress in each tube is the same when it is subject to an internal gauge pressure of 50 MPa. What are the resultant pressures at the inner and outer surfaces of both tubes and the required interference? Both tubes are of the same material, it having a modulus of elasticity of 200 GPa.

Consider first the situation when there is no internal gauge pressure for the compound cylinder. The inner cylinder will be subject to an external pressure p_s due to the shrinkage. The outer cylinder will be subject to the same pressure as an internal pressure p_s.

For the inner tube, the radial stress at its external surface is given by equation [14] as

$$\sigma_r = a_i + \frac{b_i}{r^2} = a_i + \frac{b_i}{0.030^2} = -p_s$$

and at the inner surface by

$$\sigma_r = a_i + \frac{b_i}{r^2} = a_i + \frac{b_i}{0.020^2} = 0$$

Subtracting the first equation from this gives

$$p_s = \frac{b_i}{0.020^2} - \frac{b_i}{0.030^2}$$

Hence $b_i = 7.2 \times 10^{-4} p_s$. Substituting this value in either of the equations gives $a = -1.8p_s$. Thus for the inner tube the maximum circumferential stress, at $r = 20$ mm, is given by equation [15] as

$$\sigma_c = a_i - \frac{b_i}{r^2} = -1.8p_s - \frac{7.2 \times 10^{-4} p_s}{0.020^2} = -3.60p_s$$

and the minimum circumferential stress, at $r = 30$ mm, is

$$\sigma_c = a_i - \frac{b_i}{r^2} = -1.8p_s - \frac{7.2 \times 10^{-4} p_s}{0.030^2} = -2.60p_s$$

For the outer tube, the radial stress at its internal surface is given by equation [14] as

$$\sigma_r = a_o + \frac{b_o}{r^2} = a_o + \frac{b_o}{0.030^2} = -p_s$$

and at its outer surface by

$$\sigma_r = a_o + \frac{b_o}{r^2} = a_o + \frac{b_o}{0.040^2} = 0$$

Subtracting this equation from the previous one gives

$$-p_s = \frac{b_o}{0.030^2} - \frac{b_o}{0.040^2}$$

Hence $b_o = -2.06 \times 10^{-3} p_s$. Substituting this value in one of the above equations gives $a_o = 1.29p_s$. The maximum circumferential stress is given, at $r = 30$ mm, by equation [15] as

$$\sigma_c = a_o - \frac{b_o}{r^2} = 1.29p_s + \frac{2.06 \times 10^{-3} p_s}{0.030^2} = +3.58p_s$$

and the minimum circumferential stress, at $r = 40$ mm, by

$$\sigma_c = a_o - \frac{b_o}{r^2} = 1.29p_s + \frac{2.06 \times 10^{-3} p_s}{0.040^2} = +2.58p_s$$

Now consider the stresses due to the internal pressure. We can consider the combined tube as effectively a single tube with the radial stress, at the inner surface with $r = 20\,\text{mm}$, being given by equation [14] as

$$\sigma_r = a + \frac{b}{r^2} = a + \frac{b}{0.020^2} = -50 \times 10^6$$

and at the outer surface, where $r = 40\,\text{mm}$, by

$$\sigma_r = a + \frac{b}{r^2} = a + \frac{b}{0.040^2} = 0$$

Subtracting this from the previous equation gives

$$-50 \times 10^6 = \frac{b}{0.020^2} - \frac{b}{0.040^2}$$

Hence $b = -2.67 \times 10^4$. Substituting this value in one of the equations gives $a = 1.67 \times 10^7$. Thus, due to the internal pressure, the circumferential stress will be, at $r = 20\,\text{mm}$,

$$\sigma_c = a - \frac{b}{r^2} = 1.67 \times 10^7 + \frac{2.67 \times 10^4}{0.020^2} = 83.5\,\text{MPa}$$

at $r = 30\,\text{mm}$,

$$\sigma_c = a - \frac{b}{r^2} = 1.67 \times 10^7 + \frac{2.67 \times 10^4}{0.030^2} = 46.4\,\text{MPa}$$

and at $r = 40\,\text{mm}$,

$$\sigma_c = a - \frac{b}{r^2} = 1.67 \times 10^7 + \frac{2.67 \times 10^4}{0.040^2} = 33.4\,\text{MPa}$$

For the pressure stress plus the shrinkage stress to have the same maximum value for both tubes we must have

$$83.5 \times 10^6 - 3.60p_s = 46.4 \times 10^6 + 3.58p_s$$

Thus $p_s = 5.17\,\text{MPa}$. The resultant circumferential stresses at the inner and outer surfaces of both tubes are

Inner tube, inner surface $83.5 - 3.60 \times 5.17 = 64.9\,\text{MPa}$
Inner tube, outer surface $46.4 - 2.60 \times 5.17 = 33.0\,\text{MPa}$
Outer tube, inner surface $46.4 + 3.58 \times 5.17 = 64.9\,\text{MPa}$
Outer tube, outer surface $33.4 + 2.58 \times 5.17 = 46.7\,\text{MPa}$

The interference is given by equation [24] as

$$\delta = r(\epsilon_{co} - \epsilon_{ci})$$

with the strains being the shrinkage strains at $r = 30\,\text{mm}$ and given by equations [22] and [23] as

$$\epsilon_{ci} = \frac{\sigma_{ci}}{E_i} - v_i \frac{\sigma_{ri}}{E_i}$$

$$\epsilon_{co} = \frac{\sigma_{co}}{E_o} - v_o \frac{\sigma_{ro}}{E_o}$$

The above stresses are the stresses due to shrinkage. Since $\sigma_{ri} = \sigma_{ro} = p_s$, then

$$\delta = r\left(\frac{\sigma_{ci}}{E} - \frac{\sigma_{co}}{E}\right)$$

$$= \frac{0.030}{200 \times 10^9}(-2.60 \times 5.17 - 3.58 \times 5.17) \times 10^6$$

Hence $\delta = 4.79 \times 10^{-6}$ m.

9.5.1   Compound shafts

Compound shafts formed by sleeves being shrunk on to solid shafts and problems involving hubs shrunk on to solid shafts can be tackled in the same way as the compound tubes.

For a solid shaft, the radial stress is given by equation [14] as

$$\sigma_r = a + \frac{b}{r^2}$$

at the centre of the shaft $r = 0$. Since the stress cannot be infinite at the centre then b must be zero. Thus $\sigma_r = a$. This means that the radial stress is constant throughout the shaft. Hence $\sigma_r = a = -p_s$, where p_s is the shrinkage pressure. The circumferential stress is given by equation [15] as

$$\sigma_c = a - \frac{b}{r^2}$$

and since $b = 0$ then $\sigma_c = a = \sigma_r = -p_s$. The circumferential stress is compressive and equal to the radial stress, both stresses being constant throughout the shaft.

The sleeve, or hub, can be considered as a thick cylindrical tube under an internal pressure of p_s.

Example

A bronze bush with a wall thickness of 25 mm is to be shrunk on to a steel shaft of radius 50 mm. What interference is required between the bush and shaft if there is to be a shrinkage interface pressure of 60 MPa? The steel has a modulus of elasticity of 200 GPa and Poisson's ratio 0.3, the bronze a modulus of elasticity of 100 GPa and a Poisson's ratio of 0.3.

Since $\sigma_c = a = \sigma_r = -p_s$, then the circumferential and radial stresses in the shaft are $-60\,\text{MPa}$. Now, considering the bush as a thick-walled tube, equation [14] gives

$$\sigma_{rb} = a_b + \frac{b_b}{r^2}$$

At $r = 50\,\text{mm}$

$$\sigma_{rb} = a_b + \frac{b_b}{0.050^2} = -60 \times 10^6$$

At $r = 75\,\text{mm}$

$$\sigma_{rb} = a_b + \frac{b_b}{0.075^2} = 0$$

Subtracting this from the previous equation gives

$$-60 \times 10^6 = \frac{b_b}{0.050^2} - \frac{b_b}{0.075^2}$$

Hence $b_b = -2.7 \times 10^5$. Substituting this value in either of the equations gives $a_b = 4.8 \times 10^7$. The circumferential stress for the bush at radius $50\,\text{mm}$ is thus given by equation [15] as

$$\sigma_{cb} = a_b - \frac{b_b}{r^2} = 4.8 \times 10^7 + \frac{2.7 \times 10^5}{0.050^2} = 156\,\text{MPa}$$

The interference is given by equation [24] as

$$\delta = r(\epsilon_{cb} - \epsilon_{cs})$$

with the strains being the shrinkage strains at $r = 50\,\text{mm}$ and given by equations [22] and [23] as

$$\epsilon_{cs} = \frac{\sigma_{cs}}{E_s} - v_s \frac{\sigma_{rs}}{E_s}$$

$$\epsilon_{cb} = \frac{\sigma_{cb}}{E_b} - v_b \frac{\sigma_{rb}}{E_b}$$

Since Poisson's ratio for the bronze has the same value as that for the steel, then

$$\delta = 0.050\left(\frac{156 \times 10^6}{100 \times 10^9} + \frac{60 \times 10^6}{200 \times 10^9}\right)$$

and so $\delta = 9.3 \times 10^{-5}\,\text{m}$.

Problems

(1)   A thin cylindrical pipe has an internal diameter of $200\,\text{mm}$ and a wall thickness of $4\,\text{mm}$. What will be the circumferential and longi-

tudinal stresses produced in the pipe when there is an internal gauge pressure of 2.4 MPa?

(2)   A compressed air cylinder has an internal diameter of 200 mm. What wall thickness will be required for a gauge pressure of 15 MPa, if the steel used for the cylinder has a maximum allowable stress of 100 MPa?

(3)   A vertical cylindrical tank with an internal diameter of 3.0 m, a wall thickness of 3 mm and a height of 20 m, is open at its top and filled with water, density 1000 kg/m^3. What will be the maximum circumferential stress?

(4)   A compressed air cylinder has an internal diameter of 2.0 m and is made of plates of thickness 12.5 mm. The efficiency of the longitudinal joints is 85% and of the circumferential joints 50%. What is the maximum permissible air pressure if the tensile stress in the cylinder plating is limited to 100 MPa?

(5)   A cylindrical boiler has a diameter of 2.0 m and is constructed from steel plate with a riveted joint. If the efficiency of the longitudinal joint is 80%, what thickness of plate is required if the maximum working stress is to be 60 MPa when the gauge pressure is 0.8 MPa?

(6)   A cylindrical pressure vessel has an internal diameter of 1.5 m and is made of plate of thickness 10 mm. The efficiency of the longitudinal joint is 75% and of the circumferential joints 50%. The plate has a tensile strength of 400 MPa. What is the maximum internal pressure to which the vessel should be used if there is to be a factor of safety of 5?

(7)   A cylindrical boiler, diameter 2.0 m, is constructed from plate of thickness 15 mm and has a single riveted longitudinal lap joint. The joint has rivets of diameter 20 mm with a pitch of 40 mm. What is the maximum internal gauge pressure for the boiler if the maximum allowable tensile stress for the boiler plate is 100 MPa and the shear strength of the rivets is 80 MPa? Assume that the mode of failure of the joint is by shear at the rivets.

(8)   A spherical gas container has an internal radius of 1.5 m and is subject to an internal pressure of 200 kPa. What is the minimum wall thickness for the container if the maximum tensile stress for the material is 40 MPa?

(9)   A spherical pressure vessel has an internal diameter of 400 mm and a wall of thickness 5 mm. What is the maximum pressure that can be used if the yield stress of the material is 100 MPa and a safety factor of 2 is to be used?

(10)   A steel cylindrical pressure vessel has an internal diameter of 200 mm and walls of thickness 5 mm. What is the increase in diameter of the

container when there is an internal gauge pressure of 4 MPa? The steel has a modulus of elasticity of 200 GPa and a Poisson's ratio of 0.3.

(11) A closed steel tube of internal diameter 100 mm and wall thickness 15 mm is subject to an internal gauge pressure of 2 MPa. What will be the percentage change in internal volume of the tube? The tube material has a modulus of elasticity of 200 GPa and Poisson's ratio of 0.3.

(12) A thin cylindrical shell with flat ends has a length 2.0 m, diameter 1.0 m and wall thickness 25 mm. It is filled with water at atmospheric pressure. What volume of water will need to be pumped into the vessel to obtain a gauge pressure of 10 MPa? The wall material has an elastic modulus of 200 GPa and a Poisson's ratio of 0.3. The water has a bulk modulus of 2.1 GPa.

(13) The vessel considered in problem 12 is modified by being closed with hemispherical ends. What now will be the volume of water needed?

(14) A cylinder has an internal diameter of 160 mm and wall thickness 40 mm and is subject to an internal pressure of 16 MPa and an external pressure of 6 MPa. What are the circumferential stresses at the inner and outer surfaces?

(15) The cylinder of a hydraulic ram has an internal diameter of 60 mm and is required to withstand an internal gauge pressure of 40 MPa. What wall thickness is required if the maximum tensile stress is to be 80 MPa and the maximum shear stress 50 MPa?

(16) A cylinder has an internal diameter of 300 mm, a wall thickness of 50 mm and a length of 900 mm. What will be the change in internal diameter when it is subject to an internal gauge pressure of 15 MPa? The modulus of elasticity is 200 GPa and Poisson's ratio is 0.3.

(17) What wall thickness is required for a cylinder of internal diameter 150 mm if the circumferential stress is not to exceed 125 MPa when there is an internal gauge pressure of 40 MPa?

(18) A cylindrical pressure vessel has an inner diameter of 250 mm and a wall thickness of 75 mm. What is the maximum internal pressure that can be used if the maximum is determined by a limiting tensile stress of 280 MPa with a safety factor of 2?

(19) A cylindrical pressure vessel has an inner diameter of 150 mm and a wall thickness of 30 mm. What will be the circumferential stresses at the inner and outer surfaces when the cylinder is subject to an internal gauge pressure of 40 MPa?

(20) A thick cylindrical tube has an external diameter k times its internal diameter and is subject to an internal gauge pressure. How is the

ratio of the maximum to minimum circumferential stress related to k?

(21) A steel cylindrical tube of internal radius 50 mm and external radius 75 mm has another cylindrical tube of the same material and having an external diameter of 100 mm shrunk on to it. What will be the interference required if the compound cylinder when subject to an internal gauge pressure of 100 MPa has the same maximum circumferential stresses in both cylinders? The modulus of elasticity of the steel is 200 GPa.

(22) A steel cylindrical tube of 75 mm internal radius and external radius 100 mm has another cylindrical tube of the same material and having an external diameter of 125 mm shrunk on to it. If there is a radial shrinkage stress of 17.5 MPa at the interface between the two cylinders, what will be the maximum tensile circumferential stress in the outer cylinder?

(23) A steel cylindrical tube of internal radius 50 mm and external radius 75 mm has another cylindrical tube of the same material and external radius 100 mm shrunk on to it. If there is a radial shrinkage stress of 20 MPa at the interface between the two tubes, what will be the maximum circumferential stress in the compound tube when it is subject to an internal gauge pressure of 60 MPa?

(24) A bronze bush of external diameter 75 mm is to be shrunk on to a steel shaft of radius 25 mm. What is the required interference between the two if there is to be an interface pressure of 40 MPa? For the steel the modulus of elasticity is 200 GPa, for the bronze 100 GPa. Both have a Poisson's ratio of 0.3.

Chapter 10

Linear and angular motion

10.1 Linear motion

Consider a particle moving along a straight line so that at some time t its displacement along the line from some reference point is s. If in a time interval δt the displacement increases by δs then the average velocity of the particle during that time interval is defined as being $\delta s/\delta t$. As the time interval considered is reduced, so the average velocity approaches the velocity at an instant. Thus, in the limit, the instantaneous velocity v is

$$v = \frac{ds}{dt} \qquad\qquad [1]$$

If the velocity of the particle changes by δv in a time interval δt then the average acceleration is defined as being $\delta v/\delta t$. As the time interval considered is reduced, so the average acceleration approaches the acceleration at an instant. Thus, in the limit, the instantaneous acceleration a is

$$a = \frac{dv}{dt} \qquad\qquad [2]$$

It is sometimes useful to express acceleration as

$$a = \frac{dv}{dt} = \frac{dv}{ds} \times \frac{ds}{dt} = \frac{dv}{ds} \times v$$

Example

An element in a machine accelerates from rest in a straight line such that its velocity v in m/s changes with time t in s according to $v = 0.8t^2 + 0.2t$. What is the acceleration and distance covered after $2\,$s?

Using equation [2],

$$a = \frac{dv}{dt} = \frac{d}{dt}(0.8t^2 + 0.2t) = 1.6t + 0.2$$

Thus after $2\,$s the acceleration is $3.4\,$m/s^2.
 The distance covered can be obtained by using equation [1].

$$v = \frac{ds}{dt} = 0.8t^2 + 0.2t$$

$$\int_0^s ds = \int_0^t (0.8t^2 + 0.2t)\, dt$$

$$s = \frac{0.8}{3}t^3 + \frac{0.2}{2}t^2$$

Thus when $t = 2\,\text{s}$ then $s = 2.1 + 0.4 = 2.5\,\text{m}$.

10.1.1   Motion under constant acceleration

For a particle moving in a straight line with a constant acceleration such that the velocity changes from u to v in a time t then

$$a = \frac{v - u}{t}$$

and so

$$v = u + at \tag{3}$$

The average velocity during this time t is $\frac{1}{2}(u + v)$ and thus

$$\text{average velocity} = \frac{s}{t} = \tfrac{1}{2}(u + v) = \tfrac{1}{2}[u + (u + at)]$$

$$s = ut + \tfrac{1}{2}at^2 \tag{4}$$

If we square both sides of equation [3] then

$$v^2 = (u + at)^2 = u^2 + 2uat + a^2t^2 = u^2 + 2a(ut + \tfrac{1}{2}at^2)$$

Hence

$$v^2 = u^2 + 2as \tag{5}$$

A common example of motion in a straight line with constant acceleration is that of a freely falling object or an object thrown vertically upwards, if air resistance is neglected. The falling object has the constant acceleration of about $9.8\,\text{m/s}^2$, the typical value for the acceleration due to gravity. The object thrown upwards has a constant retardation of $-9.8\,\text{m/s}^2$.

Example

A car accelerates from rest at $2.5\,\text{m/s}^2$, travels with a constant velocity of $90\,\text{km/h}$ and then decelerates at $3.0\,\text{m/s}^2$ before coming to rest. If the total distance travelled is $500\,\text{m}$, for what distance is the car travelling with constant velocity?

The velocity of $90\,\text{km/h}$ is $90 \times 10^3/3600 = 25\,\text{m/s}$. For the initial acceleration, equation [5] gives

$$v^2 = u^2 + 2as$$
$$25^2 = 0 + 2 \times 2.5 s_1$$

Hence $s_1 = 125$ m. For the deceleration, equation [5] gives

$$0 = 25^2 - 2 \times 3.0s_2$$

Hence $s_2 = 104$ m. Thus the distance travelled with constant velocity was $500 - 125 - 104 = 271$ m.

Example

A stone is thrown vertically upwards with an initial velocity of 10 m/s^2. What will be the greatest height reached?

At the greatest height the velocity of the stone will have decreased to zero. Thus, using equation [5]

$$v^2 = u^2 + 2as$$
$$0 = 10^2 - 2 \times 9.8s$$

Hence the greatest height reached s is 5.1 m.

10.1.2   Force and motion

When an object is acted on by a resultant force then an acceleration occurs (Newton's law 2, see section 1.2), the acceleration a being given by

$$F = ma \qquad\qquad [6]$$

where m is the mass of the object. Thus a force of 100 N acting on a mass of 20 kg will produce an acceleration in the direction of the force of $100/20 = 5$ m/s^2.

Suppose now that a force F acts on an object and that in exactly the opposite direction there is a resistive force R. The resultant force acting on the object will be $F - R$ and thus the acceleration is given by

$$F - R = ma \qquad\qquad [7]$$

In many situations the resistive force may be a function of the velocity of the object.

Example

A car tows a caravan of mass 1000 kg in a straight line with an acceleration of 0.2 m/s^2. If the resistance to motion is a constant 200 N, what is the force exerted by the car through the tow bar?

The resultant force acting on the caravan and which gives it the acceleration is $F - 200$ N, where F is the force exerted through the tow bar. Thus

$$F - 200 = 1000 \times 0.2$$

Hence $F = 400$ N.

Example

Figure 10.1(a) shows a pulley system involving two masses m and M connected by a light cord passing over the pulley. What is the acceleration a of the masses and the tension in the cord? Assume that the pulley is frictionless and that the moment of inertia of the pulley may be neglected.

Figure 10.1(b) shows the free body diagrams for each of the masses. Thus

$$T - mg = ma$$
$$Mg - T = Ma$$

Adding the two equations gives

$$Mg - mg = Ma + ma$$

Hence

$$a = \frac{(M - m)g}{M + m}$$

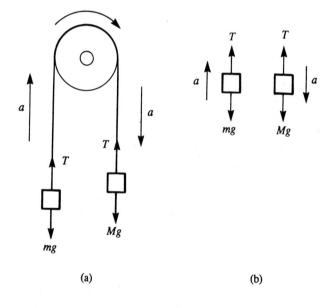

(a)   (b)

Fig. 10.1 Example.

Example

A car of mass 1000 kg starts from rest and climbs a hill which has a uniform gradient of 1 in 20 with an acceleration of 0.20 m/s². If the resistances to motion amount to 200 N, what will be the tractive effort exerted by the car?

Figure 10.2 shows the situation, the weight mg of the car having been resolved into a component down the incline and at right angles to it. The resultant force accelerating the car up the incline is thus $F - mg \sin \theta - R$, where R is the resistance to motion and F the tractive effort. Thus

$$F - mg \sin \theta - R = ma$$

A gradient of 1 in 20 means that for every 20 m up the incline a vertical height of 1 m is gained. Thus $\sin \theta = 1/20$. Hence

$$F - 1000 \times 9.81 \times (1/20) - 200 = 1000 \times 0.20$$

Hence $F = 890.5$ N.

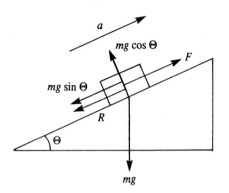

Fig. 10.2 Example.

Example

A stone of mass m is thrown vertically upwards with an initial velocity u. The air resistance at a velocity v is mkv^2, where k is a constant. What time is taken by the stone to reach its greatest height?

The resultant force on the stone in its upward motion is $-mg - mkv^2$ and thus

$$-mg - mkv^2 = ma$$

Since $a = dv/dt$, then

$$-(g + kv^2) = \frac{dv}{dt}$$

Rearranging this equation and integrating between the limits of velocity u at $t = 0$ and velocity 0 at the greatest height time of T, then

$$\int_0^T dt = -\int_u^0 \frac{dv}{g + kv^2}$$

$$T = \sqrt{(1/kg)} \tan^{-1} - [u\sqrt{(k/g)}]$$

10.1.3 Friction

The term frictional force is used to describe a force that arises when two bodies are in contact and that opposes relative motion between them. If two objects are at rest and a force is applied to move one relative to the other, then motion may not occur. This is because there is no resultant force, the applied force being balanced by an opposite and equal frictional force. The frictional force is always acting in such a direction as to oppose relative motion and always tangential to the surfaces in contact. Increasing the applied force results in a value being reached at which motion just starts. When this occurs the frictional force has reached its maximum value, this being termed the *limiting frictional force*. The resultant force acting on the object is thus the applied force minus the limiting frictional force. The *coefficient of static friction* μ_s is the ratio of the limiting frictional force F to the normal reaction R. Thus

$$F = \mu_s R \qquad\qquad [8]$$

When motion is occurring there is still a frictional force acting, this being determined by the *coefficient of kinetic friction* μ_k. This coefficient is generally lower than the static coefficient.

$$F = \mu_k R \qquad\qquad [9]$$

Example

What is the acceleration produced when a force of 100 N acts horizontally on a block of mass 5 kg which rests on a horizontal surface if the coefficient of kinetic friction is 0.6?

The reaction force is 5×9.81 N and thus the frictional force is $0.6 \times 5 \times 9.81$ N. Hence the resultant force acting on the object is

$$100 - 0.6 \times 5 \times 9.81 = ma = 5a$$

Hence $a = 14.1 \text{ m/s}^2$.

10.2 Curvilinear motion

So far in this chapter when there has been a change in velocity it has been assumed that the direction of motion remained unchanged, a change in velocity being just a change in magnitude. However, velocity is a vector quantity, having both magnitude and direction. Changes in velocity may thus be (a) a change in magnitude only, (b) a change in direction only, (c) a change in both magnitude and direction. The term speed is used for just the magnitude of the velocity. An object can thus have a constant speed but a changing velocity if the magnitude of the velocity remains constant but the direction changes.

Whenever there is a change in velocity there is an acceleration and thus, as a consequence of Newton's second law of motion, there must be a resultant force acting. The acceleration is always in the direction of the force. If there is no resultant force acting on an object then there is no acceleration and consequently it continues in motion with an unchanged velocity (Newton's first law).

For motion in a curvilinear path, i.e. a curved path on a plane, the velocity at a point A on the path is in a direction given by the tangent to the curved path at that point (figure 10.3(a)). The direction of the velocity at a point B on the path is given by the tangent to the curved path at that point. We can represent the two velocities by arrow-headed straight lines, the length of an arrow being proportional to the magnitude of a velocity and its direction to the direction of the velocity (see section 1.1 for a discussion of vectors). The change in velocity between points A and B is the difference in velocities between those points, taking into account the vector nature of velocity. Thus velocity A plus the change in velocity equals velocity B. We can add two vectors by means of the parallelogram law (see section 1.1). Figure 10.3(b) shows the parallelogram.

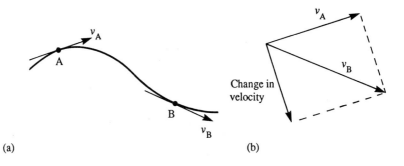

(a) (b)

Fig. 10.3 Curvilinear motion.

Example

A car travelling along a straight road with a velocity of 20 m/s turns a right-angled corner and proceeds in the new direction with a velocity of 20 m/s. What is the change in velocity?

Figure 10.4 shows the parallelogram of velocities for the situation. Hence the change in velocity has a magnitude, using Pythagoras, of $\sqrt{(20^2 + 20^2)} = 28.3$ m/s in a direction of $-135°$ from the initial direction.

10.2.1 Projectiles

Consider a projectile which has an initial velocity u at an angle θ to the horizontal (figure 10.5). The velocity can be resolved into two components, $u\cos\theta$ in a horizontal direction and $u\sin\theta$ in a vertical direction. In the

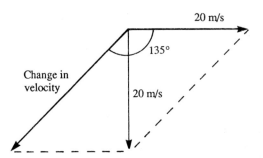

Fig. 10.4 Example.

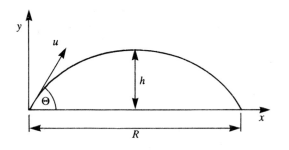

Fig. 10.5 Projectile path.

horizontal direction, if we neglect any consideration of air resistance, there is no force acting on the projectile. Thus its velocity component in that direction must remain unchanged. Hence the horizontal distance x covered in a time t is given by

$$x = ut \cos \theta \qquad [10]$$

In the vertical direction there is a retardation of $-g$ due to the force of gravity. Thus the vertical distance y covered in the time t is given by

$$y = ut \sin \theta - \tfrac{1}{2}gt^2 \qquad [11]$$

Equation [10] gives $t = x/u \cos \theta$ and thus substituting this value in equation [11] gives

$$y = x \tan \theta - \frac{xg}{2u^2 \cos^2 \theta} \qquad [12]$$

This equation describes the path of the projectile.

At the highest point h of the trajectory the vertical component of the velocity must be zero. Thus

$$0 = u \sin \theta - gt_h$$

where t_h is the time to the greatest height. Thus $t_h = (u \sin \theta)/g$. For a projectile moving over a horizontal ground then the time taken to go to the greatest height and then back down to earth is $2t_h$. This is thus the total time the projectile is in the air and hence, using equation [10], the horizontal distance covered in that time, the so-called range R, is

$$R = \frac{2u^2 \sin\theta \cos\theta}{g} = \frac{u^2 \sin 2\theta}{g} \qquad [13]$$

Example

A stream of water emerges from a jet which is set at an angle of 30° to the horizontal with a velocity of 2 m/s. What will be the greatest height above the jet reached by the water and the horizontal distance travelled before it reaches the horizontal plane of the jet?

The vertical component of the velocity is $20\sin 30°$ and the horizontal component $20\cos 30°$. The greatest height is reached when the vertical velocity is zero. Thus

$$v^2 = u^2 + 2as$$
$$0 = (2\sin 30°)^2 - 2gh$$

Hence

$$h = \frac{4\sin^2 30°}{2 \times 9.81} = 51.0\,\text{mm}$$

The horizontal distance R travelled is in twice the time taken to reach the greatest height. This time t_h is the time taken for the vertical velocity component to become zero and is thus given by

$$v = u + at$$
$$0 = 2\sin 30° - \tfrac{1}{2} \times 9.81 t_h$$

Hence $t_h = 0.204\,\text{s}$. Hence

$$R = 2 \times 0.204 \times 2\cos 30° = 707\,\text{mm}$$

10.2.2 Circular motion with constant speed

Consider a particle of mass m rotating with constant speed in a circular path of radius r (figure 10.6(a)). At point A the velocity is v in the indicated direction. At point B, a short time Δt later, the velocity has the same magnitude v but a different direction. The change in velocity Δv can be obtained from the parallelogram of velocities (figure 10.6(b)). Thus, since we are only considering a small angular change,

$$\Delta v = 2v\sin\tfrac{1}{2}\theta \approx v\theta$$

The direction of this change in velocity is towards the centre of the circle. The velocity changes by Δv in a time interval of Δt, thus there is an acceleration directed towards the centre of the circle of

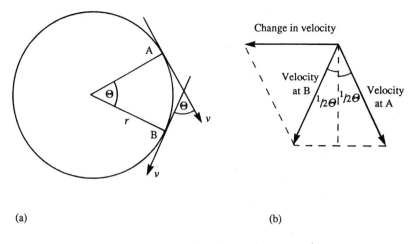

Fig. 10.6 Circular motion.

$$a = \frac{\Delta v}{\Delta t} = \frac{v\theta}{\Delta t}$$

But in the time Δt the particle has moved round the arc of the circle from A to B. Thus the magnitude of the velocity must be arc AB/Δt. Hence, since arc AB = $r\theta$,

$$v = \frac{r\theta}{\Delta t}$$

Thus

$$a = \frac{v(v\Delta t/r)}{\Delta t} = \frac{v^2}{r} \qquad [14]$$

This acceleration is directed towards the centre of the circle and is called the *centripetal acceleration*.

The centripetal acceleration must be provided by a force acting in the same direction. Thus there is a centripetal force of

$$F = ma = \frac{mv^2}{r} \qquad [15]$$

According to Newton's third law of motion there must be an opposite and equal force acting on some other object. In this case, the centripetal force is balanced by a *centrifugal force* which acts in an outward direction on the pivot around which the circular motion is occurring.

Example

What is the centripetal acceleration of a train travelling round a curve of radius 1 km at a speed of 20 m/s?

Using equation [14]

$$a = \frac{v^2}{r} = \frac{20^2}{1000} = 0.40 \, \text{m/s}^2$$

Example

What is the maximum speed with which a car can turn a horizontal corner of radius *r* if skidding is not to occur?

The reaction force for the car is *mg* and thus the frictional force is *μmg*. It is this force which provides the centripetal force and prevents sliding. Thus the maximum speed *v* is when

$$F = \frac{mv^2}{r} = \mu mg$$

$$v = \sqrt{(\mu mg)}$$

Example

At what angle should a corner of radius *r* be banked if a car is to round it without relying on frictional forces to provide the centripetal force?

Figure 10.7 shows the situation. The centripetal force is provided by the horizontal components of the reaction force, i.e. $R \sin \theta$. Thus

$$R \sin \theta = \frac{mv^2}{r}$$

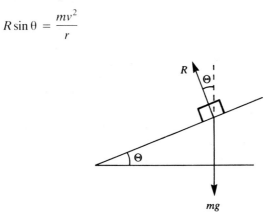

Fig. 10.7 Example.

But we must also have the vertical component of the reactive force equal to *mg*.

$$R \cos \theta = mg$$

Thus dividing the two equations gives

$$\tan \theta = \frac{v^2}{rg}$$

10.2.3 Curvilinear motion in general

Consider a particle moving in a curved path with a changing speed. The velocity of the particle is thus changing as a result of a changing magnitude and a changing direction. The changing magnitude means that there must be an acceleration tangential to the path. The magnitude of this acceleration is related to the path position s and time t by $a = d^2s/dt^2$. The changing direction means that there must be an acceleration at right angles to the path. The magnitude of this acceleration is v^2/r, where r is the radius of curvature of the path at the point concerned. Since acceleration is a vector quantity, the resultant acceleration is obtained from the tangential and radial accelerations by using the parallelogram of vectors.

10.3 Relative velocity

The displacements, velocities and accelerations considered so far in this chapter have all been measured from a stationary reference point. In considering the motion of, say, a car then the displacements, velocities and accelerations have all been considered from the point of view of an observer who is standing still in the road. However, there are situations where it is useful to consider the motion of a point relative to another point which has a different motion. For example, the motion of the car could be considered relative to an observer moving in another car.

Consider two points A and B whose displacements from a fixed point O are d_{AO} and d_{BO} (figure 10.8). A useful way of considering the problem of determining the displacement of B relative to A, i.e. d_{BA}, is to move point A to the origin by adding to it a displacement $-d_{AO}$. We must thus also add to the other displacement $-d_{AO}$. The result of this addition then gives the displacement of B relative to A.

Fig. 10.8 Relative displacement.

We can apply a similar technique to determine the velocity of one point relative to another. Consider two points A and B in motion, with A having a velocity v_{AO} relative to some stationary origin O and B a velocity v_{BO} relative to the same stationary origin. To find the velocity of B relative to A, we make A stationary by adding to its velocity the velocity $-v_{AO}$. We then add the same velocity to v_{OB} to give the velocity of B relative to A (figure 10.9).

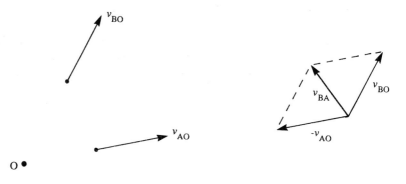

Fig. 10.9 Relative velocity.

Example

A train travelling at a constant velocity of 30 m/s crosses a bridge over a road, the road being at an angle of 45° to the rail track. A car is travelling along the road with a velocity of 20 m/s. What is the velocity of the car relative to the train?

To find the velocity of the car relative to the train we 'stop' the train by adding to it a velocity of -30 m/s in the track direction. Adding the same velocity to that of the car gives the vector diagram shown in figure 10.10. The velocity of the car relative to the train v_{CT} can be obtained from the vector diagram by the use of the cosine rule (see section 1.1.2). Thus

$$v_{CT}^2 = 20^2 + 30^2 - 2 \times 20 \times 30 \cos 45°$$

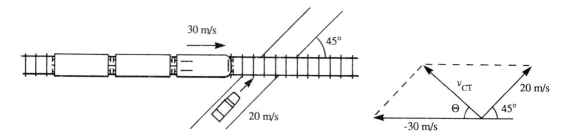

Fig. 10.10 Example.

Hence $v_{CT} = 21.2$ m/s in a direction which makes an angle θ with the track. θ can be obtained using the sine rule (see section 1.1.2) as

$$\frac{20}{\sin \theta} = \frac{21.2}{\sin 45°}$$

Thus $\theta = 41.8°$.

10.4   Angular motion

Consider a line, fixed at one end and rotating in a circular path. The angle swept out by the line is termed the angular displacement. The unit used for the angle is the radian, one complete rotation of $360°$ being equal to 2π rad. If there is an angular displacement of $\delta\theta$ in a time δt then the average angular velocity in that time is $\delta\theta/\delta t$. In the limit as the time interval tends to zero then the angular velocity ω at an instant is

$$\omega = \frac{d\theta}{dt} \qquad [16]$$

If f revolutions of the rotating line are made per second then the angular displacement per second is $2\pi f$ and so the average angular velocity in a second is

$$\omega = 2\pi f \qquad [17]$$

Angular acceleration α is the rate of change of angular velocity and so can be written as

$$\alpha = \frac{d\omega}{dt} \qquad [18]$$

An alternative and sometimes useful way of writing equation [18] is as

$$\alpha = \frac{d\omega}{dt} = \frac{d\omega}{d\theta} \times \frac{d\theta}{dt} = \frac{d\omega}{d\theta} \times \omega$$

Example

A flywheel, starting from rest, takes 4 s to achieve a rotation of 5 rev/s. What is the average angular acceleration during that time?

The change in angular velocity is from 0 to $2\pi f = 2\pi \times 5$ rad/s in 4 s. Hence the average angular acceleration is

$$\alpha = \frac{10\pi - 0}{4} = 7.85 \text{ rad/s}$$

10.4.1 Motion with constant angular acceleration

For constant angular acceleration α, when the angular velocity changes from ω_0 to ω in a time t then

$$\alpha = \frac{\omega - \omega_0}{t}$$

Hence

$$\omega = \omega_0 + \alpha t \qquad [19]$$

The average angular velocity in a time t is $\frac{1}{2}(\omega_0 + \omega)$ and so, if the angle swept out in that time is θ,

$$\frac{\theta}{t} = \frac{\omega_0 + \omega}{2} = \frac{\omega_0 + (\omega_0 + \alpha t)}{2} = \omega_0 + \frac{1}{2}\alpha t$$

Hence

$$\theta = \omega_0 t + \frac{1}{2}\alpha t^2 \qquad [20]$$

Since $\omega = \omega_0 + \alpha t$, then

$$\omega^2 = (\omega_0 + \alpha t)^2 = \omega_0^2 + 2\alpha(\omega_0 + \frac{1}{2}\alpha t^2)$$
$$\omega^2 = \omega_0^2 + 2\alpha\theta \qquad [21]$$

Equations [19], [20] and [21] are similar to those for linear motion [3], [4] and [5].

Example

A flywheel starts from rest and rotates with a constant angular acceleration of $2\,\text{rad/s}^2$. What will be the angular velocity after 10 revolutions?

The angular displacement in 10 revolutions is $10 \times 2\pi\,\text{rad}$. Thus, using equation [21]

$$\omega^2 = \omega_0^2 + 2\alpha\theta$$
$$= 0 + 2 \times 2 \times 10 \times 2\pi$$

Hence $\omega = 15.9\,\text{rad/s}$

10.4.2 Relationship between linear and angular motion

The distance covered s by the end of a radius moving round the arc of a circle of radius r is related to the angular displacement θ by $s = r\theta$. If the radius line rotates with an average angular velocity ω then in time t the angle covered will be ωt. Thus $s = r\omega t$. But s/t is the speed v of the end of the radius round the circumference of the circular path. Thus

$$v = r\omega \qquad [22]$$

If the end of the radius has an initial tangential velocity u, when the angular velocity is ω_0, and this changes to v, with angular velocity ω, in a time t, then the tangential acceleration a is

$$a = \frac{v - u}{t} = \frac{r\omega - r\omega_0}{t}$$

But $(\omega - \omega_0)/t = \alpha$. Hence

$$a = r\alpha \qquad\qquad [23]$$

Example

A bicycle has wheels of diameter 650 mm and is being ridden on level ground at 2.5 m/s. What is the angular velocity of each wheel?

Using equation [22]

$$v = 2.5 = r\omega = 0.325\omega$$

Hence $\omega = 7.7$ rad/s

10.5   Torque and angular motion

Consider the rotation of a rigid body about a fixed axis, through O in figure 10.11, with a constant angular acceleration α. A small particle of mass δm in the body a distance r from O will have a linear acceleration a, given by equation [23], of $r\alpha$ in a direction at right angles to the direction of r. Because there is an acceleration there must be a resultant force F acting on the particle. Thus

$$F = \delta m \times a = \delta m \times r\alpha$$

The turning moment of this force about the axis through O is

$$\text{moment} = Fr = r^2\alpha\,\delta m$$

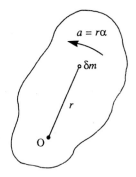

Fig. 10.11 Rotation of a rigid body.

The total torque T due to all the elements of mass in the body will be

$$T = \int r^2 \alpha \, dm$$

This is written as

$$T = I\alpha \qquad [24]$$

where I is called the *moment of inertia* of the body and is given by

$$I = \int r^2 \, dm \qquad [25]$$

The moment of inertia has the basic unit of $kg\,m^2$.

Example

A flywheel has a moment of inertia of $100\,kg\,m^2$ and is rotating at $120\,rev/min$. What braking torque is required to bring it to rest in $10\,s$?

The initial angular velocity is $2\pi \times 120/60 = 12.6\,rad/s$. Thus, using equation [19],

$$\omega = \omega_0 + \alpha t$$
$$0 = 12.6 + 10\alpha$$

Thus $\alpha = -1.26\,rad/s^2$. The retarding torque is thus given by equation [24] as

$$T = I\alpha = 100 \times 1.26 = 126\,N\,m$$

10.5.1 Radius of gyration

The moment of inertia of a body about an axis is the sum of all the elemental masses constituting the body with each multiplied by the square of their distance from the axis. The sum of all the individual elemental masses is equal to the total mass m of the body. Thus we can imagine all this mass to be concentrated at a point a distance k from the axis such that

$$mk^2 = \int r^2 \, dm$$

Then the moment of inertia I can be expressed as

$$I = mk^2 \qquad [26]$$

The distance k is called the *radius of gyration*.

Example

A wheel of mass $10\,kg$ has a radius of gyration of $200\,mm$ and is subject to a turning moment of 4 N m. What will be its angular velocity 3 s after starting from rest?

The moment of inertia $I = mk^2 = 10 \times 0.200^2 = 0.40\,\text{kg m}$. Thus

$$T = 4 = I\alpha = 0.40\alpha$$

and so the angular acceleration α is $10\,\text{rad/s}^2$. The angular velocity after 4 s is given by equation [19] as

$$\omega = \omega_0 + \alpha t$$
$$= 0 + 10 \times 3 = 30\,\text{rad/s}$$

10.5.2 *Moment of inertia*

The moment of inertia of a body is given by equation [25] as

$$I = \int r^2\,dm$$

Consider a uniform disc of radius r and mass m. The mass per unit area is thus $m/\pi r^2$. The moment of inertia about an axis through the disc centre can be found by considering the disc to be made up of elemental masses in the form of rings of radius x and thickness δx, with x taking values between 0 and r (figure 10.12). Such a ring has an area of $2\pi x\,\delta x$ and so a mass of

$$\delta m = \frac{m}{\pi r^2}\,2\pi x\,\delta x = \frac{2m}{r^2}x\,\delta x$$

The moment of inertia of the ring is thus

$$I = \frac{2m}{r^2}x^3\,\delta x$$

Thus the moment of inertia of the entire disc is

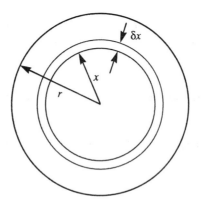

Fig. 10.12 Moment of inertia of a disc.

$$I = \int_0^r \frac{2m}{r^2} x^3 \, dx$$

$$= \tfrac{1}{2}mr^2 \tag{27}$$

Figure 10.13 shows the moments of inertia of some common shapes. The moment of inertia of composite shapes about some axis can be determined by adding together the moments of inertia of the constituent parts about the same axis.

The value of a moment of inertia for a body depends on the axis about

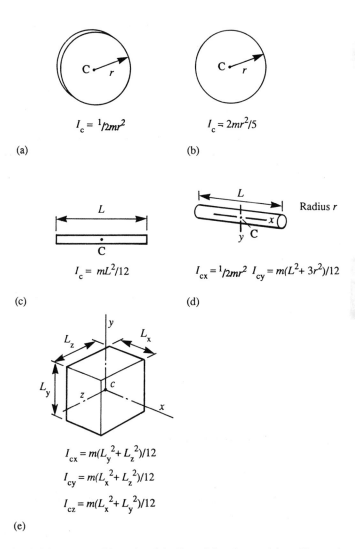

(a) $I_c = \tfrac{1}{2}mr^2$

(b) $I_c = 2mr^2/5$

(c) $I_c = mL^2/12$

(d) $I_{cx} = \tfrac{1}{2}mr^2 \quad I_{cy} = m(L^2 + 3r^2)/12$

(e)
$$I_{cx} = m(L_y^2 + L_z^2)/12$$
$$I_{cy} = m(L_x^2 + L_z^2)/12$$
$$I_{cz} = m(L_x^2 + L_y^2)/12$$

Fig. 10.13 Moments of inertia, (a) disc, (b) sphere, (c) uniform slender rod, (d) cylinder, (e) rectangular block.

which it is specified. If the moment of inertia for an object about an axis through its centroid is

$$I_{xc} = \int r^2 \, dm$$

then about a parallel axis a distance h from the original axis it is

$$I_x = \int (r + h)^2 \, dm = \int r^2 \, dm + 2h \int r \, dm + h^2 \int dm$$

But the integral of $r \, dm$ about an axis through the centroid is zero and the integral of dm is the total mass m. Thus

$$I_x = I_{xc} + mh^2 \tag{28}$$

This relationship is known as the *parallel axis theorem*.

Example

Determine the moment of inertia about an axis through its centre of a uniform ring of mass m with an internal diameter r and external diameter R.

The ring can be considered to be a composite shape consisting of a disc of radius R minus a disc of radius r. Thus, since the mass per unit area is $m/\pi(R^2 - r^2)$,

$$I = \tfrac{1}{2}[\pi R^2 m/\pi(R^2 - r^2)]R^2 - \tfrac{1}{2}[\pi r^2 m/\pi(R^2 - r^2)]r^2$$
$$= \tfrac{1}{2} m \, (R^2 + r^2)$$

Example

The moment of inertia of a thin uniform rod of mass m and length L about an axis at right angles to the rod and through its centroid is $mL^2/12$. What is the moment of inertia of the rod about an axis at right angles to the rod and at one end of it?

Using the parallel axis theorem,

$$I_x = I_{xc} + mh^2 = \frac{mL^2}{12} + m \left(\frac{L}{2} \right)^2 = \frac{mL^2}{3}$$

10.6   Balancing of rotating masses

A shaft is said to be in *static balance* if, for off-centre masses, there is no resultant force in any direction due to the weights of the masses and the algebraic sum of their turning moments about the shaft axis is zero, i.e. the centre of gravity of the assembly lies on the axis of the shaft. Thus, for the off-centre masses shown in figure 10.14, taking moments about O

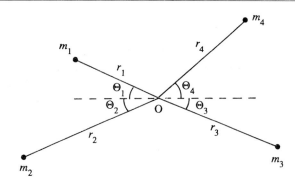

Fig. 10.14 Off-centre masses.

$$m_1gr_1\cos\theta_1 + m_2gr_2\cos\theta_2 - m_3gr_3\cos\theta_3 - m_4gr_4\cos\theta_4 = 0$$

Thus, in general

$$\Sigma mgr\cos\theta = 0 \qquad\qquad [29]$$

A vector diagram can be drawn with the sides representing the vectors m_1r_1, m_2r_2, m_3r_3 and m_4r_4 (the g is just a constant which multiplies all the vectors equally), at the appropriate angles θ_1, θ_2, θ_3 and θ_4, taken in order. For equilibrium, the diagram must be a closed polygon, as shown in figure 10.15.

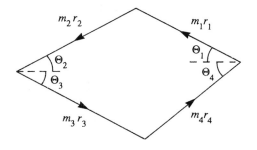

Fig. 10.15 mr polygon.

The above discussion refers to a shaft when it is stationary and not rotating. When it is rotating, each mass will experience a centripetal force of $mv^2/r = m\omega^2r$. The pivot point O will experience a centrifugal force which is opposite and equal to the centripetal force. Thus there will be centrifugal forces due to each of the off-centre masses. If all the rotating masses are in the same plane, then for *dynamic balance* we must have the vector sum of the centrifugal forces equal to zero. If we have, say, just a single off-centre mass m at distance r then to achieve dynamic balance we must have another mass M, diametrically opposite, at a distance r such that

$$MR\omega^2 = mr\omega^2$$

Since ω is the same for both masses, this means $MR = mr$. If the system is balanced for one angular speed it is balanced for every other speed.

If we have a number of off-centre masses, all in the same transverse plane, then if all the centrifugal forces are resolved into one direction

$$\Sigma m\omega^2 r \cos\theta = 0$$

Since ω is the same for all the masses attached to the shaft, then we can write

$$\Sigma mr \cos\theta = 0 \tag{30}$$

This is the same as the condition [29] for static balance. Thus if the system is dynamically balanced it is automatically statically balanced. The converse is only true when all the masses are in the same transverse plane.

A vector diagram can be drawn in which the sides represent the mr values, taken in order and at the angles of their radial lines since there are the relative directions of the centrifugal forces. If there is dynamic balance then the resulting polygon is closed.

If the off-centre masses are not all in the same transverse plane, then when the system rotates they will rotate in parallel planes and this will so give rise to centrifugal forces at different points along the shaft (figure 10.16). Thus, for dynamic balance, not only must equation [30] be true but also the turning moments about any axis and considered in any plane for the centrifugal forces acting along the length of the shaft must be zero, i.e.

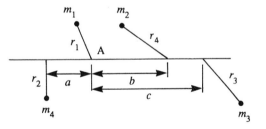

Fig. 10.16 Off-centre masses in different planes.

$$\Sigma m\omega^2 rL \cos\theta = 0$$

where L is the distance for each centrifugal force from some point along the shaft. Since the angular velocity is the same for each off-centre mass then we can write

$$\Sigma mrL \cos\theta = 0 \tag{31}$$

A vector diagram can be drawn in which the sides represent the mrL values, taken in order and at the angles of their radial lines since there are the relative directions of the centrifugal forces. If there is dynamic balance then the resulting polygon is closed.

Example

A flywheel of mass 12 kg has its centre of mass offset from the centre by 2.0 mm. At what radius should a mass of 100 g be placed so that the flywheel is in dynamic balance?

For dynamic balance we must have $MR = mr$ and so

$$12 \times 2 = 0.100r$$

and $r = 240$ mm.

Example

A shaft 2.0 m long is supported in two bearings at 200 mm from each end. The shaft carries three pulleys, one at each end and one at the middle of the length. The pulleys have the same masses, 50 kg, with the end pulleys A and C having their centre of gravities at 3.5 mm and 5.0 mm and the centre pulley B 4.5 mm from the shaft axis. The pulleys are keyed to the shaft at such angles as to give static balance. What are the relative angular settings needed to give static balance and the dynamic load on each bearing when the shaft rotates at 360 rev/min?

The mr products for the three pulleys must give a closed polygon of vectors for static balance to occur. The mr products are $50 \times 3.5 = 175$ kg mm, $50 \times 5.0 = 250$ kg mm and $50 \times 4.5 = 225$ kg mm. These when represented by lengths must form a triangle. The triangle in figure 10.17(a) has been drawn at the instant the 175 kg mm vector is horizontal. The relative angular directions of the out-of-balance forces is thus as in figure 10.17(b).

Figure 10.18(a) shows the arrangement of the bearings and pulleys along the shaft. Considering the mrL values about bearing D then they are $-50 \times 0.0035 \times 0.2 = 0.035$ kg m^2, $50 \times 0.005 \times 1.8 = 0.450$ kg m^2 and

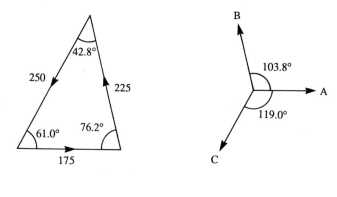

(a) (b)

Fig. 10.17 Example.

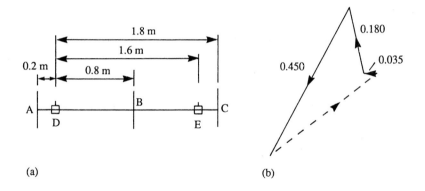

Fig. 10.18 Example.

$50 \times 0.0045 \times 0.8 = 0.180 \, \text{kg m}^2$. The minus sign is because L is to the left of bearing D for that pulley, all others being to the right. Bearing E will then have a moment which completes the polygon of vectors, as shown in figure 10.18(b). The negative mrL value is drawn in the opposite direction to the force direction indicated in figure 10.17(b). From the figure the mrL value for bearing E is thus about $0.365 \, \text{kg m}^2$. Since $L = 1.6 \, \text{m}$ then mr is $0.228 \, \text{kg m}$ and so the force is $mr\omega^2 = 0.228 \times (2\pi \times 360/60)^2 = 324 \, \text{N}$. Since there is no unbalanced force in the system, and the pulleys are statically balanced, then the force at bearing D must be opposite and equal in size to that at bearing E. In fact the forces on the bearings constitute a pure couple.

10.7 Momentum

Linear momentum of a body is defined as being the product of its mass and velocity. It is a vector quantity. When Newton's second law is written as $F = ma$, then since the acceleration $a = \mathrm{d}v/\mathrm{d}t$ we can write

$$F = m \frac{\mathrm{d}v}{\mathrm{d}t} = \frac{\mathrm{d}(mv)}{\mathrm{d}t} \qquad [32]$$

Thus the rate of change of momentum is equal to the applied force. Newton's second law can be defined as this and $F = ma$ deduced.

The product of the force and the length of time over which it acts is called the *linear impulse* and is thus, according to equation [32], equal to the change in momentum.

Newton's third law states that when two bodies interact the force experienced by one must be opposite and equal in size to the force experienced by the other body. Thus when an object of mass m_1 and velocity u_1 collides with another object of mass m_2 moving in the same straight line with velocity u_2, then on impact we must have

$$m_1 a_1 = -m_2 a_2$$

where a_1 and a_2 are the acceleration occurring at impact. If the duration of the impact is a time t and the two objects have velocities v_1 and v_2 after impact, then

$$m_1 \frac{v_1 - u_1}{t} = - m_2 \frac{v_2 - u_2}{t}$$

Thus

$$m_1 u_1 + m_2 u_2 = m_1 v_1 + m_2 v_2 \qquad [33]$$

The sum of the momentum before the collision equals the sum of the momentum after it. This is known as the *conservation of momentum*.

Example

A railway truck of mass 6000 kg when travelling at 2 m/s runs up against buffers and rebounds at 0.5 m/s. What is the force exerted by the buffers if the time of contact of the truck with the buffers is 0.5 s?

The momentum changes from 6000×2 kg m/s to -6000×0.5 kg m/s. Thus the change in momentum is 15 000 kg m/s in 0.5 s. The average rate of change of momentum is thus 30 000 kg m/s^2 and so the average force is 30 kN.

10.7.1 Angular momentum

The *angular momentum* of a particle is defined as being the moment of the linear momentum of the particle about the axis of rotation. Thus for a particle of mass m moving in a circular path of radius r with a tangential velocity v, then the moment of the linear momentum is mvr. But $v = r\omega$, where ω is the angular velocity about the axis. Thus the angular momentum is $mr^2\omega$. But mr^2 is the moment of inertia of the particle about the axis. Thus

$$\text{angular momentum} = I\omega \qquad [34]$$

Angular momentum is a vector quantity.

If the particle is acted on by a tangential force F then since the applied torque $T = Fr$ and $F = d(mv)/dt$, we can write

$$T = \frac{d(mv)}{dt} r = \frac{d(mr^2\omega)}{dt} = \frac{d(I\omega)}{dt}$$

$$= \text{rate of change of angular momentum} \qquad [35]$$

The product of the torque and the time over which it acts is called the *angular impulse* and is thus, according to equation [35], equal to the change in angular momentum.

If there is no external torque acting on an object or system of connected rigid objects then, according to equation [35], there is no change in angular momentum. The total angular momentum of the system about a fixed point remains unchanged, i.e. angular momentum is conserved. An example which illustrates this is a skater performing a spin on the toe of one skate. If the skater starts with arms and a leg outstretched and then pulls them into his or her body then the moment of inertia is reduced. Since angular momentum is conserved, the skater rotates faster.

Example

A flywheel with a moment of inertia of $0.60\,\text{kg m}^2$ is rotating at 2.0 rev/s. What is the average retarding torque needed to bring the wheel to rest in 4 s?

The change in angular momentum in bringing the wheel to rest is $I\omega = 0.60 \times (2\pi \times 2) = 7.54\,\text{kg m}^2\,\text{rad/s}$. Thus the average retarding torque is the rate of change of momentum $= 7.54/4 = 1.9\,\text{N m}$.

Example

A skater has a moment of inertia of $5.0\,\text{kg m}^2$ with both arms and one leg outstretched and rotates at 1.0 rev/s. When he or she draws the arms and leg into the body the moment of inertia drops to $0.7\,\text{kg m}^2$. What is the resulting frequency of rotation?

The initial angular moment is $I\omega = 5.0 \times 1.0 \times 2\pi\,\text{kg m}^2\,\text{rad/s}$ and thus for angular momentum to be conserved

$$0.7 \times f \times 2\pi = 5.0 \times 1.0 \times 2\pi$$

Hence the frequency of rotation $f = 7.1\,\text{rev/s}$.

10.8 Work and energy

Work W is defined as the transfer of energy occurring when the point of application of force F moves through a distance s, with

$$W = Fs \tag{36}$$

Work is a scalar quantity, and as with all forms of energy the basic unit is the joule J. The term *power* P is used for the rate of doing work. Thus

$$P = \frac{\mathrm{d}W}{\mathrm{d}t} = \frac{\mathrm{d}(Fs)}{\mathrm{d}t} = F\frac{\mathrm{d}s}{\mathrm{d}t} = Fv \tag{37}$$

The unit of power is the watt (W).

When a force is used to rotate an object (figure 10.19) then if the point of application of the tangential force F is moved through an arc of radius r and subtending an angle θ the work done is

$$W = Fs = Fr\theta = T\theta \tag{38}$$

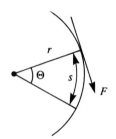

Fig. 10.19 Work done by a torque.

where T is the torque, i.e. Fr. Since power is the rate of doing work, then

$$P = \frac{d(T\theta)}{dt} = T\frac{d\theta}{dt} = T\omega \qquad [39]$$

When work is expended in slowly raising an object of mass m through a vertical height h the force mg has its point of application moved through h and so the work done is mgh. Energy has been transferred to the object and it is said to have gained *potential energy* of mgh.

$$\text{Potential energy} = mgh \qquad [40]$$

When work is expended in accelerating an object from rest to a velocity v the force F being applied to give the acceleration a is $F = ma$ and the distance s through which the point of application moves is given by $v^2 = 2as$. Thus the work done is $Fs = (ma)(\frac{1}{2}v^2/a) = \frac{1}{2}mv^2$. Energy has been transferred to the object and it is said to have gained *kinetic energy* of $\frac{1}{2}mv^2$.

$$\text{Kinetic energy} = \frac{1}{2}mv^2 \qquad [41]$$

Consider an object with an element of it having a mass δm a distance r from the axis of rotation, as in figure 10.20, and rotating about that axis with an angular velocity ω. The tangential velocity of the element is $v = r\omega$ and so the kinetic energy of the element is

$$\text{kinetic energy} = \frac{1}{2}\delta m\,(r\omega)^2 = \frac{1}{2}\omega^2 \times r^2\delta m$$

The total kinetic energy of the object is thus

$$\text{kinetic energy} = \frac{1}{2}\omega^2 \int r^2\,dm$$

But the integral is the moment of inertia I about the axis of rotation. Thus

$$\text{kinetic energy} = \frac{1}{2}I\omega^2 \qquad [42]$$

When a body has both linear motion and rotational motion then the total kinetic energy of the body is the sum of the linear and rotational kinetic energies, i.e.

$$\text{kinetic energy} = \frac{1}{2}mv^2 + \frac{1}{2}I\omega^2 \qquad [43]$$

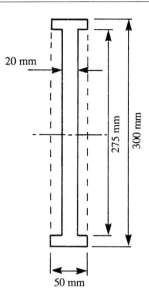

Fig. 10.20 Problem 44.

Energy is always conserved. Thus, for example, if a pulley wheel is caused to rotate by a falling weight attached to a rope wrapped round the wheel, then the gain in angular kinetic energy of the wheel must equal the work done by the falling weight, i.e. its loss in potential energy.

Example

With what velocity can a tractor of mass 600 kg pull a loaded trailer of mass 400 kg up a 1 in 12 incline if there is a total resistance to motion of 350 N and the tractor motor can develop a power of 15 kW?

The components of the weight acting down the incline are

$$(M + m)g\sin\theta = 1000 \times 9.81 \times (1/12) = 817.5\,\text{N}$$

The force being overcome in moving the load up the incline is thus $817.5 + 350 = 1167.5\,\text{N}$. Thus, using equation [37]

$$P = 15 \times 10^3 = Fv = 1167.5v$$

then $v = 12.8\,\text{m/s}$.

Example

A flywheel has a mass of 100 kg and a radius of gyration of 150 mm. It is accelerated from 100 rev/min to 250 rev/min at a constant rate in 15 s. What is the power absorbed by the flywheel in this time?

The rotational kinetic energy of the flywheel changes by

$$\text{change in KE} = \tfrac{1}{2}(mk^2)(\omega^2 - \omega_0^2)$$
$$= \tfrac{1}{2} \times 100 \times 0.150^2 (2\pi/60)^2 (250^2 - 100^2)$$
$$= 647.7\,\text{J}$$

This change in kinetic energy occurs in 15 s, as a result of work being done. Thus the power is $647.7/15 = 43.2\,\text{W}$.

Example

What is the speed of the centre of mass of a cylinder of mass m and radius r after it has rolled from rest down an inclined plane through a vertical height h without slipping?

In rolling down the incline the cylinder loses a potential energy of mgh and gains both linear and rotational kinetic energies. Thus

$$mgh = \tfrac{1}{2}mv^2 + \tfrac{1}{2}Iw^2$$

where I is the moment of inertia of the cylinder about its centre of mass and v is the speed of the centre of mass. Since $I = \tfrac{1}{2}mr^2$ (see figure 10.13) and $\omega = v/r$, then

$$mgh = \tfrac{1}{2}mv^2 + \tfrac{1}{2}(\tfrac{1}{2}mr^2)(v^2/r^2) = (3/4)mv^2$$

Thus $v = \sqrt{(1.33gh)}$.

Problems

(1) A particle starting from rest moves in a straight line with an acceleration in m/s^2 given by $a = 2t - 5$, where t is in s. What is the velocity and the distance covered after 4 s?

(2) A piston moving in a cylinder has a deceleration of $-1.6v^3$, where v is its velocity in m/s. If it starts with a velocity of 2 m/s, what is the velocity after 4 s?

(3) A particle moves so that the displacement s in metres is given by $s = 3t^2 - 3t - 2$. What will be the velocity and the acceleration when $t = 2\,\text{s}$?

(4) A particle moves in a straight line with an acceleration of $3t^2$ m/s^2 away from a reference point on the line at which it started at $t = 0$. What will be its velocity and displacement at $t = 2\,\text{s}$?

(5) A car accelerates at 2.5 m/s^2 to a velocity of 20 m/s, then travels at this constant velocity before decelarating at 4.0 m/s^2. If the total distance covered was 500 m, how long will it take to cover the distance?

(6) A car is travelling at 30 m/s when the brakes are suddenly applied

and result in a constant deceleration of $4.0 \, \text{m/s}^2$. What time is required to stop the car and what will be the distance travelling in that time?

(7)   A car has a velocity of $15 \, \text{m/s}$. How long will it take for this velocity to be doubled if it accelerates at a constant $2 \, \text{m/s}^2$?

(8)   A ball is thrown vertically upwards with a velocity of $15 \, \text{m/s}$. How long will it take before it comes back down to the thrower?

(9)   A stone is thrown vertically downwards with an initial velocity of $4 \, \text{m/s}$. How long will it take to hit the ground which is $20 \, \text{m}$ vertically below the thrower?

(10)   An aircraft of mass $4000 \, \text{kg}$ lands at $200 \, \text{km/h}$ and is brought to rest in a distance of $400 \, \text{m}$. What is the average resistance to motion due to the brakes and aerodynamic drag?

(11)   An aircraft of mass $4000 \, \text{kg}$ has a take-off speed of $250 \, \text{km/h}$ and a take-off run of $1 \, \text{km}$. If the engine delivers a thrust of $12 \, \text{kN}$, what is the average resistance to motion?

(12)   A mass of $20 \, \text{kg}$ is attached to one of a light cord which passes over a pulley and has a mass of $30 \, \text{kg}$ at the other end. Neglecting the moment of inertia of the pulley and friction, what is the acceleration of the masses?

(13)   A mass of $4 \, \text{kg}$ rests on a horizontal table and is connected by a light cord passing over a pulley wheel at the edge of the table to a mass of $6 \, \text{kg}$ which hangs vertically. What is the acceleration of the masses?

(14)   A mass of $20 \, \text{kg}$ is being pulled up, from rest, a uniform gradient of 1 in 10 with an acceleration of $0.30 \, \text{m/s}^2$. If the resistance to motion is $5 \, \text{N}$, what is the effort required?

(15)   A stone of mass m is thrown vertically upwards with an initial velocity u. The air resistance at a velocity v is mkv, where k is a constant. What time is taken by the stone to reach its greatest height?

(16)   A ball bearing of mass m is released from rest and allowed to freely fall in oil. There is a resistance to motion at a velocity v of mkv^2. After some time the ball bearing is found to be falling with a constant velocity (referred to as the terminal velocity) v_t. What is the terminal velocity?

(17)   A block rests on a plane, the angle of which can be slowly increased. When the angle reaches θ the block just begins to slide down the plane. What is the coefficient of static friction?

(18)   Packets slide down a chute which is at an angle of $20°$ to the horizontal. If the packets start at the top of the chute with zero

velocity, what will be their velocity at the bottom of the chute, its length being 3.0 m, if the coefficient of kinetic friction is 0.2?

(19) Water from a jet has a velocity of 10 m/s and is deflected through 90° by the vane of a turbine. What is the change in velocity of the water if the emerging jet has a velocity of 10 m/s in the new direction?

(20) A projectile has an initial velocity of 10 m/s at 30° to the horizontal. What is the greatest height reached and the range on the horizontal plane?

(21) A projectile is to have an initial velocity of 10 m/s and a horizontal range of 9.0 m. What is the required angle of projection?

(22) A shell is fired with a velocity of 120 m/s at 30° to the horizontal from the edge of a cliff which is 140 m above sea level. What is the time taken by the shell to reach the sea?

(23) A person standing on a smooth hillside which makes a constant angle of φ with the horizontal throws a stone down the hill with an initial velocity of u at an angle θ above the horizontal. At what distance down the slope will the stone land?

(24) What is the greatest speed with which an aircraft can move in a horizontal turn of radius 1.2 km if the acceleration experienced must not exceed 3g?

(25) A car is driven over a hump-back bridge in the form of a circular arc of radius 12 m. What is the greatest speed the car can have if it is to remain in contact with the ground?

(26) At what angle should a corner of radius 50 m be banked if a car travelling at 10 m/s is to be able to negotiate it without any side thrust on the tyres?

(27) Determine the condition for a car to be able to turn an unbanked corner of radius r without overturning. The car has a mass of m and a centre of gravity which is half way between the wheels, a distance d apart, and a height h above the ground.

(28) A ball on the end of a string of length 1 m is whirled round in a horizontal circle with the string making an angle of 45° with the vertical. What is the speed of the ball?

(29) A car is travelling along a level road at 20 m/s and rain is falling vertically with a velocity of 7 m/s. What is the velocity of the rain relative to a passenger in the car looking out of a side window?

(30) Aircraft A is flying due east at 300 km/h aircraft B due south at 400 km/h. What is the velocity of aircraft B relative to A?

(31)   What is the average angular velocity of a drive shaft if it makes 10 revolutions in 20 s?

(32)   The angular velocity of a gear wheel increases at a uniform rate from 10 rev/min to 25 rev/min in 20 s. What is the angular acceleration and angular displacement of the gear teeth in that time?

(33)   A wheel of diameter 500 mm rolls a distance of 400 mm. Through what angle has a radius rotated?

(34)   What is the linear velocity of a pulley belt if the pulley wheel, diameter 100 mm, rotates at 500 rev/min?

(35)   A wheel of diameter 300 mm rotates with a constant angular acceleration of 20 rad/s². What is the peripheral speed of the wheel after 20 revolutions, starting from rest?

(36)   A load is lifted by a rope wrapped round a pulley of diameter 500 mm. Initially the load is lifted at 0.2 m/s. Then the load is accelerated at a constant rate for 5 s until it is lifted at 1.0 m/s. What is (a) the angular acceleration of the pulley, (b) the number of turns made by the pulley during the acceleration?

(37)   A flywheel has a moment of inertia of 10 kg m². What will be the angular velocity after 5 s starting from rest if a torque of 5 N m is applied and bearing friction is equivalent to a torque of 2 N m?

(38)   A pulley with a moment of inertia of 0.50 kg m² and radius 200 mm is acted on by a tangential force of 10 N applied through a rope at its periphery. How many revolutions will the wheel make, starting from rest, before it attains an angular velocity of 20 rad/s?

(39)   An object of mass 1000 kg is lifted by means of a steel lifting cable being wound round a drum of diameter 2.5 m mounted on a horizontal shaft. The drum and shaft have a mass of 1000 kg and a radius of gyration 1.0 m. What is the torque required to give the object an upward acceleration of 0.75 m/s²?

(40)   A flywheel has a mass of 750 kg and a radius of gyration of 0.60 m. What is the torque required to increase the speed of the flywheel at a constant rate from 100 rev/min to 200 rev/min in a time of 15 s?

(41)   A pulley has a mass of 10 kg, a radius of 100 mm, and a radius of gyration of 70 mm. A rope passing over the pulley has a mass of 10 kg hanging vertically at one end and 20 kg at the other. What will be the acceleration of the 20 kg mass?

(42)   A slender uniform rod of mass 2 kg and length 1.2 m is pivoted about one end and hangs vertically. What is its initial angular acceleration when it is subject to a turning moment of 3 N m?

(43)   A hoist drum has a mass of 20 kg and a radius of 150 mm. A cable passing over the drum is used to vertically lift an object of mass

10 kg. What is the tension in that part of the cable attached to the object? Assume that the drum is uniform and that the cable does not slip.

(44) Determine the moment of inertia about an axis through its centre of a flywheel having the section shown in figure 10.20. The material used for the flywheel has a density of 7800 kg/m³.

(45) An element for machining on a lathe has a mass of 12 kg and a centre of gravity which is 140 mm from the mandrel axis. At what distance from the axis should a 5 kg mass be attached to give dynamic balance?

(46) A plate rotates about a central axis O at 300 rev/min. The plate can be considered to be formed from masses A and B of 20 kg at 500 mm radially from the axis and 15 kg at 400 mm respectively. The angle AOB is 120°. For dynamic balancing, at what distance should a mass of 20 kg be placed?

(47) A shaft of length 800 mm is supported on bearings 200 mm from each end. At the two ends are pulleys, each of mass 10 kg, and having centres of gravity 2.5 mm from the axis of rotation. At what angles should the pulleys be keyed to the shaft to give dynamic balance and what are the loads on the bearings as a result of the shaft rotating at 240 rev/min?

(48) A shaft of length 3.0 m is supported on two bearings 0.5 m from each end. The shaft carries three pulleys of mass 60 kg, one at each end and one at the middle of its length. The end pulleys have centres of gravity which are 3.5 mm and 5.0 mm respectively and the centre pulley 6.0 mm from the shaft axis. The pulleys are arranged to give static balance. What are the forces on the bearings as a result of the shaft rotating at 300 rev/min?

(49) A shaft rotates in bearings 2.0 m apart and carries a disc of mass 30 kg at 600 mm from one bearing. The centre of gravity of the disc is 2.0 mm from the axis of rotation. What are the forces on the bearings as a result of the shaft rotating at 600 rev/min?

(50) What thrust is exerted by a stationary jet engine if it has an intake of 80 kg or air per second and ejects it from the rear of the engine at a velocity of 600 m/s?

(51) What is the average recoil force experienced by an automatic rifle firing 500 rounds per minute if each has a mass of 15 g and a muzzle velocity of 800 m/s?

(52) A flywheel has a mass of 15 kg, and a radius of gyration of 120 mm, and is rotating at 6 rev/s. What is the average retarding torque needed to bring the wheel to rest in 5 s?

(53) A disc A has a mass of 50 kg, a radius of gyration of 160 mm and

rotates at 10 rev/s. Disc B has a mass of 40 kg, a radius of gyration of 120 mm, and is not rotating. Disc A and disc B are mounted on a common axis of rotation. When disc A drops on to disc B, what will be the rate of rotation when all slipping has ceased?

(54) What power must be exerted by a car for it to maintain a constant velocity of 25 m/s along a horizontal road if the total resistive force is 3 kN?

(55) What power is required by a cyclist to cycle at 2 m/s up a slope of 1 in 20 if the combined mass of rider plus cycle is 90 kg and the total frictional resistance to motion is 10 N?

(56) The total resistive force to the motion of a car of mass 600 kg is $10 + 0.3v^2$, where v is in m/s and the force in N. What power will be needed to drive the car up a gradient of 1 in 30 at a steady velocity of 20 m/s?

(57) A sphere of mass m and radius r rolls from rest down an incline of vertical height h without slipping. What is the speed of the centre of mass at the bottom?

(58) A thin rod of length L and mass m is pivoted about one end. When the rod is pulled to one side and allowed to swing, it passes through the lowest point in its path with an angular velocity ω. How high does the free end of the rod rise above its lowest position?

(59) A uniform disc of mass 30 kg and radius 200 mm can rotate about its centre and is acted on by a tangential force of 30 N applied through a rope wrapped round its circumference. How many revolutions of the disc will occur in it attaining an angular velocity of 20 rad/s starting from rest?

(60) A uniform disc of radius 200 mm and mass 2.5 kg is mounted on a central axle with a light rope wrapped round its circumference. On the end of the vertically hanging rope there is a mass of 0.5 kg. If the disc starts from rest what is (a) the angular acceleration of the disc, (b) the tension in the rope, (c) the work done on the disc in 3 s?

Chapter 11

Mechanisms

11.1 Mechanisms

A *mechanism* can be defined as a device which is used to translate a movement or displacement of one point to another point some distance away. A mechanism which is used to transmit a force is termed a *machine*.

The various parts of a mechanism are called *links*. Links may be items such as levers, cranks, pistons, sliders, shafts, connecting rods, pulleys and bearings. Each link is capable of moving relative to its neighbouring links, the types of relative motion which commonly occur being sliding, turning or screw motion. Examples of these types of motion are sliding – a piston in a cylinder, turning – a crankshaft rotating on bearings, screw motion – a nut and bolt. Two links between which there is relative motion are known as a *pair*.

As an illustration, consider a motor car engine. We can represent the main unit in such an engine as a crankshaft rotating on bearings in a fixed frame and driven by pistons moving back-and-forth in cylinders, which are part of the fixed frame. Figure 11.1 shows the form of the mechanism. It consists of four links, three of them forming turning pairs and one a sliding pair.

A mechanism is formed from an assembly of links. A group of links which is capable of relative motion but can be made rigid by the addition of a single link is called a *kinematic chain*. To act as a mechanism and transmit movement or displacement, a kinematic chain must have one of its links fixed. It is possible to obtain from one kinematic chain a number of different mechanisms by having a different link as the fixed one. The technique of obtaining different mechanisms by fixing alternative links in a chain is called *inversion*. Most mechanisms can be reduced to two basic kinematic chains,

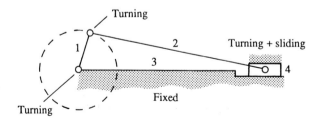

Fig. 11.1 Simple engine mechanism.

these being the *four-bar chain* (or *quadric cycle*) and the *slider–crank chain*.

11.1.1 The four-bar chain

The four-bar chain consists of four links with four turning pairs. Figure 11.2(a) shows the basic four-bar chain, links 1, 2 and 4 being movable and link 3 fixed. Figure 11.2(b) shows the four-bar chain operating as a double lever mechanism, links 1 and 4 being of such a length and constrained so that they can only oscillate backwards and forwards. Figure 11.2(c) shows the four-bar chain operating as a lever–crank system, link 1 being only able to oscillate backwards and forwards while link 4 makes complete revolutions. Figure 11.2(d) shows the four-bar chain operating as a double crank, links 1 and 4 making complete revolutions.

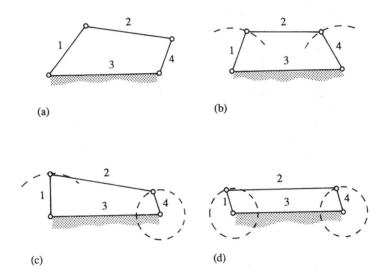

Fig. 11.2 Four-bar chains, (a) the four links, (b) a double lever, (c) a lever-crank, (d) a double crank.

Example

Identify the form of four-bar chain in the mechanism of the vehicle suspension unit illustrated in figure 11.3(a).

Figure 11.3(b) shows the suspension unit drawn as a four-bar chain. The vehicle chassis provides the fixed link 3. Links 1 and 4 can only oscillate backwards and forwards, thus the system is essentially a double lever mechanism.

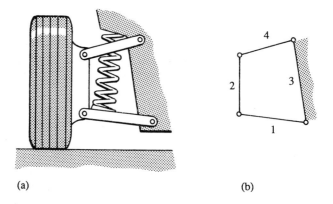

(a)   (b)

Fig. 11.3 A vehicle suspension unit.

11.1.2 *The slider–crank mechanism*

This mechanism consists of a crank, a connecting rod and a slider. Figure
11.1, in describing the mechanism of a car engine, is an example of such a
mechanism. Link 3 is fixed, link 1 is the crank which rotates, 4 is the slider
which moves relative to the fixed link and link 2 is the connecting rod.
Figure 11.4 shows an alternative version of the mechanism. This is a quick-
return mechanism. Link 3 is fixed, link 1 is the crank which rotates, 2 is the
slider which moves relative to the fixed link and link 4 is the connecting rod.
The crank rotates at constant speed, thus causing link 4 to oscillate backwards
and forwards. The angle turned by the crank to make link 4 move from one
extreme to another is greater when the crank rotates in one direction than
the other. Thus, for a constant angular crank speed, lever 4 moves more
slowly in one direction than the other.

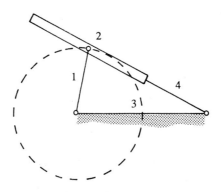

Fig. 11.4 Quick-return mechanism.

11.2   Velocity diagrams

The methods available for the analysis of the motion of links in a mechanism fall in two groups – analytical and graphical. The analytical methods can however be exceedingly complex for all but the simplest mechanism. A technique that is relatively simple and can be used with even complex mechanisms is that of velocity diagrams.

Consider an individual link AB in motion, with end B rotating with a constant angular velocity ω about end A. At some instant of time end A has a velocity v_A and end B a velocity v_B relative to the fixed frame of reference, such velocities often being termed absolute velocities (figure 11.5(a)). The velocity of B relative to A, v_{BA}, can be obtained by adding $-v_A$ to both the velocity at A and that at B, thus effectively stopping end A (see section 11.3). Figure 11.5(b) shows the resulting velocity vector diagram. Since end B is rotating with constant angular velocity about A then the relative velocity v_{BA} must be at right angles to the link BA and since $v = r\omega$, then $v_{BA} = (AB)\omega$.

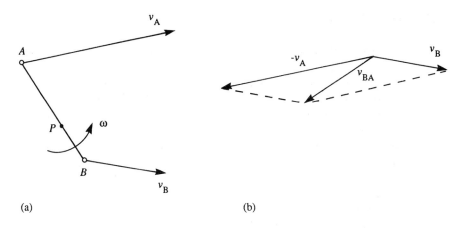

Fig. 11.5 Velocity of a rigid link.

The velocity of any intermediate point P on AB, since the velocity will be given by $r\omega$, is thus

$$\frac{v_{PA}}{v_{BA}} = \frac{AP}{AB} \qquad [1]$$

Now consider the relative velocities of the links in a mechanism. An important point to note is that, for rigid links, whatever the absolute velocities of each end of a link the relative velocity of one end with respect to the other must always be at right angles to the link. This is because there can be no relative velocity component along the length of the link. Consider the links in figure 11.6(a). Starting with a link AB for which one end A is fixed, then the velocity of end B with respect to end A is the same as the absolute velocity. If we now consider link BC, then we can use the parallelogram

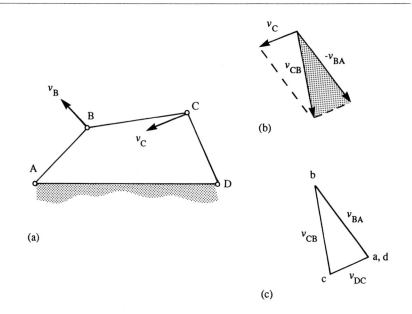

Fig. 11.6 Drawing a velocity diagram.

of vectors to determine the relative velocity of C with respect to B. It will be at right angles to the direction of the link BC. Now considering link CD, since D is fixed then the relative velocity of C with respect to D is the absolute velocity v_C of C. Hence, the relative velocity of D with respect to C, v_{DC}, is just $-v_C$. This relative velocity is at right angles to the link CD. The shaded triangle in figure 11.6(b) thus has sides which represent in size and orientation the relative velocities. This is the velocity diagram.

To draw the velocity diagram, all we have to do is draw a line to represent the relative velocity of B with respect to A (figure 11.6(c)). We can represent this velocity in magnitude and direction by a line ab. Note that the convention is used in velocity diagrams of labelling the ends of the velocity vectors by lower case letters. Also no arrows are put on the lines. Then from point b on the diagram we draw a line to represent the velocity of C relative to B, this line is then bc. From c a line representing the relative velocity of D with respect to C must end up back at point A, since both A and D are at rest. Each side of the velocity diagram is at right angles to the corresponding side in the mechanism. Note that if DEF are three points on a straight link that the velocity diagram will contain a straight line def such that de/ef = DE/EF (a consequence of equation [1]).

Example

For the four-bar chain shown in figure 11.7(a) determine the velocity of point P and the angular velocity of link CD; AB is at 60° to AD.

The first step is to draw a space diagram to scale of the mechanism. It is

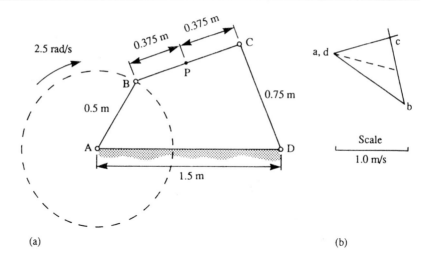

Fig. 11.7 Example.

important that the various links are drawn in their correct orientations since the angles in the velocity diagram are obtained relative to the orientations of the links. This is the diagram in figure 11.7(a).

Starting with point a, the line ab is drawn to represent v_{BA}. This is in a direction at right angles to AB and will be given by $v = r\omega$ as $0.5 \times 2.5 = 1.25$ m/s. All that is known of the velocity of C relative to B is that it must be at right angles to BC. Thus we can draw a line bc in this direction. Because A and D are fixed, then points a and d will be coincident. Relative to D all we know of the velocity at C is that it must be at right angles to DC. Thus we can draw the line dc in this direction. Where the lines drawn from d and from b intersect gives c. Figure 11.7(b) shows the resulting velocity diagram. It has been drawn to the scale indicated.

The velocity of C relative to D, obtained from measurement of the scaled velocity diagram, is 0.8 m/s. Thus the angular velocity of link CD is given by $v = r\omega$ as $0.8/0.75 = 1.1$ rad/s.

Point P, since it is half way between B and C, will be half way between b and c on the velocity diagram. The velocity of P relative to the fixed link, i.e. a,d, is thus represented by the length of ap. Hence the velocity is 0.95 m/s.

Example

For the slider–crank mechanism shown in figure 11.8(a), determine the angular velocity of the connecting rod BC.

Figure 11.8(a) shows the space diagram of the mechanism. Since the crank AB is rotating with constant angular velocity the absolute velocity of B is given by $v = r\omega$ as $0.200 \times 20 = 4.0$ m/s. Thus starting with point a for the velocity diagram, figure 11.8(b), then since A is at rest the relative velocity

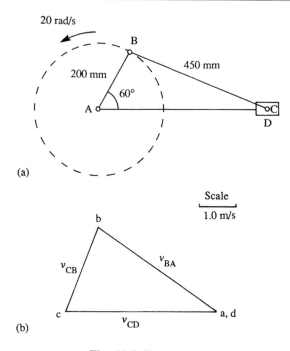

Fig. 11.8 Example.

of B, v_{BA}, is 4.0 m/s. This velocity is at right angles to AB. Thus the line ab can be drawn. All we know at this time about the velocity of C relative to B is that it will be at right angles to BC. Thus we can draw bc in this direction. The velocity of C relative to D must be along the direction of the line AD, since C is constrained to move in this direction. But D is at rest and so d is coincident with a. Hence we can draw the line from a,d to c in the direction given by DA. Thus c is located. Measurement of the length of bc gives 2.4 m/s and hence, using $v = r\omega$, an angular velocity of 2.4/0.450 = 5.3 rad/s.

Example

Determine the velocity of the slider F in the shaping tool quick-return mechanism shown in figure 11.9(a).

Figure 11.9(a) shows the space diagram for the mechanism. The angular velocity of AC is 5 rev/s and thus the velocity at C, at right angles to AC, is given by $v = r\omega$ as $0.150 \times 5 \times 2\pi = 4.7$ m/s. Thus ac can be drawn on the velocity diagram as a line equivalent to 4.7 m/s in a direction at right angles to AC. The velocity of D relative to C must be in the direction of line BDE. Thus a line can be drawn from c in this direction. The velocity of E with respect to B must be at right angles to the line BE. Since B and A are at rest then b and a are coincident. Thus a line can be drawn from a,b in this direction. Since D is a point on the line BE then d must lie on the line a,b to e. Since ad/de = AD/DE then we can now locate e. The velocity of F with

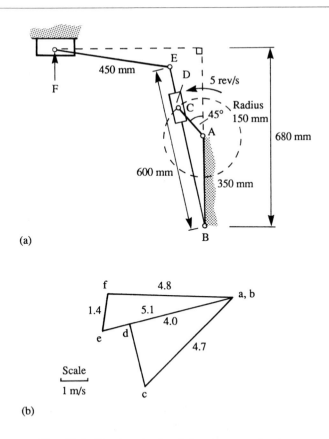

(a)

(b)

Fig. 11.9 Shaping tool quick-return mechanism.

respect to A or B must be in the horizontal direction. Thus a line can be drawn from a,b to f in this direction. The velocity of E relative to F is at right angles to FE and so a line can be drawn from e in this direction and hence locate f. The velocity of F relative to A, i.e. the absolute velocity, is given by the velocity diagram as 4.8 m/s.

11.3 Acceleration diagrams

Acceleration is a vector quantity and thus when a particle is subject to more than one acceleration then the resultant acceleration is determined by using the parallelogram of vectors. Consider a particle rotating in a circular arc. It will have a centripetal acceleration of $a = v^2/r = r^2\omega^2/r = \omega^2 r$, where r is the radius of the arc and ω the angular velocity of the particle in that path. In addition the particle may experience a tangential acceleration along its path. Thus the resultant acceleration of the particle is given by the parallelogram shown in figure 11.10.

Now consider a rigid link AB (figure 11.11(a)) with end B rotating about

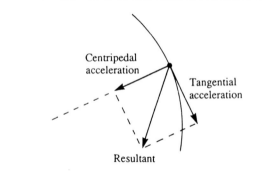

Fig. 11.10 Resultant acceleration.

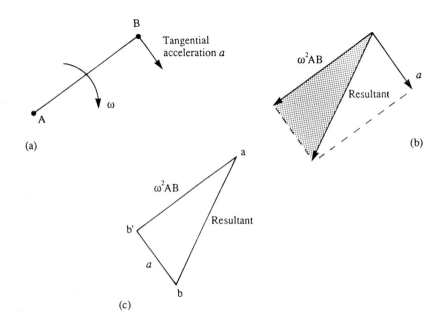

Fig. 11.11 Acceleration of a rigid link.

A with an angular velocity ω. There will be a centripetal acceleration of B relative to A of $\omega^2 AB$ directed towards A. In addition there may be a tangential acceleration, this therefore being at right angles to AB. We can draw the parallelogram of vectors for this situation in the way shown in figure 11.11(b) and so obtain the resultant acceleration of B relative to A. The shaded part of the parallelogram represents what is termed the *acceleration diagram*. What we have is an acceleration of ωAB drawn from a to b', this being the same direction as that of AB. At right angles to this, from b', we have the tangential acceleration a represented by a line b'b. The triangle is completed by the resultant drawn from a to b.

Consider a four-bar mechanism, as in figure 11.12, with link AB having an angular velocity ω which is changing with an angular acceleration α. In

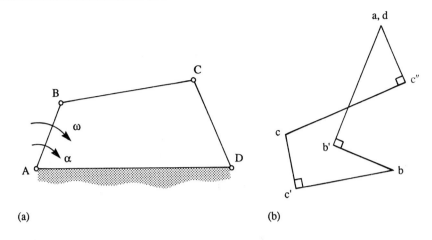

Fig. 11.12 Acceleration with a four-part mechanism.

an acceleration diagram we can represent the centripetal acceleration of B relative to A of ω AB by a line ab′ which is at the same angle as AB and with a length a scaled version of ω AB. The tangential acceleration of B relative to A is given by $a = r\alpha$ as α AB. We can represent this as line b′b. The centripetal acceleration of C relative to B is v_{CB}^2/BC and is directed to B. We can represent this on the acceleration diagram by the line bc′, the velocity v_{CB} having been obtained from a velocity diagram for the mechanism. The tangential acceleration of C relative to B is in a direction at right angles to BC but with size unknown. It can thus be represented on the acceleration diagram by a line drawn from c′ at right angles to bc′, with c being somewhere along the resulting line. The centripetal acceleration of C relative to D is given by v_{CD}^2/CD and is directed towards D. We can represent this by line dc″. The tangential acceleration of C relative to D is in a direction at right angles to CD but with size unknown. It can thus be represented on the acceleration diagram by a line drawn from c″ at right angles to dc″. The intersection of the lines through c′ and c″ gives the point c.

In determining acceleration diagrams, care must be taken to include all the acceleration terms. There can be centripetal acceleration, tangential acceleration and Coriolis acceleration. Coriolis acceleration arises when there is acceleration of a block sliding on a rotating link and is discussed in the next section.

Example

Determine the angular velocity and angular acceleration of link DC and the velocity and acceleration of point P for the four-bar mechanism shown in figure 11.13(a). Link AB rotates with an angular velocity 3 rad/s and angular acceleration 2 rad/s^2 and P is half way along link BC.

Figure 11.13(a) shows the space diagram. The velocity diagram is first

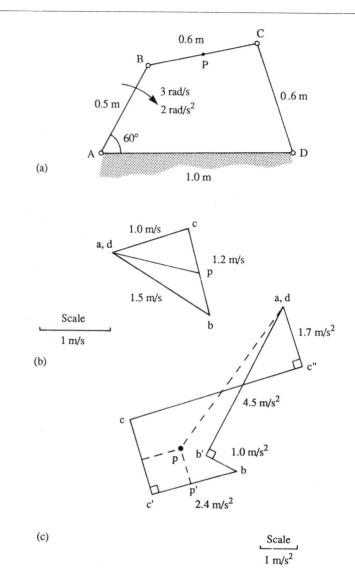

Fig. 11.13 Example.

drawn. Arm AB rotates with an angular velocity 3 rad/s and thus the tangential velocity at B is $0.5 \times 3 = 1.5$ m/s. This is the velocity of B relative to A and is drawn on the velocity diagram as ab in a direction at right angles to AB. The velocity at C relative to B is at right angles to BC, hence the line bc can be drawn in this direction. The velocity at C relative to D is at right angles to CD, hence the line dc can be drawn in this direction. The result is the velocity diagram shown in figure 11.13(b). The velocity of C relative to D is thus 1.0 m/s. Hence the angular velocity of C about D is $1.0/0.6 = 1.7$ rad/s.

Since point P is half way along link BC then p will be half way along bc. Thus the velocity diagram gives for the velocity of P relative to A, i.e. length pa, 1.2 m/s.

To draw the acceleration diagram shown in figure 11.13(c), we start by considering link AB. The centripetal acceleration of B is $r\omega^2 = 0.5 \times 3^2 =$ 4.5 m/s^2. This is drawn as the line ab′, in a direction parallel to AB. The tangential acceleration is $r\alpha = 0.5 \times 2 = 1.0$ m/s^2. This is drawn at right angles to ab′ as b′b. The resultant acceleration of AB would be ab. The centripetal acceleration of C relative to B is v_{CB}^2/r and since the velocity diagram gives v_{CB} as 1.2 m/s then the acceleration in the direction CB is $1.2^2/0.6 = 2.4$ m/s^2. This can then be drawn as the line bc′ on the acceleration diagram. The tangential acceleration of C relative to B is at right angles to this and thus the line c′c can be drawn in this direction. The centripetal acceleration of C relative to D is v_{CD}^2/r and since the velocity diagram gives v_{CD} as 1.0 m/s^2 then the centripetal acceleration is $1.0^2/0.6 = 1.7$ m/s^2. This is then drawn as dc″, with d being coincident with a. The tangential acceleration at C is a line drawn at right angles to dc″. This thus locates point c. The resultant acceleration at C relative to D is the line dc, the resultant acceleration at B being bc.

The angular acceleration of CD is related to the tangential acceleration at C, being a/r and thus, since $a = cc″ = 4.7$ m/s^2, then the angular acceleration is $4.7/0.6 = 7.8$ rad/s^2.

Since point P is half way along BC and the centripetal acceleration is $\omega^2 r$, then the centripetal acceleration of P relative to B will be half that of C relative to B. Thus point p′ is half way along bc′. Since the tangential acceleration at P relative to B will be half that at C relative to B, point p can be located. The absolute acceleration of P relative to A is given by pa, which is 4.8 m/s^2 in the direction given by pa.

11.3.1 *Coriolis acceleration*

Consider a block B sliding with a velocity v and linear acceleration a along a link AB which is rotating about A with an angular velocity ω and angular acceleration α (figure 11.14(a)). Suppose that in time δt the block increases its linear velocity along the link to $v + \delta v$, moving out from a radial distance r to $r + \delta r$, and in the same time the link AB rotates through an angle $\delta\theta$ with the angular velocity increasing from ω to $\omega + \delta\omega$.

Consider the radial velocity. This changes from v in direction AB to $v + \delta v$ in a direction at angle $\delta\theta$ to AB. The change in velocity in this time can be obtained by a parallelogram of vectors, as in figure 11.14(b). To a reasonable approximation we can take the change in velocity to be $v\delta\theta$ which is tangential to the rotation and δv in the radial direction.

Consider the angular velocity. This changes from ω to $\omega + \delta\omega$ in time δt. This is a change in the tangential velocity from ωr to $(\omega + \delta\omega)(r + \delta r)$ while the direction of the tangents changes by $\delta\theta$. The change in velocity in this time can be obtained by a parallelogram of vectors, as in figure 11.14(c). To a reasonable approximation we can take the change in velocity to be $-\omega r\delta\theta$ which is in a radial direction and

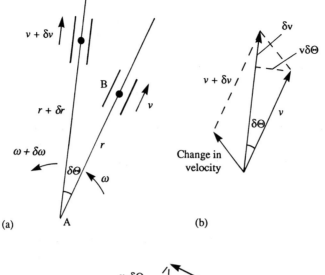

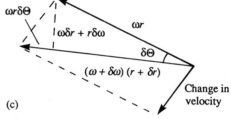

Fig. 11.14 Block sliding on a rotating link.

$$(\omega + \delta\omega)(r + \delta r) - \omega r \approx \omega\delta r + r\delta\omega$$

in the tangential direction.

The total change of velocity of B relative to A in the radial direction in a time δt is thus

$$\text{radial change in velocity} = \delta v - \omega r\delta\theta$$

Thus

$$\text{radial acceleration} = \frac{\delta v}{\delta t} - \omega r\frac{\delta\theta}{\delta t}$$

Since $a = \delta v/\delta t$ and $\omega = \delta\theta/dt$, then

$$\text{radial acceleration} = a - \omega^2 r \qquad [2]$$

The radial acceleration is thus the outward acceleration of the block minus the centripetal acceleration relative to A.

The total change in velocity of B relative to A in the tangential direction in a time δt is

$$\text{tangential change in velocity} = v\delta\theta + \omega\delta r + r\delta\omega$$

Thus

$$\text{tangential acceleration} = v\,\frac{\delta\theta}{\delta t} + \omega\,\frac{\delta r}{\delta t} + r\,\frac{\delta\omega}{\delta t}$$

Since $\omega = \delta\theta/\delta t$, $v = \delta r/\delta t$ and $\alpha = \delta\omega/\delta t$, then

$$\text{tangential acceleration} = v\omega + \omega v + r\alpha$$
$$= \alpha r + 2v\omega \qquad\qquad [3]$$

The tangential acceleration is thus the sum of the tangential acceleration αr due to just the angular acceleration and the term $2v\omega$ which represents the tangential acceleration of the block resulting from its velocity v along AB. This is called the *Coriolis acceleration*.

Example

For the mechanism shown in figure 11.15(a), determine the angular acceleration of link CD when link AE rotates at a constant angular velocity of 2 rad/s.

The tangential velocity of B relative to A is given by $v = r\omega$ as $2\sqrt{(0.4^2 + 0.4^2)} = 1.13$ m/s. This is represented in size and direction by ab in the velocity diagram in figure 11.15(b). The velocity of C relative to A will be in the direction of AE, thus the direction of bc on the velocity diagram will be at right angles to ab. The velocity of C relative to D will be at right angles to CD, thus the direction of cd is horizontal. Hence c can be located on the velocity diagram. This gives $v_{CB} = 1.13$ m/s and $v_{CD} = 1.60$ m/s.

The radial acceleration of B relative to A is the centripetal acceleration given by $v_{BA}^2/r = 1.13^2/\sqrt{(0.4^2 + 0.4^2)} = 2.26$ m/s^2 and is represented on the acceleration diagram by ab′. The tangential acceleration of B relative to A has two components, a sliding component and a Coriolis component (equation [3]). Since there is no angular acceleration of AB then there is no sliding acceleration component. The Coriolis component is given by $2v_{AB}\omega = 2 \times 1.13 \times 2 = 4.52$ m/s^2. This then gives b′b on the acceleration diagram (figure 11.15(c)). The acceleration of C relative to B is parallel to the direction AE and thus gives the direction of the line bc. The centripetal acceleration of C relative to D is $v_{CD}^2/r = 1.60^2/0.4 = 6.4$ m/s^2 and is represented by dc″ on the acceleration diagram. The tangential acceleration is given by c″c at right angles to dc″.

The acceleration diagram thus gives for the tangential acceleration of C relative to D, i.e. cc″, 0.5 m/s^2. Hence the angular acceleration is $a/r = 0.5/0.4 = 1.3$ rad/s^2.

11.4 Reciprocating mechanisms

A rotating crank driving a connecting rod to move a piston, i.e. a slider, in a cylinder, as in figure 11.1, is a mechanism that is of common application and

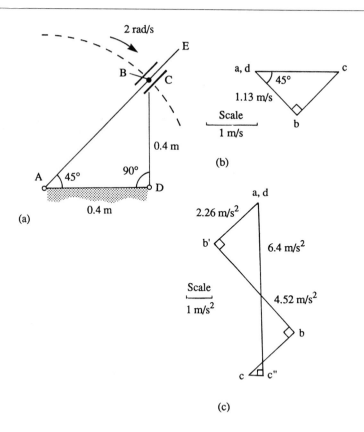

Fig. 11.15 Example.

is often referred to as a reciprocating mechanism. It describes, for example, the basic principles of the mechanism of the internal combustion engine.

In describing such a mechanism the term *inner dead centre* is used to describe the orientation of the crankshaft when it is pointing directly towards the piston and *top dead centre* when it is at right angles to this direction.

11.4.1 Velocity and acceleration diagrams

Consider the reciprocating mechanism shown in figure 11.16(a) with a crankshaft rotating with an angular velocity ω. The tangential velocity of the crankshaft is $v_c = \omega \times AB$. Thus in the velocity diagram (figure 11.16(b)) ab has a scaled length of $\omega \times AB$ in the direction at right angles to AB. The velocity of the piston C is at right angles to CB and thus cb is in this direction. The direction ac is parallel to AC. The triangle abc is similar to the triangle ABD (rotate the velocity diagram through 90°). Thus

$$\frac{v_p}{v_c} = \frac{ac}{ab} = \frac{AD}{AB}$$

[4]

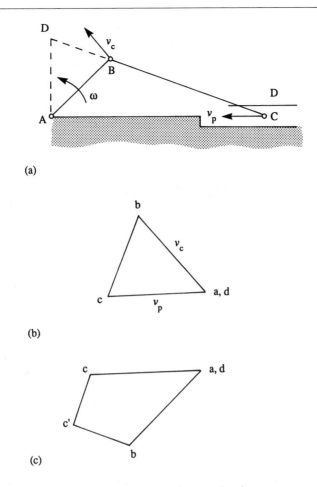

Fig. 11.16 Reciprocating mechanism.

The centripetal acceleration of B relative to A is $\omega^2 \times$ AB and is represented on the acceleration diagram by ab, in a direction parallel to AB. The centripetal acceleration of C relative to B is $v_{BC}^2/$BC and is represented by bc′ in a direction parallel to BC. The tangential acceleration at C is represented by c′c. The acceleration of C with respect to D is in the direction of AC. Thus the acceleration diagram is as shown in figure 11.16(c).

11.4.2   *Analytical analysis*

The velocity and acceleration of the piston in the cylinder can be obtained analytically by deriving an equation describing how the displacement of the slider depends on the angle of the crank, then differentiating once to give the velocity and then a second time to give the acceleration.

For the system shown in figure 11.17, the displacement *x* of the piston when measured from the crank axis A is

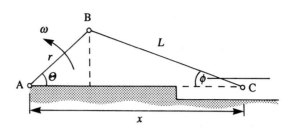

Fig. 11.17 Reciprocating mechanism.

$$x = r\cos\theta + L\cos\phi \tag{5}$$

The vertical height of B above the line AC is

$$r\sin\theta = L\sin\phi \tag{6}$$

Hence

$$r^2\sin^2\theta = L^2\sin^2\phi = L^2(1 - \cos^2\phi)$$

$$\cos^2\phi = 1 - \frac{r^2}{L^2}\sin^2\theta$$

Thus equation [5] becomes

$$x = r\cos\theta + L\sqrt{\left(1 - \frac{r^2}{L^2}\sin^2\theta\right)}$$

Since reciprocating machines generally have r/L significantly less than 1 then we can simplify the above expression (using the binomial series expansion of the square root term) to give

$$x = r\cos\theta + L - \frac{r^2}{2L}\sin^2\theta$$

$$= r\cos\theta + L - \frac{r^2}{2L}\tfrac{1}{2}(1 - \cos 2\theta)$$

Differentiating this with respect to time gives the velocity v_p of the piston C relative to A.

$$\frac{dx}{dt} = -r\sin\theta\,\frac{d\theta}{dt} - \frac{r^2}{2L}\sin 2\theta\,\frac{d\theta}{dt}$$

Since $d\theta/dt = \omega$, then

$$v_p = -\omega r\left(\sin\theta + \frac{r}{2L}\sin 2\theta\right) \tag{7}$$

Differentiating equation [7] with respect to time gives the acceleration a_p of the piston C relative to A.

$$\frac{dv_p}{dt} = -\omega r\left(\cos\theta\,\frac{d\theta}{dt} + \frac{r}{L}\cos 2\theta\,\frac{d\theta}{dt}\right)$$

Since $d\theta/dt = \omega$, then

$$a_p = -\omega^2 r\left(\cos\theta + \frac{r}{L}\cos 2\theta\right) \qquad [8]$$

We can determine the angular velocity and angular acceleration of the connecting rod BC by differentiating equation [6].

$$r\sin\theta = L\sin\phi$$

$$r\cos\theta\,\frac{d\theta}{dt} = L\cos\phi\,\frac{d\phi}{dt}$$

The angular velocity $\omega_{BC} = d\phi/dt$ and thus, since $\omega = d\theta/dt$,

$$\omega_{BC} = \frac{r\omega\cos\theta}{L\cos\phi}$$

Since ϕ is small, then $\cos\phi$ approximates to 1 and so

$$\omega_{BC} = \frac{r\omega\cos\theta}{L} \qquad [9]$$

The angular acceleration of the connecting rod α_{BC} is $d\omega_{BC}/dt$ and thus

$$\alpha_{BC} = -\frac{r\omega\sin\theta}{L}\,\frac{d\theta}{dt}$$

$$= -\frac{r\omega^2\sin\theta}{L} \qquad [10]$$

Example

An engine mechanism has a crank shaft radius of rotation of 150 mm and a connecting rod of 600 mm. If the crank rotates at 5 rad/s, what is the velocity and acceleration of the piston when the crankshaft makes an angle of 60° with the inner dead centre position?

This can be solved by either calculation or drawing the velocity and acceleration diagrams. By calculation, equation [7] gives

$$v_p = -\omega r\left(\sin\theta + \frac{r}{2L}\sin 2\theta\right)$$

$$= -5 \times 0.150\left(\sin 60° + \frac{0.150}{1.200}\sin 120°\right)$$

$$= -0.73\,\text{m/s}$$

Equation [8] gives

$$a_p = -\omega^2 r\left(\cos\theta + \frac{r}{L}\cos 2\theta\right)$$

$$= -5^2 \times 0.150\left(\cos 60° + \frac{0.150}{0.600}\cos 120°\right)$$

$$= -1.41\,\text{m/s}^2$$

11.4.3   Torque on crankshaft

Consider a crankshaft in a reciprocating engine mechanism, as in figure 11.18, with the force on the piston F_p resulting in a force at right angles to the crankshaft at the crank pin. Neglecting the inertia and gravity effects of the connecting rod and considering only the crank shaft torque due to the piston force and mass, then if we neglect frictional forces the component of the force F_p acting along the connecting rod will be $F_p/\cos\phi$ and thus at B we have a torque of

$$T = (F_p/\cos\phi)x$$

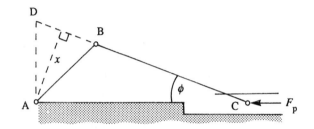

Fig. 11.18 Torque on a crankshaft.

But $\cos\phi = x/\text{AD}$. Thus

$$T = F_p \times \text{AD} \tag{11}$$

The torque can thus be determined by calculation or from a scale drawing of the mechanism.

Example

A horizontal engine has a stroke of 200 mm and a connecting rod of length 200 mm. What is the torque exerted on the crankshaft when the piston exerts a force of 4 kN and the crankshaft is at 45° to the horizontal?

Thus AB = 100 mm, BC = 200 mm and angle BAC = 45°. Figure 11.19 shows a scale diagram of the mechanism. From measurement on the diagram, AD = 98 mm. Hence, using equation [11],

$$T = F_p \times \text{AD} = 4000 \times 0.098 = 392\,\text{N m}$$

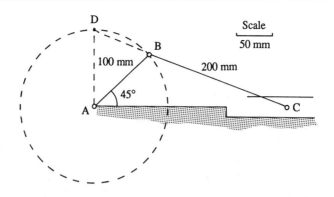

Fig. 11.19 Example.

11.4.4 *Force on piston*

The force acting on the piston is provided by a gas pressure p. Thus if A is the area of the piston then the force is pA. If the mass of the reciprocating parts, i.e. the piston plus gudgeon pin plus possibly some allowance for the mass of the connecting rod, is m then the force necessary to accelerate the parts is ma_p, where a_p is the acceleration of the piston. Thus the net force F_p transmitted is $(pA - ma_p)$ during the acceleration of the piston and $(pA + ma_p)$ during its retardation. Using equation [8] the net force is

$$F_p = pA + ma_p = pA - m\omega^2 r \left(\cos\theta + \frac{r}{L}\cos 2\theta \right) \qquad [12]$$

The above refers to a horizontal engine. If the engine is vertical then the weight mg has to be added to the piston force in the downstroke and subtracted in the upstroke.

Example

A horizontal engine has a cylinder of diameter 200 mm and a stroke of 400 mm. The connecting rod has a length of 800 mm. The reciprocating parts have a mass of 150 kg and the crank rotates at 5 rev/s. When the crank is at 45° to the inner dead centre the net piston pressure is 1 MPa. What is the force acting on the piston at that instant?

Using equation [12]

$$F_p = pA - m\omega^2 r \left(\cos\theta + \frac{r}{L}\cos 2\theta \right)$$

$$= 10^6 \times \tfrac{1}{4}\pi \times 0.200^2$$

$$- 150 \times (2\pi \times 5)^2 \times 0.200 \left(\cos 45° + \frac{0.200}{0.800}\cos 90° \right)$$

$$= 10.5\,\text{kN}$$

Alternatively the acceleration could have been obtained from an acceleration diagram and substituted in

$$F_p = pA - ma_p$$

11.4.5 Forces on cylinder and connecting rod

Consider the equilibrium of the forces acting on the piston (figure 11.20). If there are no frictional forces acting on the piston when it slides along the cylinder then there is a reaction force S acting at right angles to it, i.e. a side thrust on the cylinder. If Q is the force in the connecting rod, then resolving the forces in a vertical direction gives

$$S = Q \sin \phi$$

and horizontally

$$F_p = Q \cos \phi$$

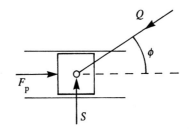

Fig. 11.20 Forces on cylinder and connecting rod.

Thus, rearranging the last equation,

$$Q = \frac{F_p}{\cos \phi} \qquad [13]$$

and dividing the two equations gives

$$S = F_p \tan \phi \qquad [14]$$

If there are frictional forces F acting on the piston then the forces are as shown in figure 11.21. The frictional force $F = \mu R$, where μ is the coefficient of friction and R the reaction force. The forces F and R can be combined to give the side thrust S acting at an angle θ_f, where

$$\tan \theta_f = \frac{F}{R} = \mu \qquad [15]$$

Example

A horizontal engine has a cylinder of diameter 120 mm and a stroke of 140 mm. The connecting rod has a length of 240 mm and the crank shaft

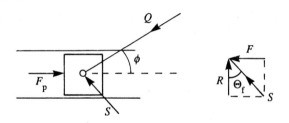

Fig. 11.21 Friction at the piston.

rotates at 20 rev/s. If the reciprocating parts have a mass of 1.2 kg, what is the resultant load on the gudgeon pin and thrust on the cylinder wall when the crank is at 30° from the inner dead centre and the net pressure on the piston is 0.8 MPa?

Using equation [12]

$$F_p = pA - m\omega^2 r\left(\cos\theta + \frac{r}{L}\cos 2\theta\right)$$

$$= 0.8 \times 10^6 \times \tfrac{1}{4}\pi \times 0.120^2$$

$$- 1.2 \times (2\pi \times 20)^2 \times 0.070\left(\cos 30° + \frac{0.070}{0.240}\cos 60°\right)$$

$$= -7.71\,\text{kN}$$

Using figure 11.17 and equation [6]

$$r\sin\theta = L\sin\phi$$

$$0.070\sin 30° = 0.240\sin\phi$$

Thus $\phi = 8.39°$. Thus the resultant force on the gudgeon pin is given by equation [13] as

$$Q = \frac{F_p}{\cos\phi} = \frac{7.71}{\cos 8.39°} = 7.79\,\text{kN}$$

The thrust on the cylinder wall is given by equation [14] as

$$S = F_p\tan\phi = 7.71\tan 8.39° = 1.14\,\text{kN}$$

11.4.6 *Inertia of connecting rod*

Any body of mass can be replaced by a dynamically equivalent system of two point masses at the ends of a rod of zero mass provided the total mass remains the same, the centre of gravity is the same and the moment of inertia of the system is unchanged. Thus if we consider the connecting rod of a reciprocating system (figure 11.22) to have a mass m, then for the equivalent system of point masses of m_1 and m_2 we must have

$$m = m_1 + m_2 \tag{16}$$

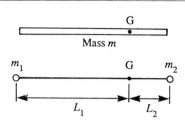

Fig. 11.22 Connecting rod and equivalent system.

If the centre of gravity of the link is L_1 from one end and L_2 from the other, then for the centre of gravity of the equivalent system with the two masses located at each end of the link we must have for an unchanged location of the centre of gravity, for moments about the centre of gravity,

$$m_1 L_1 = m_2 L_2 \qquad [17]$$

For the moment of inertia to be unchanged then, if I_G is the moment of inertia of the link about its centre of gravity

$$I_G = m_1 L_1^2 + m_2 L_2^2 \qquad [18]$$

In practice, for a connecting rod, it is usually adequate to just use equations [16] and [17]. We can thus just consider the mass of the connecting rod to be subdivided between part which is added to the piston mass and part to the crank-pin.

Example

A horizontal engine has a crank of radius 50 mm and a connecting rod of length 200 mm and mass 1.4 kg. The centre of gravity of the connecting rod is 80 mm from the crank-pin end. The piston has a mass of 1.5 kg. What will be the crankshaft torque when the crank is at 45° to the inner dead centre and the crank is rotating at 20 rev/s?

The acceleration of the piston a_c can be found from a construction of the velocity and acceleration diagrams or calculation. By calculation, equation [8] gives

$$a_p = -\omega^2 r \left(\cos\theta + \frac{r}{L} \cos 2\theta \right)$$

$$= -(2\pi \times 20)^2 \times 0.050 \left(\cos 45° + \frac{0.050}{0.400} \cos 90° \right)$$

$$= -558 \, \text{m/s}^2$$

The mass of the connecting rod can be replaced by two equivalent masses, one at the piston end and one at the crank-pin end. Thus, using the conditions given by equations [16] and [17],

$$1.4 = m_1 + m_2$$
$$m_1 L_1 = m_2 L_2$$

But $L_1 = 120\,$mm and $L_2 = 80\,$mm. Thus $m_2 = 1.5 m_1$ and so

$$1.4 = m_1 + 1.5 m_1$$

Hence $m_1 = 0.56\,$kg. Thus we can consider the total mass of the reciprocating parts to be the mass of the piston plus 0.56 kg, i.e. 2.06 kg. Hence the force at the piston $F_p = m a_p = 2.06 \times 558 = 1.15\,$kN.

Figure 11.23 shows a scale drawing of the mechanism. From that drawing $AD = 42\,$mm. Thus, using equation [11],

$$T = F_p \times AD = 1.15 \times 0.042 = 0.048\,\text{kN m} = 48\,\text{N m}$$

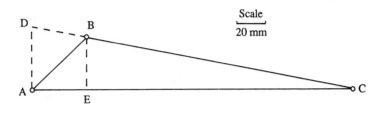

Fig. 11.23 Example.

The dead weight of the mass m_2 at B will also produce a torque. This will be $m_2 g \times AE$. Since $AE = 50 \cos 45° = 35\,$mm then this torque is $0.84 \times 9.8 \times 0.035 = 0.29\,$N m. To the accuracy with which the 48 N m torque was obtained, this contribution is insignificant.

Problems

(1) Determine the angular velocity of CD for the four-bar chain shown in figure 11.24.

(2) Determine the velocities of the sliders C in figure 11.25(a) and (b).

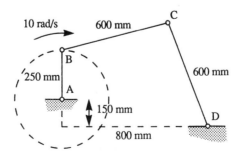

Fig. 11.24 Problem 1.

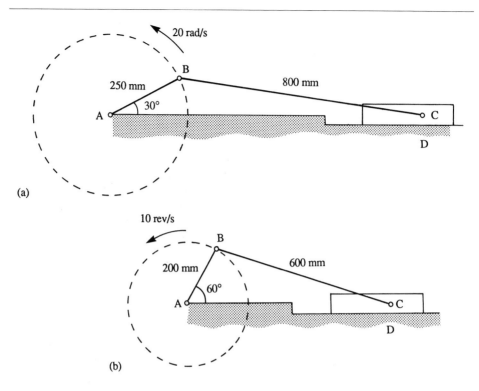

(a)

(b)

Fig. 11.25 Problem 2.

(3) Determine the velocity of point C in the quick-return mechanism shown in figure 11.26.

(4) Determine the velocity of the slider F in the quick-return mechanism shown in figure 11.27.

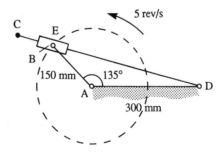

Fig. 11.26 Problem 3.

(5) Determine the angular velocity and angular acceleration of the arm DC in the mechanism described by figure 11.28.

(6) A collar slides along a rod with a velocity of 2 m/s and an acceleration of 3 m/s². The rod is rotating at 3 rad/s and with an angular

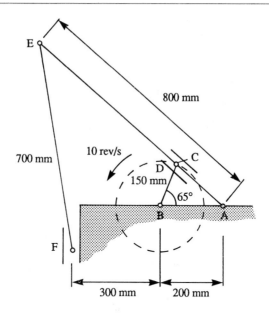

Fig. 11.27 Problem 4.

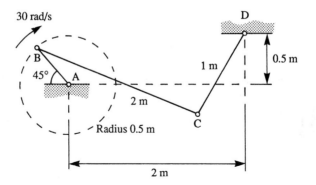

Fig. 11.28 Problem 5.

acceleration of 2 rad/s². If the rod is at an angle of 60° to the horizontal and the collar is a distance of 0.2 m along the rod, what is the velocity and acceleration of the collar?

(7)   A mechanism is of the form shown in figure 11.15(a), with CD = 0.2 m, angle CDA = 60° and the angular velocity of AB = 25 rad/s. What is the angular acceleration of the link CD?

(8)   For the mechanism shown in figure 11.29, what is the acceleration of C when AB has an angular velocity of 10 rad/s?

(9)   Determine the angular acceleration of link DC in figure 11.28, problem 5.

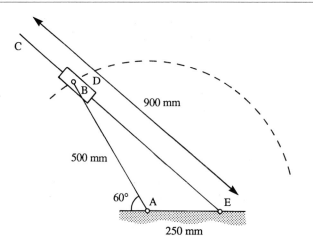

Fig. 11.29 Problem 8.

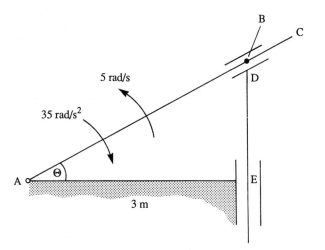

Fig. 11.30 Problem 10.

(10) For the mechanism shown in figure 11.30, link DE is constrained by guides to move vertically, being driven by a crank ABC, with an angular velocity of 5 rad/s and an angular acceleration of -35 rad/s^2, and a sliding block at D. What is the acceleration of DE?

(11) Derive analytically equations for the velocity and acceleration of a point P situated a distance p from the piston along the connecting rod of a reciprocating mechanism, as illustrated in figure 11.17.

(12) An engine mechanism has a crank shaft radius of rotation of 120 mm and a connecting rod of 400 mm. If the crank rotates at 4 rev/s, what is the velocity and acceleration of the piston when the crankshaft makes an angle of 30° with the inner dead centre position?

(13)   A horizontal engine has a stroke of 200 mm and a connecting rod of length 400 mm. What is the torque exerted on the crankshaft when the piston exerts a force of 4 kN and the crankshaft is at 45° to the horizontal?

(14)   A horizontal engine has a stroke of 100 mm and a connecting rod of length 240 mm. What is the torque exerted on the crankshaft when the piston exerts a force of 10 kN and the crankshaft is at 30° to the horizontal?

(15)   A horizontal engine has a cylinder of diameter 200 mm and a stroke of 300 mm. The connecting rod has a length of 600 mm. The reciprocating parts have a mass of 180 kg and the crank shaft rotates at 6 rev/s. What is the force acting on the piston when the crankshaft is at 45° to the inner dead centre and the net pressure on the piston is 1 MPa?

(16)   A horizontal engine has a cylinder of diameter 100 mm and a stroke of 140 mm. The connecting rod has a length of 250 mm and the crankshaft rotates at 30 rev/s. If the reciprocating parts have a mass of 1.2 kg, what is the resultant load on the gudgeon pin and thrust on the cylinder wall when the crank is at 20° from the inner dead centre and the net pressure on the piston is 0.8 MPa?

(17)   A horizontal steam engine has a cylinder of diameter 200 mm and a stroke of 600 mm. The connecting rod has a length of 1200 mm and the crankshaft rotates at 4 rev/s. If the reciprocating parts have a mass of 100 kg, what is the resultant load on the gudgeon pin, the thrust on the cylinder wall and the crankshaft torque when the crank is at 60° from the inner dead centre and the net pressure on the piston is 0.5 MPa?

(18)   A vertical engine has a cylinder of diameter 260 mm and a stroke of 450 mm. The connecting rod has a length of 900 mm and the crankshaft rotates at 6 rev/s. If the reciprocating parts have a mass of 180 kg, what is the crankshaft torque when the crank is at 45° from the inner dead centre and the net pressure on the piston is 1 MPa?

(19)   A horizontal engine has a stroke of 140 mm and a connecting rod of length 250 mm and mass 1.5 kg. The centre of gravity of the rod is 80 mm from the crank-pin end. The piston has a mass of 1.8 kg. What will be the crankshaft torque when the crank is at 45° to the inner dead centre and rotating at 25 rev/s?

(20)   A horizontal engine has a stroke of 200 mm and a connecting rod of length 350 mm and mass 120 kg. The centre of gravity of the rod is 150 mm from the crank-pin end. The piston has a mass of 100 kg. What will be the crankshaft torque when the crank is at 45° to the inner dead centre and rotating at 4 rev/s?

Chapter 12

Turning moment diagrams

12.1 Crank effort diagrams

The torque on the crankshaft of a reciprocating engine depends on the position of the crank and the pressure in the cylinder. Since these vary through the working cycle, the torque also varies throughout the cycle. A graph of the crankshaft torque plotted against the crank angle gives, what is termed, a *turning moment diagram* or *crank–effort diagram*.

Figure 12.1 shows an example of such a diagram for a single cylinder four-stroke engine. The working cycle consists of the gas pressure rising as a result of the engine firing, i.e. the gas exploding, and forcing the piston outwards, then the gas being exhausted as the piston moves back, followed by a suction stroke when the gas is sucked into the cylinder, followed by a compression stroke which then leads to firing and a repeat of the cycle. The torque varies from large and positive during the firing or working stroke to small and negative during the exhaust and compression stroke and small and positive during the suction stroke. The torque can be made more even by increasing the number of cylinders to four and arranging the firing strokes to be at 180° intervals. In a multi-cylinder engine, the total torque for any crankshaft position is the algebraic sum of the torques exerted by the various cranks. Figure 12.2 shows a turning moment diagram for a four cylinder four-stroke engine.

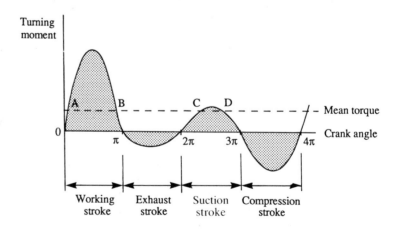

Fig. 12.1 Turning moment diagram for single cylinder four-stroke engine.

316

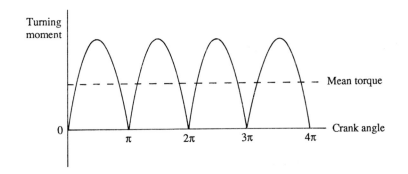

Fig. 12.2 Turning moment diagram for a four cylinder four-stroke engine.

Turning moment diagrams can be drawn for electric motors or indeed any engine or motor drive. Electric motors give an inherently smoother torque output than petrol or steam engines. The torque depends on the angle of the rotor, varying as each pole on the rotor passes the poles and gaps in the stator. The turning moment diagram thus tends to consist of a high frequency ripple superimposed on a constant value.

The average height on the turning moment diagram is the mean torque developed. Since the work done by a torque T in producing an angular rotation θ is $T\theta$ (equation [38], chapter 10) then the area under the turning moment diagram is the work done, e.g. the shaded area in figure 12.1.

12.2   Fluctuations of speed and energy

If the mean speed of rotation of the crank shaft is to remain constant then the mean torque delivered by an engine must just equal the resisting torque of the load. When the torque produced by the engine is greater than that required by the load then the engine crank shaft will accelerate. If the torque produced by the engine is less than that required by the load then the engine will decelerate. Thus, for the engine giving the turning moment diagram in figure 12.1, if the resisting torque is equal to the mean torque then between A and B the engine will accelerate, between B and C decelerate, between C and D accelerate, and between D and the onset of the next cycle at A decelerate. There will thus be fluctuations in speed. Between A and B there is acceleration and so the speed increases, reaching a maximum value at B. Then there is deceleration between B and C and so the minimum speed is attained at C. Points of maximum speed thus occur at B and D and minimum speed at C and A.

The kinetic energy of rotation of the system is $\frac{1}{2}I\omega^2$ (equation [40], chapter 10), where ω is the angular velocity and I the total moment of inertia of all the rotating parts, i.e. the engine, the load and flywheel (see later for discussion of flywheels). Thus the points of maximum and minimum excess energy coincide with the points of maximum and minimum speed. These thus occur where the engine torque curve cuts the mean torque line.

If ω_{max} is the maximum angular velocity and ω_{min} the minimum, then the excess energy available between points of minimum and maximum speeds is

$$\Delta E = \tfrac{1}{2}I\,(\omega_{max}^2 - \omega_{min}^2) \qquad [1]$$

ΔE is termed the *energy fluctuation*. The larger the moment of inertia, the smaller the difference between the maximum and minimum speeds.

Equation [1] can be rearranged to give

$$\Delta E = I(\omega_{max} - \omega_{min})\,\tfrac{1}{2}(\omega_{max} + \omega_{min})$$

If the fluctuation in speed is either side of the mean speed ω_{mean} then to a reasonable approximation $\tfrac{1}{2}(\omega_{max} + \omega_{min}) = \omega_{mean}$. Thus

$$\Delta E = I\omega_{mean}(\omega_{max} - \omega_{min}) \qquad [2]$$

The ratio of the maximum fluctuation of speed in a cycle to the mean speed is termed the *coefficient of fluctuation of speed A*. Thus

$$A = \frac{\omega_{max} - \omega_{min}}{\omega_{mean}} \qquad [3]$$

Hence equation [2] can be written as

$$\Delta E = AI\omega_{mean}^2 \qquad [4]$$

The ratio of the maximum fluctuation in energy in a cycle to the work done per cycle W is termed the *coefficient of fluctuation of energy B*. Thus

$$B = \frac{I\omega_{mean}\,(\omega_{max} - \omega_{min})}{W} = \frac{AI\omega_{mean}^2}{W} \qquad [5]$$

Example

A single cylinder four-stroke engine has a mean speed of 300 rev/min when running against a constant load torque. The turning moment diagram has areas above and below the mean torque line of working cycle: $+7000\,J$, exhaust stroke: $-1800\,J$, suction stroke $+500\,J$, compression stroke: $-5700\,J$. If the speed varies between 295 and 305 rev/min, what is the moment of inertia of the system?

This example can be considered in relation to the turning moment diagram given in figure 12.1. At point B the excess energy is $+7000\,J$. At C the excess energy is $+7000 - 1800 = +5200\,J$. At D the excess energy is $+7000 - 1800 + 500 = +5700\,J$. Then by A in the next cycle the excess energy $= +7000 - 1800 + 500 - 5700 = 0\,J$. The fluctuation in energy is the difference between the maximum and minimum energy values, in this case between the values at B and C. Thus $\Delta E = 7000 - 5200 = 1800\,J$. Hence, using equation [1],

$$\Delta E = \tfrac{1}{2}I(\omega_{max}^2 - \omega_{min}^2)$$
$$1800 = \tfrac{1}{2}I(305^2 - 295^2)\,(2\pi/60)^2$$

Hence $I = 54.7\,\text{kg}\,\text{m}^2$.

12.2.1 Machines offering variable resistance

In the above discussion it has been assumed that the engine had a constant resisting torque. For such machines as presses this is not the case. With an electric motor operated press the motor delivers an essentially constant input torque while the resisting torque is fluctuating. The press output torque is likely to be high during the pressing operation and virtually zero during the time between pressings. Figure 12.3 illustrates such a situation.

During the pressing part of the operation the motor has to deliver a higher torque than the mean input torque with the result that the motor slows down. During the remainder of the cycle the motor is delivering more torque than is required and so speeds up. There are thus fluctuations in speed and energy.

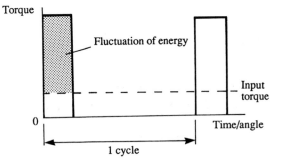

Fig. 12.3 Torque–time/angle diagram.

12.3 Flywheels

In order to reduce the speed variations of an engine a flywheel can be added. This adds inertia to the system and thus, as equation [1] indicates, reduces the difference between the maximum and minimum speeds. The flywheel is storing energy during the periods of excess torque and releasing it during the periods when the torque is deficient.

Example

An engine has a turning moment diagram with the areas above and below the mean torque line of $-50\,\text{mm}^2$, $+120\,\text{mm}^2$, $-100\,\text{mm}^2$, $+150\,\text{mm}^2$, $-80\,\text{mm}^2$, $+70\,\text{mm}^2$ and $-110\,\text{mm}^2$, with the turning moment diagram scales being for the torque $1\,\text{mm} \equiv 100\,\text{N}\,\text{m}$ and for the crank angle $1\,\text{mm} \equiv 2°$. What is the smallest mass a flywheel of radius of gyration $250\,\text{mm}$ can have, when running at $600\,\text{rev/min}$, if the total fluctuation of speed is not to exceed 2% of the mean?

Labelling the turning moment diagram in the same way as figure 12.1, then the areas up to the various points are: at $A = -50\,\text{mm}^2$, up to $B = -50 + 120 = +70\,\text{mm}^2$, up to $C = +70 - 100 = -30\,\text{mm}^2$, up to $D = -30 + 150 = +120\,\text{mm}^2$, up to $E = +120 - 80 = +40\,\text{mm}^2$, up to $F = +40 + 70 = +110\,\text{mm}^2$, and up to the end of the cycle $+110 - 110 = 0\,\text{mm}^2$. The maximum energy excess is thus between D and A. Since 1 mm on the torque scale represents 100 N m and 1 mm on the angle scale $2 \times 2\pi/360 = 0.0349\,\text{rad}$, then $1\,\text{mm}^2$ represents an energy of $100 \times 0.0349 = 3.49\,\text{J}$. Thus the energy excess is

$$\Delta E = (120 + 50) \times 3.49 = 593.3\,\text{J}$$

Since $(\omega_{max} - \omega_{min})$ is to be at the most $0.02\omega_{mean}$, then using equation [2]

$$\Delta E = I\omega_{mean}\,(\omega_{max} - \omega_{min})$$
$$593.3 = I \times (600 \times 2\pi/60)\,(0.02 \times 600 \times 2\pi/60)$$

Hence the required moment of inertia is $7.51\,\text{kg m}^2$. Since $I = mk^2 = m \times 0.250^2$ then $m = 120\,\text{kg}$.

Example

A motor driven punching machine used to stamp out shapes from sheet metal has a motor which exerts a constant torque of 600 N m on a flywheel which has a mean speed of 2 rev/s. The punch operation lasts 0.2 s and is carried out once every second. What is the resisting torque, assumed to be constant during the pressing operation, and the moment of inertia required of the flywheel if the speed fluctuation is not to exceed 2% of the mean?

Figure 12.4 shows the torque–time/angle diagram for the machine. The mean torque is 600 N m and during the punching operation there is excess energy required. During one complete working cycle the area above the mean torque line must equal the area under it. These areas are shown shaded in the figure. Thus, since the angle rotated is proportional to the time,

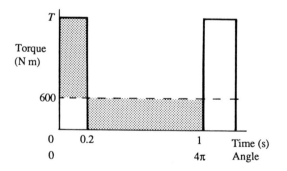

Fig. 12.4 Example.

$$(T - 600) \times 0.2 = 600 \times (1 - 0.2)$$

Hence $T = 3000\,\text{N m}$.

Since the angle rotated during $0.2\,\text{s}$ is $2 \times 2\pi \times 0.2 = 2.5\,\text{rad}$, the excess energy is

$$\Delta E = (3000 - 600) \times 2.5 = 6000\,\text{J}$$

The speed fluctuation is not to exceed 2% of the mean, i.e. $0.02 \times 2 \times 2\pi\,\text{rad/s}$. Hence, equation [2] gives

$$\Delta E = I\omega_{\text{mean}}\,(\omega_{\text{max}} - \omega_{\text{min}})$$
$$6000 = I \times 2 \times 2\pi \times 0.02 \times 2 \times 2\pi$$

Hence $I = 1900\,\text{kg m}^2$.

Example

During the working stroke of a single cylinder four-stroke engine $10\,\text{kJ}$ of work is done by the expanding gas on the piston and during the compression stroke the work done on the gas is $5\,\text{kJ}$. The energy involved in the exhaust and suction strokes is negligible. If the mean speed is $300\,\text{rev/min}$ and the fluctuation of speed is not to exceed 2% of the mean, what is the minimum power that the engine is required to deliver and the moment of inertia required for the flywheel?

For a single cylinder four-stroke engine the full working cycle involves the crank rotating through 4π radians (see figure 12.1). Thus the net work done during the full working cycle is given by $\omega = T\theta$ as

$$W = 10\,000 - 5000 = T \times 4\pi$$

Hence the mean torque is $398\,\text{N m}$. The engine power is given by

$$P = T\omega = 398 \times (300 \times 2\pi/60) = 12.5\,\text{kW}$$

The average work done per stroke, each stroke being an angle of $\pi\,\text{rad}$, is given by $T\theta = 398 \times \pi = 1.25\,\text{kJ}$. Thus the excess energy at the end of the working stroke is $10 - 1.25 = 8.75\,\text{kJ}$ and at the end of the compression stroke is $10 - 1.25 - 5 = 3.75\,\text{kJ}$. Thus the fluctuation of the energy is $8.75 - 3.75 = 5\,\text{kJ}$. Hence, using equation [2],

$$\Delta E = I\omega_{\text{mean}}\,(\omega_{\text{max}} - \omega_{\text{min}})$$
$$5000 = I \times 5 \times 2\pi \times 0.02 \times 5 \times 2\pi$$

Hence $I = 253\,\text{kg m}^2$.

Problems

(1) An engine has a turning moment diagram which has areas above and below the mean torque line of $+25\,\text{J}$, $-60\,\text{J}$, $+4\,\text{J}$, $-30\,\text{J}$, $+85\,\text{J}$, $-24\,\text{J}$. If the rotating parts have a total moment of inertia of $0.8\,\text{kg m}^2$,

what is the coefficient of fluctuation of speed when the engine runs at 30 rev/s?

(2) An engine has a turning moment diagram which has areas above and below the mean torque line of $+390\,mm^2$, $-200\,mm^2$, $+260\,mm^2$, $-310\,mm^2$, $+130\,mm^2$, $-230\,mm^2$, $+170\,mm^2$ and $-210\,mm^2$ and runs at 2 rev/s. The scales used for the diagram were for torque 1 mm $\equiv 400\,N\,m$ and angle 1 mm $\equiv 2°$. What moment of inertia flywheel is required if the fluctuation in speed is to be restricted to 2% of the mean speed?

(3) An engine has a turning moment diagram which has areas above and below the mean torque line of $-10\,mm^2$, $+128\,mm^2$, $-86\,mm^2$, $+102\,mm^2$, $-98\,mm^2$, $+72\,mm^2$, $-114\,mm^2$, $+86\,mm^2$ and $-76\,mm^2$ and runs at 400 rev/min. The scales used for the diagram were for torque 1 mm $\equiv 500\,N\,m$ and angle 1 mm $\equiv 5°$. What moment of inertia flywheel is required if the fluctuation in speed is to be restricted to 1.5% of the mean speed?

(4) An engine has a power of 90 kW and a mean speed of 100 rev/min. The flywheel, between its maximum and minimum speeds, takes up 20% of the work done in the cylinders per revolution. What is the moment of inertia of the flywheel if the speed fluctuation is kept within 2% of the mean speed?

(5) The turning moment diagram for a four-stroke engine has the scales of torque 1 mm $\equiv 500\,N$ and angle 1 mm $\equiv 10°$. The fluctuation of energy between the crank positions for maximum and minimum speeds corresponds to an area on the turning moment diagram of $10\,mm^2$. If, when the engine is running at a mean speed of 1500 rev/min, the fluctuation of speed is not to exceed 0.5%, what is the necessary moment of inertia of the flywheel?

(6) An engine has a torque which is zero at crank angles of 0° and 180°. From 0° to 30° the torque increases at a uniform rate to reach a maximum of 300 N m. This torque then remains constant until the crank angle reaches 60°, then it decreases at a constant rate to zero at 180°. This cycle is repeated in each half revolution. If the mean speed of the engine is 5 rev/s, what is the required moment of inertia for the flywheel if the speed fluctuation is not to exceed 1.5%?

(7) An electric motor is used to drive a power press which makes steel pressings from a metal sheet. The motor runs at a mean speed of 900 rev/min, each pressing takes 1 s and the time between a pressing and the next one is 4 s. The energy required for each pressing is 1000 J. What is the smallest power motor that can be used and what size of flywheel should be used if the speed fluctuation is not to exceed 5% of the mean?

(8) An electric motor is used to drive a power press which makes steel

pressings from a metal sheet. The motor runs at a mean speed of 50 rev/s. The torque required is 10.0 N m for 0.2 s, followed by 1.0 N m for 0.3 s with this sequence then being repeated. What is the minimum power required of the motor and the moment of inertia required for the flywheel if the speed fluctuations are to be restricted to 1.5%?

(9)   An engine has three cylinders with cranks at 120°. For each cylinder the torque increases at a constant rate from zero at 0° to 100 N m at 60° before decreasing to zero at 180° and then remaining at zero for the rest of the rotation to 360°. What is the mean torque?

(10)   A two-stroke engine has three cylinders with cranks at 120°. For each cylinder the torque increases at a constant rate from zero to a maximum during a rotation of 90° and then falls at the same constant rate to zero over the next 90°. It then remains at zero for the remainder of the revolution. The engine develops 10 kW per cylinder when running at an average speed of 400 rev/min. What moment of inertia will be required of a flywheel if the total speed variation is not to exceed 1% of the mean?

(11)   A four stroke engine has six cylinders with cranks at 120°. For each cylinder the torque increases at a constant rate from zero to a maximum of 1.5 kN m during a rotation of 60° and then falls at a constant rate to zero over the next 120°. It then remains at zero for the rest of the cycle. The engine has a mean speed of 30 rev/s. What moment of inertia will be required of a flywheel if the total speed variation is to be kept within 1% of the mean?

Chapter 13

Geared systems

13.1 Definitions

Gears are used to transmit rotary motion and power from one shaft to another. While there are other methods of doing this, e.g. belt drives (see chapter 14), gears have the advantages of providing positive drive without slip and permit high torques to be transmitted. They can also be used between shafts which can be parallel or inclined to one another. Figure 13.1 shows some examples of gears.

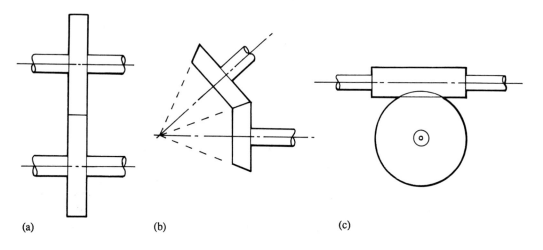

(a) (b) (c)

Fig. 13.1 Examples of gears for use with (a) parallel axes, (b) axes inclined to one another, (c) non-intersecting axes.

Gears for use with parallel shaft axes, as for example in figure 13.1(a), may have axial teeth with the teeth being cut along axial lines parallel to the shaft (figure 13.2(a)) and are then called *spur gears*, or helical teeth with the teeth being cut on a helix (figure 13.2(b)) and are then called *helical gears*. Helical gears have the advantage that there is a gradual engagement of any individual tooth and consequently there is a reduction in noise, a smoother drive, and generally prolonged life. However, a consequence of the inclination of the teeth is that there will be an axial force component on the shaft bearing. This can be overcome by using double helical gears with two sets of teeth back to back and cut with opposite inclinations (figure 13.2(c)).

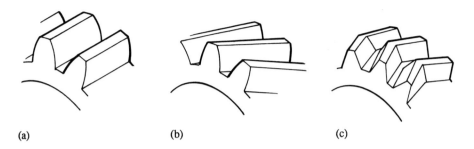

Fig. 13.2 (a) Axial teeth, (b) helical teeth, (c) double helical teeth.

Gears for use with shafts inclined to one another are termed *bevel gears* when the lines of the shafts intersect, as for example in figure 13.1(b). The teeth may be cut straight or curved, these giving the same benefits as helical gears. The terms *skew* or *spiral* gears are used for gears when the shafts are non-intersecting. A special form of such a gear is the *worm and wheel* (figure 13.1(c)), this term being generally used when one of the wheels, the worm, is very small and has relatively few teeth.

When two gears are in mesh the larger gear is often called the *spur* or *crown wheel* or just *wheel* and the smaller one the *pinion*.

13.1.1   Spur gears

Consider two parallel shafts, each with a disc which is so arranged that the edges of the two discs press together. When one shaft, and its disc, rotates then because of friction the other disc and shaft are constrained to rotate. Such an arrangement would not be able to transmit much power since slipping would readily occur. However if teeth are put on each disc we have an arrangement which prevents slipping. The two teethed discs in such an arrangement are called spur gears.

Figure 13.3 shows some of the terms used with spur gears. The *pitch circle diameters* are the diameter of the discs, without the teeth, which would

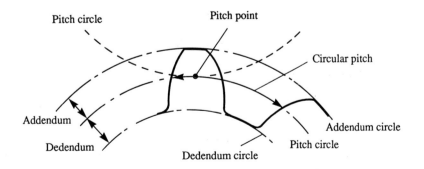

Fig. 13.3 Spur gear terms.

transmit the same velocities by friction as the gear wheels. If the pitch circle diameters for the wheels in a pair are d_A and d_B, then since the teeth will be the same size

$$\frac{d_A}{d_B} = \frac{t_A}{t_B}$$

where t_A is the number of teeth on wheel A and t_B the number on wheel B. If there are 50 teeth on wheel A and 100 teeth on wheel B then wheel A must rotate through 2 revolutions in the same time as B rotates through 1. Thus the ratio of the angular velocities of the two wheels is the inverse ratio of the teeth numbers.

$$\frac{d_A}{d_B} = \frac{t_A}{t_B} = \frac{\omega_B}{\omega_A} \tag{1}$$

where ω_A is the angular velocity of wheel A and ω_B that of wheel B. The *pitch point* is the point of contact between the two pitch circles.

The *circular pitch p* is the distance measured along the pitch circle between a point on one tooth and the corresponding point on the next tooth.

$$p = \frac{\pi d_A}{t_A} = \frac{\pi d_B}{t_B} \tag{2}$$

The *diametral pitch P* is the number of teeth per millimetre of pitch circle diameter.

$$P = \frac{t_A}{d_A} = \frac{t_B}{d_B} = \frac{\pi}{p} \tag{3}$$

The *module m* is the number of millimetres of pitch circle diameter per tooth.

$$m = \frac{d_A}{t_A} = \frac{d_B}{t_B} = \frac{1}{P} = \frac{p}{\pi} \tag{4}$$

The part of the tooth above the pitch circle is called the *addendum* and that part below the pitch circle the *dedendum*. Standard gear teeth have an addendum of $1/P = m$, and a dedendum of $1.25/P = 1.25m$. Figure 13.4 shows the profile of a standard tooth.

Example

The axes of the two wheels in a spur gear are 370 mm apart. If one wheel rotates four times faster than the other, what are the diameters of the pitch circles?

We must have $\frac{1}{2}(d_A + d_B) = 370$ mm. Equation [1] gives

$$\frac{d_A}{d_B} = \frac{\omega_B}{\omega_A} = 4$$

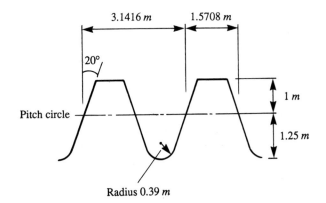

Fig. 13.4 Standard tooth.

Thus $d_A = 4d_B$ and so $5d_B = 740\,\text{mm}$ and $d_B = 148\,\text{mm}$. This then gives $d_A = 592\,\text{mm}$.

13.2   Gear trains

The term *gear train* is used to describe a series of gear wheels which are used to transmit rotational motion from an input shaft to an output shaft. For any pair of meshing gear wheels, whatever the type, the ratio of the angular speeds of the wheels is proportional to the inverse ratio of the number of teeth on the wheels. The term *gear ratio G* is used for the ratio of the angular speeds.

$$G = \frac{\omega_A}{\omega_B} = -\frac{t_B}{t_A} \qquad [5]$$

The minus sign is often included in such an equation to indicate that the angular velocity of wheel A is in the opposite direction to that of B.

In the case of spur gears on parallel shafts, the pitch circle diameters are proportional to the number of teeth. Thus the gear ratio is proportional to the inverse ratio of the pitch circle diameters (see equation [1]). This relationship is not valid for skew and worm gears.

13.2.1   Simple trains

A gear train in which each shaft carries only one wheel is called a simple gear train. Figure 13.5 shows a simple gear train involving three gear wheels. The overall gear ratio G is the ratio of the angular velocities at the input and output shafts and is thus ω_A/ω_C. Hence

$$G = \frac{\omega_A}{\omega_C} = \frac{\omega_A}{\omega_B} \times \frac{\omega_B}{\omega_C}$$

$$= \left(\frac{-t_B}{t_A}\right)\left(\frac{-t_C}{t_B}\right) = \frac{t_C}{t_A} \qquad [6]$$

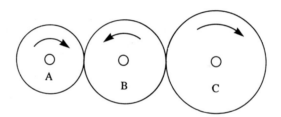

Fig. 13.5 Simple gear train.

The gear ratio is independent of the size of the intermediate wheel. The presence of the intermediate wheel does however result in a change in angular direction of the output wheel. The intermediate wheel is often called an *idler*.

13.2.2 Compound trains

A gear train in which two wheels are mounted on a common shaft is called a compound gear train. Figure 13.6 shows two examples of compound gear trains. The two wheels mounted on the same shaft rotate with the same angular velocity. Thus $\omega_B = \omega_C$. The overall gear ratio G, which is ω_A/ω_D, is thus given by

$$G = \frac{\omega_A}{\omega_D} = \frac{\omega_A}{\omega_B} \times \frac{\omega_B}{\omega_C} \times \frac{\omega_C}{\omega_D} = \frac{\omega_A}{\omega_B} \times \frac{\omega_C}{\omega_D}$$

$$= \left(\frac{-t_B}{t_A}\right)\left(\frac{-t_D}{t_C}\right) = \frac{t_B t_D}{t_A t_C} \tag{7}$$

The form of compound gear train shown in figure 13.6(b) is called a *reverted* or *co-axial* train. With such an arrangement the output shaft can be brought into line with the input shaft. For such a requirement, not only does equation [7] have to hold but also

$$r_A + r_B = r_C + r_D \tag{8}$$

where r_A is the pitch circle radius of A, r_B that of B, etc. Since the module (equation [4]) for gear wheels A and B is $m_{AB} = d_A/t_A = d_B/t_B$ and for gear wheels C and D $m_{CD} = d_C/t_C = d_D/t_D$ then equation [8] can be written as

$$\tfrac{1}{2}m_{AB}t_A + \tfrac{1}{2}m_{AB}t_B = \tfrac{1}{2}m_{CD}t_C + \tfrac{1}{2}m_{CD}t_D \tag{9}$$

An alternative way of obtaining the ouput shaft in line with the input shaft is the epicyclic gear train.

The typical motor car gear box has the input shaft in line with the output shaft and uses the arrangement shown in figure 13.7. The pairs of wheels CD, EF and GH are to allow different gear ratios to be selected. One of the wheels D, F or G is connected to the output shaft when the gear is selected. Typical gear ratios used are first gear 4 to 1, second gear 2.5 to 1, third gear 1.5 to 1, and fourth gear 1 to 1.

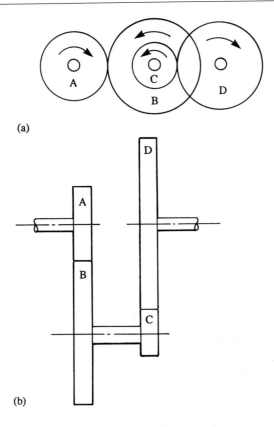

(a)

(b)

Fig. 13.6 Compound gear trains.

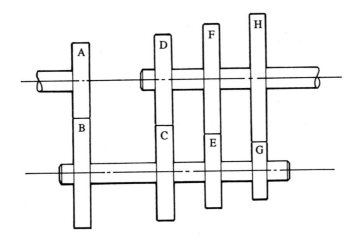

Fig. 13.7 Typical car gear box.

Example

A crane has to lift a load at a speed of 5 m/min by means of a rope which is wound round a drum of 0.80 m diameter. The hoist motor runs at 400 rev/min. What will be the number of teeth required on the wheels of a compound gear train of the form shown in figure 13.6(a), to reduce the motor speed to the required speed if the maximum gear ratio that can be used for a pair of spur wheels is 5 and the minimum number of teeth on a wheel is to be 15?

The angular velocity of the drum is given by $v = r\omega$ as $5/0.40 = 12.5$ rad/min. The angular velocity of the motor is $400 \times 2\pi = 2513.3$ rad/s. Thus the overall gear ratio required is $2513.3/12.5 = 201.1$. If each pair of wheels in the train gave the same gear ratio then, with the maximum ratio of 5,

$5^n = 201.1$

where n is the number of pairs of wheels. This gives $n = 3.3$. Hence taking n to be 4 means a gear ratio for each pair of wheels of

$G^4 = 201.1$

and so $G = 3.77$. We could thus take each pair to have 15:57 teeth (or even closer to the required ratio 17:64), with four pairs being used.

Example

A reverted gear train, as in figure 13.6(b), is to be used to connect two shafts X and Z in the same straight line through an intermediate shaft Y. The wheels connecting X and Y have a module of 2 and those connecting Y and Z a module of 3. The speed of the output shaft Z is to be about but not less than 1/10 th that of the input shaft X. The two pinions each have 24 teeth. Suggest possible number of teeth on the wheels and the distance of shaft Y from X and Z.

The requirement is, using the symbols given in figure 13.6, for $\omega_D \geq 0.1\omega_A$ with $t_A = t_C = 24$. Thus, using equation [7],

$$G = \frac{\omega_A}{\omega_D} = \frac{t_B t_D}{t_A t_C}$$

$$\frac{t_B t_D}{24 \times 24} \geq 10 \tag{10}$$

Because the input and output shafts are to be in line we must have (equation [8])

$$r_A + r_B = r_C + r_D$$

which when written in terms of the modules leads to (equation [9])

$$\tfrac{1}{2}m_{AB}t_A + \tfrac{1}{2}m_{AB}t_B = \tfrac{1}{2}m_{CD}t_C + \tfrac{1}{2}m_{CD}t_D$$

Thus

$$2 \times 24 + 2t_B = 3 \times 24 + 3t_D$$

We need values which will satisfy both this equation and equation [10]. Possible values which will do this are $t_D = 61$ and $t_B = 104$.

The distance between the centres of the shafts X and Y is $r_A + r_B$. In terms of the module this is

$$\text{distance} = \tfrac{1}{2}m_{AB}t_A + \tfrac{1}{2}m_{AB}t_B = \tfrac{1}{2} \times 2 \times 24 + \tfrac{1}{2} \times 2 \times 104$$
$$= 128 \, \text{mm}$$

13.3   Epicyclic gear trains

In the epicyclic gear one or more wheels is carried on an arm which can rotate about the main axis of the train. Such wheels are called *planets* and the wheel around which the planets revolve is called the *sun*.

Figure 13.8 shows an elementary epicyclic train. The centres of the planet wheel P and the sun wheel S are linked by an arm. If the sun wheel is fixed then rotation of the arm around the sun causes the planet wheel P to rotate. In order to determine the amount of this rotation a technique that can be used is to first imagine the link to be fixed while wheel S rotates through $+1$ revolution. This causes wheel P to rotate through t_s/t_p revolutions. Then we imagine the gears to be locked solid and give a rotation of -1 revolution to all the wheels and the arm about the axis through S. Now if we had taken S to be fixed with the arm rotating then the result is identical with the sum of the above two operations. Thus while the arm rotates through -1 revolution the planet gear rotates through $-(1 + t_s/t_p)$. Table 13.1 summarises the effects on the wheels.

Table 13.1

Operation	Arm	Rotation S	P
1. Fix arm and rotate S by 1 rev.	0	$+1$	$-t_s/t_p$
2. Give all -1 rev.	-1	-1	-1
Adding 1 and 2	-1	0	$-(1 + t_s/t_p)$

Figure 13.9 shows an epicyclic gear train in which A is an annulus having internal teeth and there are three planets which can rotate about pins through the arms. The planets mesh externally with the sun wheel S and internally with the annulus. There are usually three or four planets in order to improve the dynamic characteristics. The same technique can be used to establish the relative motion of the wheels and arm. First the arm is considered fixed and the annulus is rotated through $+1$ revolution. Then we consider the gears to be locked solid and all given -1 rotation about the train axis.

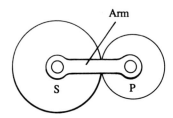

Fig. 13.8 Simple epicyclic train.

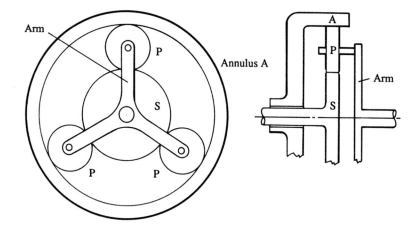

Fig. 13.9 Epicyclic train.

Table 13.2 summarises the situation. Thus -1 revolution of the arm, with the annulus fixed, results in a revolution for S of $-1-t_A/t_S$ and for P of $-1+t_A/t_P$.

Table 13.2

Operation	Rotation			
	Arm	A	S	P
1. Fix arm and rotate A by $+1$ rev.	0	$+1$	$-t_A/t_S$	$+t_A/t_P$
2. Give all -1 rev.	-1	-1	-1	-1
Adding 1 and 2	-1	0	$-1-t_A/t_S$	$-1+t_A/t_P$

Example

An epicyclic train of the form shown in figure 13.9 has a fixed annulus with 100 teeth and a sun wheel with 40 teeth. What will be the number of teeth required for a planet wheel and the gear ratio between the sun and the arm?

Referring to figure 13.9, then

$$r_S + 2r_P = r_A$$

But we must have $d_S/d_P = t_S/t_P$ and $d_P/d_A = t_P/t_A$ (see equation [1]). Thus

$$\frac{r_S}{r_P} + 2 = \frac{r_A}{r_P}$$

$$\frac{t_S}{t_P} + 2 = \frac{t_A}{t_P}$$

$$t_S + 2t_P = t_A$$

Hence

$$t_P = \tfrac{1}{2}(t_A - t_S) = \tfrac{1}{2}(100 - 40) = 30$$

To determine the rotation of the arm in relation to that of the sun wheel we can first consider the position when the arm is fixed and the annulus is given a $+1$ rotation. Then we consider what happens when the gears are locked and all the wheels are given a rotation of -1 about the main axis. The reason for this is that when we add the results of the two rotations we want to end up with the annulus having zero revolutions, since it is the fixed item. Table 13.3 shows the outcomes.

Table 13.3

Operation		Rotation		
	Arm	A	S	P
1. Fix arm and rotate A by $+1$ rev.	0	$+1$	$-100/40$	$+100/30$
2. Give all -1 rev.	-1	-1	-1	-1
Adding 1 and 2	-1	0	-3.5	$+2.3$

Thus a rotation of -1 for the arm results in -3.5 for the sun wheel. The gear ratio is thus 3.5.

Example

Figure 13.10 shows an epicyclic gear train in which wheel A has 20 teeth and rotates at 20 rev/s, the fixed annular wheel C has 120 teeth and the other annular wheel E 100 teeth, and wheel D has 40 teeth. What is the speed of rotation of the annular wheel E? B and D rotate together.

Since $r_A + 2r_B = r_C$ and d_A/d_B and $d_C/d_B = t_C/t_B$, then

$$t_A + 2t_B = tC$$

Hence $t_B = \tfrac{1}{2}(120 - 20) = 50$

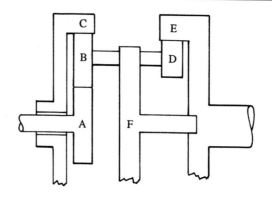

Fig. 13.10 Example.

We can now construct the table with the first condition being the arm F fixed and C given a rotation of +1. A will have a rotation of +120/20, B and D a rotation of −120/50, and E a rotation of (120/50) (40/100). Then with the gears locked the entire train is given a rotation of −1. The reason for choosing to give C a rotation of +1 and then −1 revolutions is so that when they are added they give 0. This is because C is the fixed item. Table 13.4 shows the results. Thus when A makes −7 revolutions then E makes −0.04 revolutions. Hence when A makes 20 revolution in 1 s then E makes $20 \times 0.04/7 = 0.114$ revolutions in the same direction. Thus the angular speed of E is 0.114 rev/s.

Table 13.4

| Operation | Rotation | | | | |
	A	B and D	C	E	F
1. Fix arm and rotate C by +1 rev.	−6	+2.4	+1	+0.96	0
2. Give all −1 rev.	−1	−1	−1	−1	−1
Adding 1 and 2	−7	+1.4	0	−0.04	−1

Example

The epicyclic train shown in figure 13.11 has shaft A rotating at 600 rev/min clockwise while the annulus D rotates at 300 rev/min anticlockwise. What is the angular speed of the shaft connected to arm F? Wheel A has 40 teeth, wheel B 25 teeth and annulus D 80 teeth. B and C rotate together.

Since $r_A + 2r_B = r_E$ and $d_A/d_B = t_A/t_B$ and $d_E/d_B = t_E/t_B$, then

$$t_A + 2t_B = t_E$$

Hence $t_E = 40 + 2 \times 25 = 90$. Since $r_D = r_A + r_B + r_C$ then

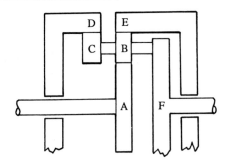

Fig. 13.11 Example.

$$t_D = t_A + t_B + t_C$$

Hence $t_C = 80 - 40 - 25 = 15$.

We can now construct the table with the first condition of the arm F fixed and wheel A rotated by $+n$ revolutions. B and C will rotate by $-(40/25)n$ revolutions, D by $-(40/25)(15/80)n$ and E by $-(40/90)n$. Then with all the gears locked the entire train is given a rotation of m. Table 13.5 shows the above results. Thus if we have A rotating at 600 rev/min then

$$600 = m + n$$

and if D rotates at -300 rev/min then

$$-300 = m - 0.3n$$

Subtracting the equations gives $n = 692.3$ and so $m = -92.3$. Hence shaft F rotates at -92.3 rev/min.

Table 13.5

Operation	Rotation				
	A	B and C	D	E	F
1. Fix arm and rotate A by $+n$ rev.	$+n$	$-1.6n$	$-0.3n$	$-0.44n$	0
2. Give all m rev.	$+m$	$+m$	$+m$	$+m$	$+m$
Adding 1 and 2	$m+n$	$m-1.6n$	$m-0.3n$	$m-0.44n$	$+m$

13.3.1 *Compound epicyclic trains*

The term compound or multi-stage is used for epicyclic trains where there are two or more co-axial simple epicyclic gear trains with members forming parts of two consecutive trains. The following example and figure 13.12 deals with a compound train involving two stages with the planet of one stage being attached to and rotating with the annulus of the second stage.

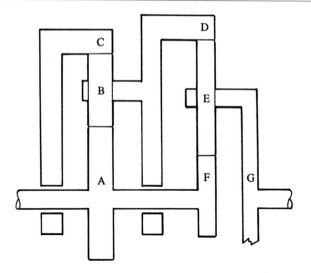

Fig. 13.12 Two stage epicyclic train.

Example

Figure 13.12 shows a two stage epicyclic train. The input shaft is connected to the sun wheels A and F, with A having 50 teeth and F 25 teeth. The output shaft is connected to the arm for the planet E. The annulus C has 90 teeth and the annulus D 125 teeth. Annulus D forms the arm for planet B. What is the rate of revolution of the output shaft when the input shaft rotates at 400 rev/min and annulus C is fixed?

Consider just the first stage. Since C is to be fixed we consider giving C a rotation of $+n$ revolutions and then all a rotation of $-n$ revolution, so that when we add the two we have C at rest. Table 13.6 is the result of considering the rotations.

Table 13.6

Operation	Rotation		
	A	C	D
1. Fix arm and rotate C by $+n$ rev.	$-1.8n$	$+n$	0
2. Give all $-n$ rev.	$-n$	$-n$	$-n$
Adding 1 and 2	$-2.8n$	0	$-n$

Now consider the second stage. No item is fixed. Table 13.7 is the result of considering the relevant rotations.

Table 13.7

Operation		Rotation	
	D	F/A	G
1. Fix arm and rotate			
D by $+m$ rev.	$+m$	$-5m$	0
2. Give all p rev.	$+p$	$+p$	$+p$
Adding 1 and 2	$m+p$	$p-5m$	$+p$

The speeds given by the two tables must be the same. Thus we must have, for D

$$-n = m + p$$

and for F/A

$$-2.8n = p - 5m$$

Thus, subtracting the equations gives $m = 0.3n$ and hence $p = -1.3n$.

If we have A rotating at 400 rev/min, then putting $-2.8n = 400$ gives for G the rotation of $+p = -1.3n = 185.7$ rev/min.

13.4 Torques on gear trains

Consider a gear system which has an input torque of T_i at an angular velocity ω_i and an output torque, i.e. the resisting torque on the output shaft, of T_o at an angular velocity ω_o.

The input torque is in the same direction as the angular rotation, the output resistive torque is however in the opposite direction to the output angular rotation. Figure 13.13(a) shows the situation when the input and output shafts rotate in the same directions and figure 13.13(b) when they rotate in opposite directions. If there is to be no angular acceleration of the system then the net torque on the gear system must be zero and so we must have a holding torque T_h which prevents the gear system as a whole rotating. The direction of this holding torque will depend on the sizes and directions of the input and output torques. Then

$$T_i + T_o + T_h = 0 \qquad [11]$$

The input power is $T_i\omega_i$ and the output power is $-T_o\omega_o$. The minus sign is because the output reaction torque acts in the opposite sense to the angular velocity ω_o. The transmission efficiency ζ is the ratio of the output power to the input power. Thus

$$\zeta = \frac{\text{output power}}{\text{input power}} = \frac{-T_o\omega_o}{T_i\omega_i}$$

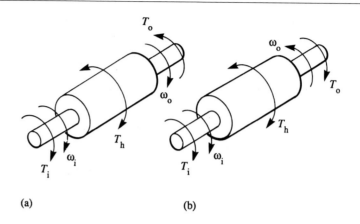

Fig. 13.13 Torques on gear trains.

Hence

$$T_o\omega_o + \zeta T_i\omega_i = 0 \tag{12}$$

Example

What is the holding torque necessary for a gear box when the input power is 10 kW at 40 rev/s, the gear ratio is a reduction of 2 to 1 and the efficiency is 98%?

The input power is equal to $T_i\omega_i$. Hence

$$10 \times 10^3 = T_i \times 2\pi \times 40$$

Thus $T_i = 39.8$ N m. The output angular speed is $40/2 = 20$ rev/s. Hence, using equation [12],

$$T_o\omega_o + \zeta T_i\omega_i = 0$$
$$T_o \times 20 \times 2\pi + 0.98 \times 10\,000 = 0$$

Hence $T_o = -78.0$ N m. Since the sum of the torques on the system must be zero, the holding torque is thus given by (equation [11])

$$T_i + T_o + T_h = 0$$
$$39.8 - 78.0 + T_h = 0$$

Hence $T_h = 38.2$ N m. The direction of the holding torque is the same as the input torque.

13.4.1 *Torque to accelerate geared systems*

With a gear system the shafts rotate in a fixed speed ratio. Thus for two shafts A and B we have the gear ratio $G = \omega_B/\omega_A$. If the angular speed of one shaft changes then the angular speed of the other shaft must change to maintain the same relationship. Thus if one has an angular acceleration α_A then it follows that the angular acceleration α_B of the other shaft must be related by $G = \alpha_B/\alpha_A$. If I_A is the moment of inertia associated with shaft A and I_B that with shaft B, then the torque on A to accelerate it at α_A is $I_A\alpha_A$ and that on B to accelerate it at α_B is $I_B\alpha_B$. The power needed to accelerate the system to give these angular accelerations must therefore be

$$P = T_A\omega_A + T_B\omega_B = I_A\alpha_A\omega_A + I_B\alpha_B\omega_B = I_A\alpha_A\omega_A + I_BG\alpha_AG\omega_A$$
$$= (I_A + G^2I_B)\alpha_A\omega_A \qquad [13]$$

If this power is produced as a result of applying a torque T_A to shaft A then the power is $T_A\omega_A$. Hence

$$T_A = (I_A + G^2I_B)\alpha_A \qquad [14]$$

The term $(I_A + G^2I_B)$ is called the *equivalent inertia* I_e referred to shaft A. Thus

$$T_A = I_e\alpha_A \qquad [15]$$

The above analysis could be extended to apply to a system in which there are n shafts geared together. Then

$$I_e = I_A + \Sigma G_x^2I_x \qquad [16]$$

where I_x is the moment of inertia of the xth shaft and G_x the ratio of the speed of the xth shaft to that of shaft A.

In, for example, problems involving hoists the moments of inertia of the various gears may be reduced to an equivalent moment of inertia on the motor shaft or drum shaft. The torque then obtained by the use of equation [14] is the torque required for the acceleration of the gears and is in addition to that required to overcome external resisting torques.

Now consider the two shafts A and B with a torque T_A applied to A and in addition a resisting torque T_B applied to B (it is resisting because it is in the direction which is in the opposite sense to that of ω_B). The net power absorbed by the system is now

$$P = T_A\omega_A - T_B\omega_B = T_A\omega_A - T_BG\omega_A = (T_A - GT_B)\omega_A$$

Hence equation [13] now becomes

$$(T_A - GT_B)\omega_A = (I_A + G^2I_B)\alpha_A\omega_A$$

and so

$$T_A - GT_B = (I_A + G^2I_B)\alpha_A \qquad [17]$$

The resistive torque acting on shaft B has thus been effectively transferred to shaft A by multiplying it by the speed ratio between the two shafts.

Example

Consider the hoist system shown in figure 13.14. A motor rotates wheel A, which is geared to wheel B and hence lifts the load by a rope wrapping

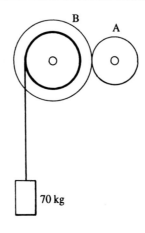

Fig. 13.14 Example.

round the drum attached to wheel B. What motor torque is required to lift a load of 70 kg with an acceleration of $1.0 \, \text{m/s}^2$? The drum has a radius of 200 mm, wheel B has 100 teeth and wheel A 20 teeth. The moment of inertia of the motor shaft and its wheel is $0.8 \, \text{kg} \, \text{m}^2$ and the moment of inertia of the rope drum and wheel is $10 \, \text{kg} \, \text{m}^2$.

The angular acceleration of wheel B is that of the drum which is $\alpha_B = a/r = 1.0/0.200 = 5.0 \, \text{rad/s}^2$. The angular acceleration of wheel A is $\alpha_B/G = 5.0/(20/100) = 25.0 \, \text{rad/s}^2$. The equivalent moment of inertia of the geared system referred to A is

$$I_e = I_A + G^2 I_B = 0.8 + (20/100)^2 \times 10 = 1.2 \, \text{kg} \, \text{m}^2$$

Hence the torque required to accelerate the gears is

$$T = I_e \alpha_A = 1.2 \times 25.0 = 30 \, \text{N} \, \text{m}$$

The torque required on B to accelerate the load is that due a net force of $(mg + ma)$ at radius r, i.e. a torque of

$$T = (mg + ma)r$$
$$= (70 \times 9.8 + 70 \times 1.0) \times 0.200 = 151.2 \, \text{N} \, \text{m}$$

The torque required at A is thus $151.2 \times (20/100) = 30.2 \, \text{N} \, \text{m}$, being obtained from $T_B \omega_B = T_A \omega_A$. Thus the total torque required is $60.2 \, \text{N} \, \text{m}$.

Example

A hoist is used to lift a load of 200 kg by means of a rope wound round a 1.0 m diameter drum. The drum has a mass of 30 kg and a radius of gyration of 450 mm with a bearing friction torque on the drum shaft of 75 N m. The drum is driven by the gear train shown in figure 13.15 and has an angular speed one-quarter that of the intermediate shaft of the compound gear train. The rotating parts of the intermediate shaft have a moment of inertia

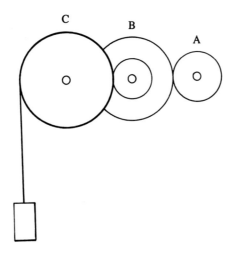

Fig. 13.15 Example.

of $2 \, \text{kg m}^2$ and experience a frictional torque of $20 \, \text{N m}$. The rotating parts of the motor have a moment of inertia of $0.8 \, \text{kg m}^2$ and experience a frictional torque of $4 \, \text{N m}$. If the motor exerts a torque of $60 \, \text{N m}$, what is the gear ratio between the intermediate shaft and the motor for the maximum acceleration of the load?

The equivalent moment of inertia of the gear system at the motor shaft is given by equation [16] as

$$I_e = I_A + \Sigma G_x^2 I_x$$
$$= 0.8 + (\omega_C/\omega_A)^2 \times 30 \times 0.450^2 + (\omega_B/\omega_A)^2 \times 2$$
$$= 0.8 + (\omega_C/\omega_B)^2 (\omega_B/\omega_A)^2 \times 30 \times 0.450^2 + (\omega_B/\omega_A)^2 \times 2$$
$$= 0.8 + 2.38G^2$$

where $G = \omega_B/\omega_A$.

The angular acceleration α_C of the drum is $a/r = a/0.5$, where a is the linear upward acceleration of the load. The angular acceleration of the intermediate shaft α_B is $4\alpha_C = 8a$. The angular acceleration of the motor shaft α_A is $\alpha_B/G = 8a/G$. Thus the torque required to accelerate the gear train is

$$T = I_e\alpha_A = (0.8 + 2.38G^2)(8a/G)$$

The equivalent frictional torque at the intermediate shaft of that at the drum shaft is given by $T_C\omega_C = T_B\omega_B$ as $T_B = 75 \times \frac{1}{4}$. The torque of $75 \times \frac{1}{4} + 20$ at B has then an equivalent at A of $(75 \times \frac{1}{4} + 20)G$ and so the total equivalent torque at the motor shaft is

$$T_e = 4 + (20 + 75 \times \tfrac{1}{4})G = 4 + 38.75G$$

The tension in the rope supporting the load is $m(g + a)$, where a is the linear acceleration upwards of the load. The torque due to this on the drum shaft C is $m(g + a)r$. The torque due to this on the intermediate shaft is $m(g + a)r \times \frac{1}{4}$. The torque on the motor shaft due to this is thus

$$m(g + a)r \times \tfrac{1}{4}G = 200(9.81 + a) \times 0.5 \times \tfrac{1}{4}G = 25(9.81 + a)G$$

Thus the total torque acting on the motor shaft is

$$60 = (0.8 + 2.38G^2)\,(8a/G) + 4 + 38.75G + 25(9.81 + a)G$$
$$= a(6.4/G + 29.76G) + 4 + 284G$$

Hence

$$a = \frac{56 - 284G}{29.76G + 6.4/G} = \frac{56G - 284G^2}{29.76G^2 + 6.4}$$

For maximum acceleration $da/dG = 0$. This means

$$(29.76G^2 + 6.4)\,(56 - 568G) - (56G - 284G^2) \times 59.52G = 0$$
$$- 1666G^2 - 3635G + 358.4 = 0$$

Hence $G = 0.0945$. This means that the angular speed of the motor shaft is $1/0.0945 = 10.6$ times that of the intermediate shaft.

13.4.2   Acceleration of combined linear and rotational system

Consider a motor vehicle which has a total mass m, an engine with rotational parts having a moment of inertia I_e and a wheels-axle system with a moment of inertia I_w. The linear kinetic energy of the vehicle is $\frac{1}{2}mv^2$ when it is moving with a linear velocity v. The rotational kinetic energy of the rotating parts in the vehicle is, when referred to the wheels, given by equation [16] as

$$\tfrac{1}{2}(I_w + G^2 I_e)\omega_w^2$$

Since we have $v = r\omega_w$, where r is the radius of a wheel and ω_w the angular velocity, then the total kinetic energy can be written as

$$\tfrac{1}{2}(mr^2 + I_w + G^2 I_e)\omega_w^2$$

Thus the quantity within the brackets can be considered to be the equivalent inertia of the vehicle when taking into account both the linear and rotational motion and referred to the road wheels.

$$\text{Equivalent moment of inertia} = mr^2 + I_w + G^2 I_e \tag{18}$$

Example

A motor vehicle of total mass 1000 kg has road wheels of effective diameter 700 mm. The moments of inertia of the four road wheels with the rear axle is $7.0\,\text{kg m}^2$. The moment of inertia of the engine rotational parts is $0.5\,\text{kg m}^2$. The engine torque can be assumed to be constant at 160 N m. The tractive resistance is $100 + v^2\,\mathrm{N}$ when the vehicle speed is $v\,\mathrm{m/s}$. What is the gear ratio, engine to back axle, to provide maximum acceleration when the car is travelling up a slope of 1 in 10 at a speed of $10\,\mathrm{m/s^2}$?

Using equation [17], the net torque acting on the wheels is

$$\text{net torque} = G \times \text{engine torque} - \text{resistance} \times \text{wheel radius}$$

The resistance is provided by the tractive resistance of $100 + 10^2 = 200\,\mathrm{N}$ and the component of the car weight acting down the slope, namely $1000 \times 9.81 \times (1/10) = 981\,\mathrm{N}$. Hence

$$\text{net torque} = 160G - (200 + 981) \times 0.350 = 160G - 413.4$$

The equivalent moment of inertia of the system when referred to the wheels is given by equation [18] as

$$I = mr^2 + I_w + G^2 I_e$$
$$= 1000 \times 0.350^2 + 7 + 0.5G^2 = 129.5 + 0.5G^2$$

Thus the angular acceleration α_w of the wheels is given by

$$T = I\alpha_w$$
$$160G - 413.4 = (129.5 + 0.5G^2)\alpha_w$$

But $\alpha_w = a/r$, where a is the linear acceleration. Thus

$$160G - 413.4 = (129.5 + 0.5G^2)\,(a/0.350)$$

Hence

$$a = \frac{56G - 144.7}{129.5 + 0.5G^2}$$

For maximum acceleration $da/dG = 0$. Hence

$$(129.5 + 0.5G^2)56 - (56G - 144.7)1.0G = 0$$

Thus

$$28G^2 - 144.7G - 7252 = 0$$

Hence $G = 18.9$.

Example

A motor vehicle of total mass $800\,\mathrm{kg}$ has road wheels of effective diameter $600\,\mathrm{mm}$. The moments of inertia of the wheels–axle system is $6\,\mathrm{kg\,m^2}$ and of the engine rotational parts $0.5\,\mathrm{kg\,m^2}$. The engine delivers a constant torque of $140\,\mathrm{N\,m}$ and drives the axle through a 4 to 1 reduction. The resistance to motion is given by $150 + 0.8v^2\,\mathrm{N}$, where the speed v is in m/s. What is the acceleration of the vehicle along a horizontal road at a speed of $12\,\mathrm{m/s}$ and the time taken to accelerate from 12 to $20\,\mathrm{m/s}$?

Using equation [17], the net torque acting on the wheels is

$$\text{net torque} = G \times \text{engine torque} - \text{resistance} \times \text{wheel radius}$$

The resistance is provided by the tractive resistance of $150 + 0.8v^2$. Hence the net torque is

$$\text{net torque} = 140 \times 4 - (150 + 0.8v^2)0.300 = 515 - 0.24v^2$$

The equivalent moment of inertia of the system when referred to the wheels is given by equation [18] as

$$I = mr^2 + I_w + G^2 I_e$$
$$= 800 \times 0.300^2 + 6 + 0.5 \times 4^2 = 86 \, \text{kg m}^2$$

Thus the angular acceleration α_w of the wheels is given by

$$T = I\alpha_w$$
$$515 - 0.24v^2 = 86\alpha\omega = 86(a/0.300)$$

Thus, when $v = 12 \, \text{m/s}$, the acceleration $a = 1.68 \, \text{m/s}^2$.

Since $a = dv/dt$, the above expression for the acceleration can be written as

$$515 - 0.24v^2 = 286.7 \frac{dv}{dt}$$

Hence the time taken to accelerate from 12 to 20 m/s is given by

$$\int_0^t dt = \int_{12}^{20} \frac{286.7}{515 - 0.24v^2} dv = \frac{286.7}{0.24} \int_{12}^{20} \frac{1}{2146 - v^2} dv$$

$$= \frac{286.7}{0.24 \times 2 \times 46.3} \int_{12}^{20} \left(\frac{1}{46.3 + v} + \frac{1}{46.3 - v} \right) dv$$

$$t = 12.9 \left[\ln(46.3 + v) - \ln(46.3 - v) \right]_{12}^{20}$$
$$= 5.08 \, \text{s}$$

Problems

(1) The axes of the two wheels in a spur gear are 250 mm apart. If one wheel rotates three times faster than the other, what are the diameters of the pitch circles?

(2) The axes of the two wheels in a spur gear are to be 200 mm apart with one wheel rotating about four times faster than the other. If the module is to be 2.5, what are the number of teeth on each wheel and what is the velocity ratio achieved?

(3) A gear box with the input and output shafts in the same line is required to have a drive ratio of approximately 4 to 1. The gears are to have a module of 3 and the intermediate shaft is to have its centre parallel to the input and output shafts but offset by

67.5 mm. If no gear is to have less than 15 teeth, suggest suitable teeth numbers.

(4)   Two shafts X and Z are to be in the same straight line and geared through an intermediate parallel shaft Y in the way shown in figure 13.6. Wheels A and B have a module of 2 and wheels C and D a module of 4. The number of teeth on any wheel is not to be less than 15 and the speed of shaft Z is to be about but not greater than 1/10 th that of X. The ratio of the reduction at each pair of wheels is to be the same. What are the numbers of teeth on each wheel and the distance of the intermediate shaft Y from shafts X and Z?

(5)   A four speed gear box, of the form shown in figure 13.8, is required to have the nominal drive ratios of 4:1, 2.5:1, 1.5:1 and 1:1 with the input/output shaft axis 72 mm from the layshaft. If all the gears are to have a module of 3 and no wheel is to have less than 15 teeth, suggest suitable teeth numbers.

(6)   A four speed gear box, of the form shown in figure 13.8, is required to have the nominal drive ratios of 3.9:1, 2.3:1;, 1.5:1; and 1:1 with the input/output shaft axis 90 mm from the layshaft. If all the gears are to have a module of 4 and no wheel is to have less than 15 teeth, suggest suitable teeth numbers.

(7)   An epicyclic train of the form shown in figure 13.9 has a fixed annulus with 80 teeth and a sun wheel with 40 teeth. What will be the number of teeth required for a planet wheel and the gear ratio between the sun and the arm?

(8)   The epicyclic train shown in figure 13.16 has shaft A rotating at 20 rev/s. Wheel B has 27 teeth and wheel C 30 teeth, wheel B meshing with wheel D which has 24 teeth and is keyed to the output shaft and wheel C meshing with the fixed wheel E which has 21 teeth. Wheels B and C are fixed together. What is the angular speed of the output shaft?

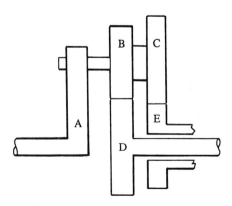

Fig. 13.16 Problem 8.

(9) The epicyclic train shown in figure 13.17 has shaft A stationary and arm B rotating at 300 rev/min. Wheel A meshes with wheel C. Wheels C and D are fixed together, wheel C having 20 teeth and wheel D 40 teeth. Wheel D meshes with wheel E which has 30 teeth. What is the rate of revolution of the shaft connected to E?

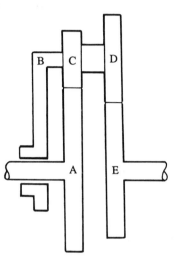

Fig. 13.17 Problem 9.

(10) The epicyclic train shown in figure 13.11 has A rotating at 800 rev/min clockwise and F rotating at 200 rev/min anticlockwise with the numbers of teeth being A 30, B 20 and D 80. What is the speed of rotation of annulus E?

(11) The two stage epicyclic train shown in figure 13.12 has the drive shaft on which wheels A and F are mounted rotating at 1000 rev/min. Wheel A has 26 teeth, wheel F 31 teeth. If the annulus C has 88 teeth and the annulus D 83 teeth, what is the speed of revolution of G?

(12) The two stage epicyclic train shown in figure 13.12 has the drive shaft on which wheels A and F are mounted rotating at 2500 rev/min. Wheel A has 19 teeth, wheel F 23 teeth. If the annulus C has 83 teeth and the annulus D 79 teeth, what is the speed of revolution of G?

(13) The two stage epicyclic train shown in figure 13.18 has shaft A rotating at 1000 rev/min. Wheel A has 24 teeth, annulus C 66 teeth, wheel E 28 teeth and annulus F 62 teeth. If annulus F is fixed, what is the rate of revolution of the output shaft D?

(14) What is the holding torque necessary for a gear box when the input

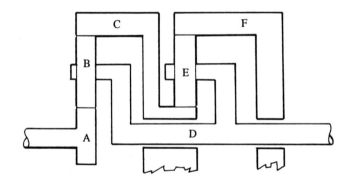

Fig. 13.18 Problem 13.

power is 2 kW at 20 rev/s, the gear ratio is a reduction of 4 to 1 and the efficiency is 98%?

(15)   What is the torque on the driven shaft and the torque on the gear box casing for a gear box which has an input power of 30 kW at + 60 rev/s and gives an output shaft rotation of + 10 rev/s if losses can be neglected?

(16)   An epicyclic gear train has an input power of 4 kW at 5 rev/s and produces a gear ratio reduction of 4 to 1. What is the torque transmitted if the efficiency is 95%?

(17)   An epicyclic gear train of the form shown in figure 13.9 has an arm which makes one revolution for every five revolutions of the sun wheel. If a torque of 20 N m is applied to the shaft carrying the sun wheel, what torque will be required to keep the annulus stationary?

(18)   For the hoist system shown in figure 13.14, a motor rotates wheel A, which is geared to wheel B and hence lifts the load by a rope wrapping round the drum attached to wheel B. What motor torque is required to lift a load of 100 kg with an acceleration of 2.0 m/s²? The drum has a radius of 200 mm, wheel B has 100 teeth and wheel A 25 teeth. The moment of inertia of the motor shaft and its wheel is 0.8 kg m² and the moment of inertia of the rope drum and wheel is 12 kg m².

(19)   For the hoist system shown in figure 13.15, the drum has a radius of 310 mm, a moment of inertia of 65 kg m² and a frictional torque of 250 N m. The intermediate shaft and wheel has a moment of inertia of 9 kg m² and a frictional torque of 43 N m. The motor shaft and wheel has a moment of inertia of 2 kg m² and a frictional torque of 20 N m. The motor shaft wheel has 36 teeth, the intermediate wheels 150 and 50 teeth, and the drum wheel 180 teeth. What is the torque required at the motor shaft to raise a load of 3000 kg at 1.5 m/s²?

(20)   A motor vehicle of mass 800 kg has road wheels of diameter 600 mm.

The engine delivers a constant torque and rotates at n times the angular velocity of the wheels. The engine has rotational parts having a moment of inertia of $0.4 \, \text{kg} \, \text{m}^2$ and the wheels–axle system has a moment of inertia of $6.0 \, \text{kg} \, \text{m}^2$. Ignoring all external resistances what value of n will give the greatest linear acceleration on a horizontal road?

(21)   A motor vehicle of total mass $800 \, \text{kg}$ has road wheels of effective diameter $600 \, \text{mm}$. The moment of inertia of the four road wheels with the rear axle is $5.0 \, \text{kg} \, \text{m}^2$. The moment of inertia of the engine rotational parts is $0.5 \, \text{kg} \, \text{m}^2$. The engine torque can be assumed to be constant at $160 \, \text{N} \, \text{m}$ and there is friction torque on the engine shaft of $20 \, \text{N} \, \text{m}$. The total axle friction torque is $25 \, \text{N} \, \text{m}$. The tractive resistance is $150 + v^2 \, \text{N}$ when the vehicle speed is $v \, \text{m/s}$. What is the gear ratio, engine to back axle, to provide maximum acceleration when the car is travelling along a horizontal road at a speed of $10 \, \text{m/s}^2$?

(22)   A motor vehicle of total mass $900 \, \text{kg}$ has road wheels of effective diameter $600 \, \text{mm}$. The moment of inertia of the four road wheels with the rear axle is $6.0 \, \text{kg} \, \text{m}^2$ and the moment of inertia of the engine rotational parts is $0.6 \, \text{kg} \, \text{m}^2$. The gear ratios provided are $4{:}1$, $2.5{:}1$, $1.5{:}1$ and $1{:}1$. If the engine torque is constant at $150 \, \text{N} \, \text{m}$, what is the maximum acceleration of the car on each gear? At what gear ratio would the acceleration be a maximum?

(23)   A motor vehicle of total mass $1000 \, \text{kg}$ has road wheels of effective diameter $720 \, \text{mm}$. The moment of inertia of the four road wheels with the rear axle is $7.0 \, \text{kg} \, \text{m}^2$ and the moment of inertia of the engine rotational parts is $0.4 \, \text{kg} \, \text{m}^2$. The engine delivers a constant torque of $150 \, \text{N} \, \text{m}$ with the frictional torque on the engine shaft being $20 \, \text{N} \, \text{m}$ and that on the axle $25 \, \text{N} \, \text{m}$. The tractive resistance is $200 + v^2 \, \text{N}$, where v is the speed in m/s. What will be the time taken by the vehicle to accelerate from (a) 0 to $10 \, \text{m/s}$ and (b) 10 to $20 \, \text{m/s}$?

(24)   A motor vehicle of total mass $900 \, \text{kg}$ has road wheels of effective diameter $600 \, \text{mm}$. The moment of inertia of the four road wheels with the rear axle is $8.5 \, \text{kg} \, \text{m}^2$ and the moment of inertia of the engine rotational parts is $0.8 \, \text{kg} \, \text{m}^2$. The engine delivers a constant torque of $135 \, \text{N} \, \text{m}$. The transmission efficiency is 90% and the tractive resistance is $130 + 0.8v^2 \, \text{N}$, where v is in m/s. What will be the gear ratio to give maximum acceleration up a gradient of 1 in 20 when the vehicle is travelling at $20 \, \text{m/s}$?

Chapter 14
Friction clutches, bearings and belt drives

14.1 Friction

Consider a simple problem, that of pushing a box across the floor. Because of friction, a certain force has to be applied before slip starts. This is because up to that force the friction force is opposite and equal to the applied force and there is no resultant force. There is however a maximum, a so-called limiting, value for the frictional force. Slip does not occur until this force has been reached. Only then is there a resultant force.

The greater the coefficient of friction between the two surfaces in contact the greater the value of this limiting frictional force (see section 11.1.3). Thus if the requirement is for as little a force as possible to be needed to move the box over the floor then the requirement is for a low coefficient of friction. If the requirement is that the box does not move relative to the floor then the requirement is for a high coefficient of friction.

With clutches and belts the basic requirement is that there is no slip between two surfaces when a torque or force is applied to one. A high coefficient of friction is thus required. The situation in these cases is rather like pushing the box when it is resting on a trolley. The aim is that the box together with the trolley moves and that the box does not slide across the surface of the trolley. With bearings the basic requirement is that frictional torques and forces are kept to a minimum. A low coefficient of friction is thus required or better still, the surfaces are kept apart by means of a lubricant.

14.2 Friction clutches

A clutch is a device for connecting and disconnecting the drive between two coaxial shafts. Friction clutches may be in the form of flat plates or conical surfaces. But whatever the form, the principle is the same. Connecting the two coaxial shafts involves bringing together two surfaces and holding them together. The requirement is that when one side of the clutch is made to rotate then the junction behaves as a solid coupling and transmits torque to the other shaft.

14.2.1 Plate clutches

Consider the forces occurring when two flat circular plates are in contact

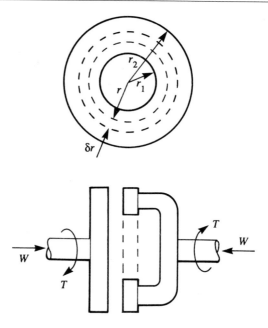

Fig. 14.1 Forces for flat surfaces pressed together.

and pressed together by an axial force W, as illustrated in figure 14.1. For a ring segment of the surfaces in contact at radius r and thickness δr, if the pressure exerted by one plate on the other is p then the axial force on the ring is the product of the pressure and the area over which it is acting and so

axial force on ring $= p2\pi r\delta r$

Thus the total axial force on the two surfaces in contact is

$$W = 2\pi \int_{r_1}^{r_2} pr\,dr \qquad [1]$$

The frictional force F is related to the normal reaction force R by $F = \mu R$ (equation [8], chapter 11). Thus the frictional force on the ring segment is

frictional force $= \mu p2\pi r\delta r$

This frictional force is tangential to the ring. Thus the moment of tangential force acting on the ring is

moment of tangential force $= (\mu p2\pi r\delta r)r$

Hence the total torque T on the shaft is

$$T = 2\pi\mu \int_{r_1}^{r_2} pr^2\,dr \qquad [2]$$

The integrals in equations [1] and [2] can be evaluated if some assumptions are made regarding the variation of the pressure p with radius r. Two cases are considered, one where it is assumed that the pressure is constant and does not vary with radius and one where it is assumed that there is uniform wear over the contact area. The uniform pressure assumption is particularly relevant to unworn clutches, while the uniform wear assumption seems to be more appropriate to worn clutches.

For *uniform pressure*, equation [1] thus gives

$$W = 2\pi \int_{r_1}^{r_2} pr\,dr = 2\pi p \int_{r_1}^{r_2} r\,dr = \pi p(r_2^2 - r_1^2) \qquad [3]$$

Equation [2] gives

$$T = 2\pi\mu \int_{r_1}^{r_2} pr^2\,dr = 2\pi\mu p \int_{r_1}^{r_2} r^2\,dr = \frac{2}{3}\pi\mu p(r_2^3 - r_1^3) \qquad [4]$$

Eliminating p between equations [3] and [4] give for the relationship between the torque and the load

$$T = \frac{2\mu W}{3}\left(\frac{r_2^3 - r_1^3}{r_2^2 - r_1^2}\right) \qquad [5]$$

For *uniform wear*, we assume that the wear is proportional to the product of the pressure and the velocity and so, since the velocity is proportional to the radius,

$$pr = k$$

where k is a constant. Equation [1] thus becomes

$$W = 2\pi \int_{r_1}^{r_2} pr\,dr = 2\pi k \int_{r_1}^{r_2} dr = 2\pi k(r_2 - r_1) \qquad [6]$$

Equation [2] becomes

$$T = 2\pi\mu \int_{r_1}^{r_2} pr^2\,dr = 2\pi\mu k \int_{r_1}^{r_2} r\,dr = \pi\mu k(r_2^2 - r_1^2) \qquad [7]$$

Eliminating k from equations [6] and [7] gives

$$T = \tfrac{1}{2}\mu W(r_2 + r_1) \qquad [8]$$

The torque given by equation [8] is less than that given by equation [5]. If however, r_1/r_2 is much greater than $\tfrac{1}{4}$ then the difference between the two equations is small. It should also be recognised that there is a significant degree of uncertainty concerning the value of the coefficient of friction.

The discussion above involved just two surfaces coming into contact. The construction used however for the single plate clutch doubles the area of frictional surfaces in contact by using both sides of each plate. A consequence of this is that the torque given by equations [5] and [8] has to be doubled. Figure 14.2 shows the basic form of the clutch used in motor vehicles. The centre plate is able to move along the splines on the end of one shaft. When the clutch is engaged the friction liners press against both sides of the centre plate and hence it rotates and so does the splined shaft. When the clutch is

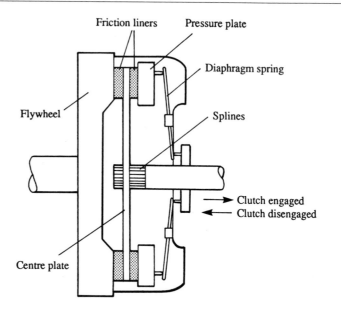

Fig. 14.2 Diaphragm spring flat plate clutch.

disengaged the pressure forcing the friction liners against the centre plate is removed. This pressure which forces the liners against the centre plate is provided by a diaphragm spring.

In a multi-plate clutch, not only are both sides of the plates used but also there are a number of such plates attached to each shaft. The total torque is thus increased by a factor of $2n$ where n is the number of plates. Multi-plate clutches have the advantage of being more compact than single plate clutches.

Example

Two coaxial shafts A and B are connected by a single plate friction clutch of internal diameter 120 mm and external diameter 200 mm, both sides of the plate being used. Shaft A is rotating at a constant angular speed of 200 rev/min and B is initially stationary. Shaft B has an effective moment of inertia of 20 kg m². When the clutch is engaged it takes 5 s for B to reach full speed. If the coefficient of friction is 0.3 and the uniform pressure condition can be assumed, what is the required spring force?

The angular acceleration of B is $(200 \times 2\pi/60)/5 = 4.19 \, \text{rad/s}^2$. Thus the torque acting on B is $I\alpha = 20 \times 4.19 = 83.8 \, \text{N m}$. Thus equation [5] gives, when it is assumed that both sides of each plate are involved,

$$T = 83.8 = 2 \times \frac{2\mu W}{3} \left(\frac{r_2^3 - r_1^3}{r_2^2 - r_1^2} \right)$$

$$= 2 \times \frac{2 \times 0.3W}{3} \left(\frac{0.100^3 - 0.060^3}{0.100^2 - 0.060^2} \right)$$

Hence $W = 1710\,\text{N}$.

Example

A multi-plate clutch has four pairs of frictional surfaces, each of internal diameter 120 mm and external diameter 200 mm. If the coefficient of friction is 0.3 and the uniform pressure condition can be assumed, what is the required spring force when the plates transmit 20 kW at 1400 rev/min?

Since the power is $T\omega$, then the torque is

$$T = \frac{20 \times 10^3 \times 60}{1440 \times 2\pi} = 132.6\,\text{N m}$$

For uniform pressure, equation [5] gives, for four pairs of surfaces,

$$T = 132.6 = 4 \times \frac{2\mu W}{3} \left(\frac{r_2^3 - r_1^3}{r_2^2 - r_1^2} \right)$$

$$= 4 \times \frac{2 \times 0.3W}{3} \left(\frac{0.100^3 - 0.060^3}{0.100^2 - 0.060^2} \right)$$

Hence $W = 1353\,\text{N}$.

14.2.2 Cone clutches

A cone clutch has a single pair of conical friction faces, as illustrated in figure 14.3. Consider the forces on a ring segment of radius r and radial thickness δr. If p is the normal pressure between the surfaces, then the normal force is $p \times 2\pi r \delta r / \sin \alpha$. The axial component of this force is

$$\text{axial force} = (p \times 2\pi r \delta r / \sin \alpha) \sin \alpha = p \times 2\pi r \delta r$$

Thus the total axial force W is

$$W = 2\pi \int_{r_1}^{r_2} pr\,\mathrm{d}r \tag{9}$$

The frictional force on the ring element is μ multiplied by the normal reaction force and is thus $\mu p 2\pi r \delta r / \sin \alpha$. The moment of this force about the axis is $(\mu p 2\pi r \delta r / \sin \alpha)r$. Thus the torque transmitted is

$$T = \frac{2\pi\mu}{\sin \alpha} \int_{r_1}^{r_2} pr^2\,\mathrm{d}r \tag{10}$$

Assuming *uniform pressure*, equation [9] becomes

$$W = \pi p (r_2^2 - r_1^2) \tag{11}$$

and equation [10]

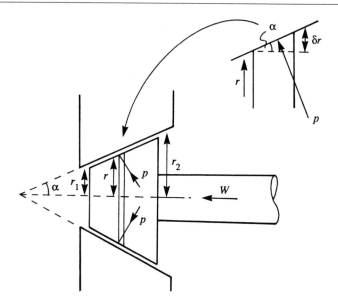

Fig. 14.3 Conical clutch.

$$T = \frac{2\pi\mu p}{3\sin\alpha}\,(r_2^3 - r_1^3) \qquad\qquad [12]$$

Eliminating p between equations [11] and [12] gives

$$T = \frac{2\mu W}{3\sin\alpha}\left(\frac{r_2^3 - r_1^3}{r_2^2 - r_2^2}\right) \qquad\qquad [13]$$

Assuming *uniform wear*, i.e. $pr = k$, equation [9] becomes

$$W = 2\pi k(r_2 - r_1) \qquad\qquad [14]$$

and equation [10]

$$T = \frac{\pi\mu k}{\sin\alpha}\,(r_2^2 - r_1^2) \qquad\qquad [15]$$

Eliminating k between equations [14] and [15] gives

$$T = \frac{\mu W}{2\sin\alpha}\,(r_1 + r_2) \qquad\qquad [16]$$

The equations for the cone clutch are the same as those for the flat plate clutch if an equivalent coefficient of friction of $\mu/\sin\alpha$ is used.

Example

A cone clutch has a bearing surface with a mean diameter of 300 mm and a cone angle of 15°. The coefficient of friction 0.3 and the axial width of conical bearing surface is 100 mm. If the maximum normal pressure is to be

50 kPa what is the greatest power that can be transmitted at 20 rev/s without the clutch slipping? Assume the constant wear condition.

Using the notation given in figure 14.3, then if w is the axial width of the bearing surface

$$r_2 - r_1 = w \sin \alpha = 100 \sin 15° = 25.88 \, \text{mm} \qquad [17]$$

Also

$$\tfrac{1}{2}(r_2 + r_1) = 150 \, \text{mm}$$

$$r_2 + r_1 = 300 \, \text{mm} \qquad [18]$$

Subtracting equation [17] from [18] gives $r_1 = 137.1 \, \text{mm}$.

For the constant wear condition then $pr = k$. The maximum value of the pressure will occur at the minimum value of the radius, i.e. r_1. Thus

$$k = 50 \times 10^3 \times 0.1371 = 6.855 \times 10^3$$

Equation [15] thus gives, using the above equation and equations [17] and [18],

$$
T = \frac{\pi \mu k}{\sin \alpha}(r_2^2 - r_1^2) = \frac{\pi \mu k (r_2 + r_1)(r_2 - r_1)}{\sin \alpha}
$$

$$
= \frac{\pi \times 0.3 \times 6.855 \times 10^3 \times 0.300 \times 0.02588}{\sin 15°}
$$

$$
= 193.8 \, \text{N m}
$$

The power is

$$P = T\omega = 193.8 \times 20 \times 2\pi = 24.4 \, \text{kW}$$

14.3 Bearings

Rotating shafts are often subject to an axial thrust and a bearing surface is required to take this thrust and maintain the shaft in its correct axial position. Relative motion between the contact surfaces is resisted by the frictional forces between the surfaces. The requirement of such a thrust bearing is to keep the friction torque to a minimum. Figure 14.4 shows four types of such bearings.

There are also situations where it is necessary to support load carrying shafts at right angles to the axis of rotation of the shaft. The term *journal bearing* is used where a shaft is supported by a concentric bearing, such a form of bearing being discussed in section 14.3.3.

The term *plain bearing* is used where the bearing surfaces concerned are plane and slide past each other. In section 14.3.4 rolling friction and ball bearings are discussed.

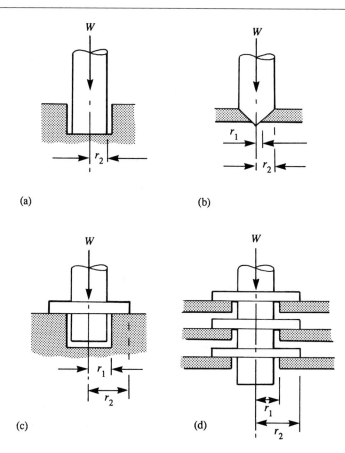

Fig. 14.4 Plain bearings, (a) footstep, (b) conical, (c) collar, (d) multi-
collar.

14.3.1 Dry plain bearings

Consider plain bearings when dry, i.e. no fluid lubrication (see 14.3.2 and
14.3.5). In such a situation the coefficient of friction between the surfaces
can be assumed to be constant and the frictional torque is given by the same
equations as were developed for friction clutches. As with friction clutches
the axial load W produces a pressure at the contacting surfaces. The relationship between the torque and the axial load will depend on the distribution
of pressure across the contacting surfaces. If the pressure is assumed to be
uniform across the surfaces then the uniform pressure equation [5] can be
used for the footstep and collar bearings equation [13]. If constant wear is
assumed then for the footstep and collar bearings equation [8] can be used,
and for the conical bearing equation [16].

Example

A vertical shaft of diameter 60 mm runs in a footstep bearing at 600 rev/min. The pressure between the bearing surfaces may be assumed to be constant at 120 kPa. The coefficient of friction between the surfaces is 0.05. What is the power loss due to friction?

The axial load W is

$$W = pA = p\pi r_2^2 = 120 \times 10^3 \times \pi \times 0.030^2 = 339.3 \, \text{N}$$

The frictional torque T is given by equation [5] as

$$T = \frac{2\mu W}{3} \left(\frac{r_2^3 - r_1^3}{r_2^2 - r_1^2} \right)$$

and since $r_1 = 0$, then

$$T = \frac{2 \times 0.05 \times 339.3 \times 0.030}{3} = 0.339 \, \text{N m}$$

Thus the power loss is

$$P = T\omega = 0.339 \times 2\pi \times 10 = 21.3 \, \text{W}$$

Example

A flat collar bearing has an internal diameter of 50 mm and an external diameter of 80 mm. The pressure exerted on the bearing surfaces may be assumed to be constant at 120 kPa. The coefficient of friction between the surfaces is 0.1. What is the power loss at 360 rev/min in overcoming friction at the bearing?

The axial load W is

$$W = pA = p\pi(r_2^2 - r_1^2)$$
$$= 120 \times 10^3 \times \pi(0.040^2 - 0.025^2) = 367.6 \, \text{N}$$

The frictional torque T is given by equation [5] as

$$T = \frac{2\mu W}{3} \left(\frac{r_2^3 - r_1^3}{r_2^2 - r_1^2} \right)$$
$$= \frac{2 \times 0.1 \times 367.6}{3} \left(\frac{0.040^3 - 0.025^3}{0.040^2 - 0.025^3} \right) = 1.216 \, \text{N m}$$

The power loss is thus

$$P = T\omega = 1.216 \times 2\pi \times 6 = 45.8 \, \text{W}$$

Example

A multi-collar bearing has flat collars with external diameters of 120 mm and internal diameters of 80 mm. The pressure exerted on the bearing surfaces may be assumed to be constant at 200 kPa. If the greatest thrust to be taken by a bearing is 5 kN, how many collars will be required?

The bearing surface of one collar has an area of

$$\text{area } A = \pi(0.060^2 - 0.040^2) = 6.283 \times 10^{-3} \, \text{m}^2$$

The total load carried by n collars is

$$W = npA = n \times 200 \times 10^3 \times 6.283 \times 10^{-3} = 1257n$$

But the maximum load is 5 kN, hence

$$1257n = 5 \times 10^3$$

Hence $n = 3.98$ and the number of collars required is 4.

14.3.2 Lubricated plain bearings

With dry friction a constant value for the coefficient of friction may be assumed. However, if the bearing is lubricated this is not the case.

Consider the situation where the moving surfaces are separated by a continuous film of lubricant, such an arrangement being referred to as *film* or *fluid lubrication*. In such a case the resistance to motion arises solely from the viscosity of the lubricant. The relative movement of two parallel surfaces results in the fluid between them being subject to shear (figure 14.5). As a consequence a velocity gradient is produced, the velocity gradient being the viscous strain rate. The viscous resistance per unit area is the viscous stress. According to Newton's law of viscous flow the shear stress is proportional to the viscous strain rate, the constant of proportionality being called the *coefficient of viscosity* η.

$$\text{Viscous resistance/unit area} = \eta \, \frac{dv}{dx} \qquad [19]$$

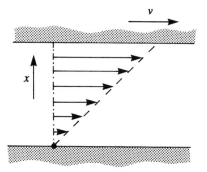

Fig. 14.5 Viscous flow.

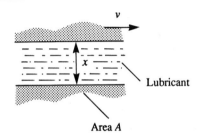

Fig. 14.6 Parallel bearings.

The coefficient of viscosity has the units of $\mathrm{N\,s/m^2}$ and depends on the temperature.

For *parallel bearings* (figure 14.6) we have two flat surfaces, area A, between which there is a constant thickness layer of lubricant, then if we assume that the velocity gradient is uniform we have $dv/dx = v/x$ and equation [19] gives

$$\text{viscous resistance force } F = \eta\,\frac{v}{x}\,A$$

The power loss as a result of this viscous resistance is

$$P = Fv = \frac{\eta v^2 A}{x} \tag{20}$$

Consider a *collar bearing* of external radius r_2 and internal radius r_1 where the shaft is supported on a film of lubricant of thickness x (figure 14.7). The velocity of the collar at a radius r is $v = r\omega$, where ω is the angular velocity of the shaft. Thus if we consider an annular element of radius r and thickness δr the viscous resistance force F is given by equation [19] as

$$F = \eta\,\frac{dv}{dx}\,A = \eta\,\frac{r\omega}{x}\,2\pi r\,\delta r$$

Hence the torque on the element is

$$\text{torque on element} = \eta\,\frac{r\omega}{x}\,2\pi r\,\delta r \times r = \frac{2\pi\eta\omega}{x}\,r^3\,dr$$

The total torque T on the collar is thus

$$T = \frac{2\pi\eta\omega}{x}\int_{r_1}^{r_2} r^3\,dr = \frac{\pi\eta\omega}{2x}\,(r_2^4 - r_1^4) \tag{21}$$

The power loss is thus

$$P = T\omega = \frac{\pi\eta\omega^2}{2x}\,(r_2^4 - r_1^4) \tag{22}$$

For a *footstep bearing* of radius r_2 (see figure 14.4(a)) equations [21] and [22] become, with $r_1 = 0$,

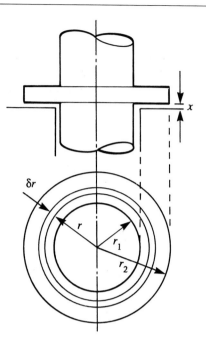

Fig. 14.7 Collar bearing.

$$T = \frac{\pi \eta \omega r_2^4}{2x} \tag{23}$$

$$P = \frac{\pi \eta \omega^2 r_2^4}{2x} \tag{24}$$

The above assumes that the viscous resistance with the footstep bearing is due entirely to the lubricant between the base of the shaft and its housing and that between the sides of the shaft and the housing is ignored.

Example

A collar bearing has an inner diameter of 60 mm and an outer diameter of 120 mm and runs in its housing on a film of lubricant of thickness 0.3 mm. If the viscosity of the lubricant is 0.1 N s/m², what is the power loss at the bearing when the shaft is rotating at 1200 rev/min?

Using equation [22]

$$P = \frac{\pi \eta \omega^2}{2x} (r_2^4 - r_1^4)$$

$$= \frac{\pi \times 0.1 \times (2\pi \times 20)^2}{2 \times 0.3 \times 10^{-3}} (0.060^4 - 0.030^4) = 100.5\,\text{W}$$

14.3.3 *Journal bearings*

Consider an unlubricated journal bearing in which a shaft of radius r is supported by a concentric bearing, as in figure 14.8. If the radial load between the shaft and the bearing is W then there is a tangential frictional force of μW and hence a frictional torque of

$$T = \mu W r \tag{25}$$

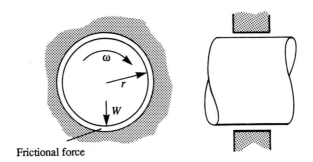

Frictional force

Fig. 14.8 Journal bearing.

The power loss due to friction, when the shaft has an angular velocity ω, is thus

$$P = T\omega = \mu W r \omega \tag{26}$$

For a lubricated journal bearing we will make the simplifying assumption that the radial clearance c is constant (in reality the lubricant film will have a varying thickness). Then the viscous resistance force is, according to equation [19],

$$F = \eta \frac{\mathrm{d}v}{\mathrm{d}x} A = \eta \frac{r\omega}{c} 2\pi r L = \frac{2\pi\eta\omega r^2 L}{c}$$

where L is the length of the bearing. The torque is thus

$$T = Fr = \frac{2\pi\eta\omega r^3 L}{c} \tag{27}$$

The power loss is thus

$$P = T\omega = \frac{2\pi\eta\omega^2 r^3 L}{c} \tag{28}$$

Example

A journal bearing has a diameter of 50 mm and a length of 40 mm. It carries a radial load of 4 kN and the shaft runs at 1200 rev/min. What is the power loss due to (a) friction when the bearing is unlubricated and the coefficient of friction is 0.05 and (b) viscous resistance when the bearing is lubricated with oil of viscosity $0.1\,N\,s/m^2$ and there is a constant radial clearance of 0.2 mm?

(a) Using equation [26]

$$P = \mu Wr\omega = 0.05 \times 4 \times 10^3 \times 0.025 \times 2\pi \times 20 = 628\,W$$

(b) Using equation [28]

$$P = \frac{2\pi\eta\omega^2 r^3 L}{c} = \frac{2\pi \times 0.1 \times (2\pi \times 20)^2 \times 0.025^3 \times 0.040}{0.2 \times 10^{-3}}$$

$$= 31\,W$$

14.3.4 Ball bearings

Friction only occurs when one body slides or tends to slide over another. If one body rolls over another then ideally there is no relative sliding at the point of contact and therefore no friction. However, in practice a pure rolling motion does not occur. The surfaces of the materials are always deformed to some extent so that the point of contact degenerates into an area of contact (figure 14.9). It is this deformation which gives rise to rolling friction. The harder the materials the less will be the deformation and consequently the smaller the area of contact and so the smaller the friction force. Think of the effort required to pull a garden roller over a lawn, the lawn deforms and the softer the lawn the greater the effort required.

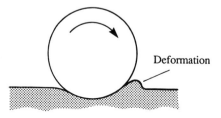

Deformation

Fig. 14.9 Rolling friction.

The replacement of sliding friction by rolling friction can result in a considerable reduction in frictional resistance. Figure 14.10 shows the basic form of a ball bearing that can be used to support a shaft. The surfaces of the shaft and the bearing housing are separated by a number of balls. Each ball has rolling contact with the adjacent surfaces, thus replacing sliding friction by rolling friction.

The advantages of ball bearings over plain journal bearings are that they are much shorter for a given load-carrying capacity, they give a lower effective coefficient of friction, usually between about 0.001 and 0.004. Although a lubricated plain bearing may give as low a coefficient of friction

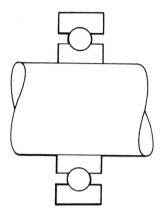

Fig. 14.10 Ball bearing.

when running, it has a much higher coefficient when starting from rest because of the squeezing out of the oil film. Ball bearings however have the disadvantages of being more expensive than plain bearings.

14.3.5   Boundary layer lubrication

When the relative speed between two surfaces is very low, or the load is high, it is not generally possible to maintain a thick layer of lubricant between the two surfaces. In such circumstances there is only a very thin layer, perhaps only a few molecules thick, of lubricant which may adhere to the surfaces. Such a form of lubrication is known as *boundary layer lubrication*. With such lubrication viscosity plays little if any part in the resistance to motion and we can effectively consider there to be just a low coefficient of friction between the surfaces. Typically this is between about 0.05 and 0.10.

14.4   Belt drives

Power can be transmitted from one shaft to another by means of a belt running between pulleys attached to the shafts. Belt drives use the friction that develops between the pulley and belt around the arc of contact in order to transmit a torque. The driving pulley transmits torque to the belt by the frictional forces between the contact surfaces of belt and pulley and the driven pulley acquires a torque as a result again of the frictional forces between the contact surfaces of belt and pulley.

The transmitted torque is due to the differences in tension that exist in the belt during operation. Figure 14.11 shows a belt system in which pulley A is the driver. The difference in tension results in a tight side and a slack side for the belt. The torque on the driving pulley A is

$$\text{torque on driving pulley} = T_1 r_A - T_2 r_A = (T_1 - T_2) r_A \qquad [29]$$

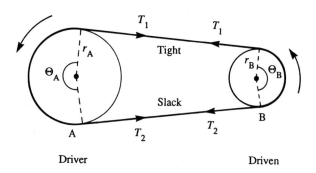

Fig. 14.11 Belt drive.

The torque acting on the driven pulley B is

$$\text{torque on driven pulley} = T_1 r_B - T_2 r_B = (T_1 - T_2) r_B \qquad [30]$$

The power transmitted is torque × angular velocity and since the angular velocity is v/r_A for pulley A and v/r_B for pulley B, v being the belt speed, then for either pulley

$$P = (T_1 - T_2) v \qquad [31]$$

The term *angle of lap* is used for the angle subtended at the centre of a pulley by the contact length of the belt with the pulley wheel. In figure 14.11 these are angles θ_A and θ_B. The rims of pulleys may be flat or V-grooved. Belts may be flat, i.e. of rectangular section, or of wedge section, i.e. V-shaped belt, or circular (these being referred to as ropes).

Compared with gear systems as a means of transmitting power between shafts, belt drives have several advantages. Belt drives are automatically protected against overload because slipping occurs if the loading exceeds the maximum tension that can be sustained by the frictional forces. The length of belt can easily be adapted to suit a wide range of shaft centre distances. Different size pulleys may be used to give a gearing effect. There are however some disadvantages. The speed ratio is generally limited to about 3 to 1 because of the need to maintain an adequate arc of contact between belt and pulley. Repeated flexing of the belt running on to and off the pulleys results in fatigue failure of the material. Belts stretch during operation and consequently alter the relationship between tensions, speed and torque.

14.4.1 *Flat pulley drives*

Consider a flat belt passing round a flat pulley so that the angle of lap is θ and the tensions in the belt are T_1 and T_2 when it is just on the point of slipping (figure 14.12). For a small element of the belt which subtends an angle of $\delta\theta$ at the pulley centre the resolved components of the forces in the radial direction are

$$\text{radial force} = (T + \delta T)\sin\tfrac{1}{2}\delta\theta + T\sin\tfrac{1}{2}\delta\theta - R$$

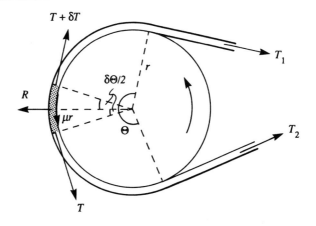

Fig. 14.12 Flat pulley drive.

R is the reaction force. Since we are considering only a small element $\delta\theta$ is very small and so $\sin\frac{1}{2}\delta\theta \approx \frac{1}{2}\delta\theta$. Thus

$$\text{radial force} = (T + \delta T)\tfrac{1}{2}\delta\theta + T\tfrac{1}{2}\delta\theta - R$$

The belt is moving in the arc of a circle and so this radial force is the centripetal force. If the mass of the belt is m per unit length, then since the length of the belt subtending an angle $\delta\theta$ is $r\delta\theta$ its mass is $mr\delta\theta$. Thus the radial force is

$$(T + \delta T)\tfrac{1}{2}\delta\theta + T\tfrac{1}{2}\delta\theta - R = \frac{(mr\delta\theta)v^2}{r}$$

where v is the belt speed. Hence, neglecting as insignificant $\delta T\delta\theta$,

$$(T - mv^2)\,\delta\theta = R \tag{32}$$

The resolved components of the forces in the tangential direction acting on the element of belt are

$$(T + \delta T)\cos\tfrac{1}{2}\delta\theta - T\cos\tfrac{1}{2}\delta\theta - \mu R = 0$$

Since $\delta\theta$ is very small then $\cos\delta\theta \approx 1$ and so

$$\delta T = \mu R \tag{33}$$

Thus $R = \delta T/\mu$ and substituting this value in equation [32] gives

$$\frac{\delta T}{T - mv^2} = \mu\delta\theta$$

Integrating over the angle of lap gives

$$\int_0^\theta \mu\,\mathrm{d}\theta = \int_{T_2}^{T_1} \frac{\mathrm{d}T}{T - mv^2}$$

$$\mu\theta = \ln\left(\frac{T_1 - mv^2}{T_2 - mv^2}\right)$$

Hence

$$\frac{T_1 - mv^2}{T_2 - mv^2} = e^{\mu\theta} \tag{34}$$

In the above equation mv^2 is sometimes referred to as the *centrifugal* or *centripetal tension*. It has the effect of reducing the tension available for transmitting power. Under low speed conditions mv^2 can be neglected to give

$$\frac{T_1}{T_2} = e^{\mu\theta} \tag{35}$$

14.4.2 Power transmission

The power transmitted by the belt is given by equation [31] as

$$P = (T_1 - T_2)v$$

which becomes, when mv^2 is not neglected and equation [34] is used,

$$P = \left(T_1 - \frac{T_1 - mv^2}{e^{\mu\theta}} - mv^2 \right)v$$

$$= (T_1 - mv^2)(1 - e^{-\mu\theta})v \tag{36}$$

and when mv^2 is neglected and equation [35] used

$$P = T_1(1 - e^{-\mu\theta})v \tag{37}$$

Example

Two pulleys, one of diameter 150 mm and the other 200 mm, are on parallel shafts with centres 600 mm apart. What is the angle of contact between a flat belt and each pulley? What power can be transmitted when the larger pulley rotates at 300 rev/min if the maximum tension in the belt is to be 1 kN and the coefficient of friction between belt and pulley is 0.25?

Figure 14.13 shows the arrangement. This gives

$$\sin\phi = \frac{25}{600}$$

Hence $\phi = 2.39°$. Then

$$\theta_A + 180° + 2\phi = 360°$$

and so $\theta_A = 175.22°$ or 3.06 rad. The angle of lap for pulley B is $360° - \theta_A = 184.78°$ or 3.23 rad.

Equation [37] gives for the power

$$P = T_1(1 - e^{-\mu\theta})v$$

For the larger pulley $v = r\omega = 0.100 \times 2\pi \times 5 = 3.14$ m/s. The maximum ten-

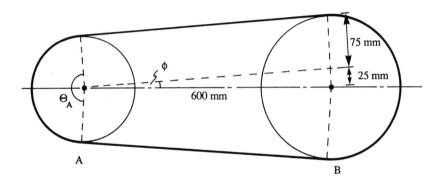

Fig. 14.13 Example.

sion $T_1 = 1\,\text{kN}$. The maximum power will depend on the angle of lap of the smaller pulley. This is because the length of belt in contact with the pulley is the smallest and slip will thus occur there before occurring at the larger pulley. Thus

$$P = 1 \times 10^3 \, (1 - e^{-0.25 \times 3.06}) \times 3.14 = 1.68\,\text{kW}$$

Example

A flat belt is required to transmit $100\,\text{kW}$ at a belt speed of $20\,\text{m/s}$ between two pulleys of diameters $250\,\text{mm}$ and $400\,\text{mm}$ when the distance between the pulley centres is $1200\,\text{mm}$. The material to be used for the belt has a maximum permissible stress of $8.0\,\text{MPa}$ and is available in thickness to width ratios of 1 to 10. The belt material has a density of $1100\,\text{kg/m}^3$ with a coefficient of friction between it and the pulleys of 0.25. What is the minimum belt thickness that can be used?

As with the previous example, a figure similar to figure 14.13 can be drawn. This then gives

$$\sin\phi = \frac{75}{1200}$$

Hence $\phi = 3.58°$ and so the angle of lap on the smaller pulley is given by

$$\theta_A + 180° + 2 \times 3.58° = 360°$$

Thus $\theta_A = 172.84° = 3.02\,\text{rad}$.

Equation [36] gives for the power

$$P = (T_1 - mv^2) \, (1 - e^{-\mu\theta})v$$

The maximum tension T_1 is $8.0 \times 10^6 \times A$, where A is the cross-sectional area of the belt. If the belt thickness is t then, since the width is $10t$, the area is $10t^2$. The mass per unit length m is $1100A = 1100 \times 10t^2$. Thus

$$P = 100 \times 10^3 = (8.0 \times 10^6 \times 10t^2 - 1100 \times 10t^2 \times 20)$$
$$(1 - e^{-0.3 \times 3.02}) \times 20$$

Hence $t = 10.3$ mm.

14.4.3 Initial tension

Let T_0 be the initial static tension in the belt. When the belt is running then the tension on the tight side becomes T_1 and that on the slack side T_2. If we assume that the belt material is elastic and obeys Hooke's law then if the overall length of the belt remains unchanged, the increase in length of the slack side must be balanced by the reduction in length of the tight side. Since the change in length of the belt material will be proportional to the change in tension we must therefore have

$$T_1 - T_0 = T_0 - T_2$$

and so

$$T_0 = \tfrac{1}{2}(T_1 + T_2) \qquad\qquad [38]$$

Example

A flat belt is installed with an initial tension of 400 N. The coefficient of friction between the belt and pulley is 0.3 and the angle of lap on the smaller pulley is 165°. The smaller pulley has a diameter of 100 mm and rotates at 600 rev/min. Determine the maximum power which the belt can transmit if it is assumed to have negligible mass.

Equation [38] gives for the relation between the initial tension and the tensions of the slack and tight belt

$$400 = \tfrac{1}{2}(T_1 + T_2)$$

The negligible mass condition means that the centripetal tension can be neglected. Thus equation [35] gives, with $\theta = 165° = 2.88$ rad,

$$\frac{T_1}{T_2} = e^{\mu\theta} = e^{0.3 \times 2.88} = 2.37$$

Hence $T_1 = 2.37 T_2$. Substituting this value in the earlier equation gives

$$400 = \tfrac{1}{2}(2.37 T_2 + T_2)$$

Hence $T_2 = 237.4$ N and $T_1 = 562.6$ N.
  The power is given by equation [31] as

$$P = (T_1 - T_2)v = (562.6 - 237.4)0.050 \times 2\pi \times 10 = 1.02 \text{ kW}$$

14.4.4 *Maximum power transmission*

The equation for the initial static tension (equation [38])

$$T_0 = \tfrac{1}{2}(T_1 + T_2)$$

can be written, by subtracting $2mv^2$ from both sides, as

$$2T_0 - 2mv^2 = T_1 - mv^2 + T_2 - mv^2$$

Substituting for $(T_2 - mv^2)$ from equation [34] gives

$$2T_0 - 2mv^2 = T_1 - mv^2 + (T_1 - mv^2)\, e^{-\mu\theta}$$

Hence

$$T_1 - mv^2 = \frac{2\,(T_0 - mv^2)}{1 + e^{-\mu\theta}}$$

The equation for the power transmitted, equation [36], can thus be written as

$$P = (T_1 - mv^2)(1 - e^{-\mu\theta})v$$
$$= 2(T_0 - mv^2)\left(\frac{1 - e^{-\mu\theta}}{1 + e^{-\mu\theta}}\right) v \qquad\qquad [39]$$

The maximum power for a given setting of T_0 is when $dP/dv = 0$. This is thus when

$$T_0 - 3mv^2 = 0$$

Since the centripetal tension is mv^2, maximum power is when the static initial tension is three times the centripetal tension. The above equation can also be written as

$$v = \sqrt{(T_0/3m)} \qquad\qquad [40]$$

The maximum belt tension occurs when $v = 0$, i.e. when the centripetal tension is zero.

Example

A flat belt having a mass of 0.2 kg/m is used to transmit power between two pulleys of diameters 100 mm and 200 mm, the distance between their centres being 1000 mm. If the maximum permissible belt tension is 500 N and the coefficient of friction is 0.3, at what belt speed will the power transmitted be a maximum?

The arrangement is similar to that shown in figure 14.13. Thus

$$\sin\phi = \frac{50}{1000}$$

Thus $\phi = 2.87°$ and so

$$\theta_A + 180° + 2 \times 2.87° = 360°$$

Hence $\theta_A = 174.26° = 3.04\,\text{rad}$. The maximum tension is determined by this angle of lap on the smaller pulley.

The maximum tension occurs when the centripetal tension is zero and is then $T_1 = 500\,\text{N}$. Thus

$$\frac{T_1}{T_2} = e^{\mu\theta} = e^{0.3 \times 3.04} = 2.49$$

Hence $T_2 = 500/2.49 = 200.8\,\text{N}$. The static initial tension is thus given by equation [38] as

$$T_0 = \tfrac{1}{2}(T_1 + T_2) = \tfrac{1}{2}(500 + 200.8) = 350.4\,\text{N}$$

The maximum belt speed is thus given by equation [40] as

$$v = \sqrt{\left(\frac{T_0}{3m}\right)} = \sqrt{\left(\frac{350.4}{3 \times 0.2}\right)} = 24.2\,\text{m/s}$$

14.4.5 V-pulley drives

Figure 14.14 shows a V-belt in the V-groove of a pulley. The V has an angle of 2α. The normal reaction in each face of an element is N. The component of these reactions at right angles to the belt, i.e. the radial reaction component R, is $2N\sin\alpha$.

$$R = 2N\sin\alpha$$

The frictional force is $2\mu N$. Hence we can write

$$\text{frictional force} = 2\mu N = \frac{\mu R}{\sin\alpha}$$

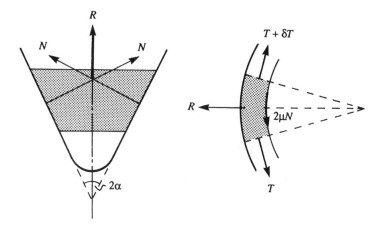

Fig. 14.14 V-pulley drive.

Thus in the equations obtained for the flat pulley drive in section 14.4.1, we just need to replace μ by $\mu/\sin \alpha$ to obtain the equations for V-pulley drives.

The same result is obtained for circular cross-section rope in a V-shaped groove.

Example

A V-belt is used to transmit power between two pulleys of diameters 100 mm and 200 mm, the distance between their centres being 1000 mm. The groove angle is 40° and the coefficient of friction 0.3. If the maximum permissible belt tension is 400 N, what power can be transmitted if the larger pulley is driven at 600 rev/min?

The arrangement is similar to that shown in figure 14.13. Thus

$$\sin \phi = \frac{50}{1000}$$

Thus $\phi = 2.87°$ and so

$$\theta_A + 180° + 2 \times 2.86° = 360°$$

Hence $\theta_A = 174.26° = 3.04$ rad. The maximum tension is determined by this angle of lap on the smaller pulley.

Using equation [35] with μ replaced by $\mu/\sin \alpha$ gives

$$\frac{T_1}{T_2} = e^{\mu\theta/\sin \alpha} = e^{0.3 \times 3.04/\sin 20°} = 14.39$$

The maximum tension will be T_1. Thus $T_2 = 400/14.39 = 27.8$ N. The power transmitted is given by equation [31] as

$$P = (T_1 - T_2)v = (400 - 27.8)\, 0.100 \times 2\pi \times 10 = 2.34 \text{ kW}$$

Problems

(1) Two coaxial shafts A and B are connected by a single plate friction clutch of internal diameter 120 mm and external diameter 220 mm, both sides of the plate being used. Shaft A is rotating at a constant angular speed of 240 rev/min and B is initially stationary. Shaft B has an effective moment of inertia of 12 kg m². When the clutch is engaged it takes 4 s for B to reach full speed. If the coefficient of friction is 0.3 and the uniform pressure condition can be assumed, what is the required spring force and pressure?

(2) The clutch of a motor vehicle has a plate with an internal diameter of 200 mm and an external diameter of 360 mm, both sides of the plate being used. What is the spring force and the power transmitted at 360 rev/min when there is a uniform pressure of 160 kPa and the coefficient of friction is 0.3?

(3)   A multi-plate clutch has four pairs of frictional surfaces, each of internal diameter 100 mm and external diameter 200 mm. If the coefficient of friction is 0.3 and the uniform pressure condition can be assumed, what is the required spring force when the plates transmit 30 kW at 420 rev/min?

(4)   A multi-plate clutch has four pairs of frictional surfaces, each of internal diameter 120 mm and external diameter 240 mm. The total spring force is 1.2 kN. If the coefficient of friction is 0.3 and the uniform wear condition can be assumed, what is the maximum power that can be transmitted at 1200 rev/min?

(5)   What will be the time taken to accelerate a shaft of moment of inertia 20 kg m^2 from rest to a speed of 250 rev/min through a single plate clutch of internal diameter 250 mm and external diameter 400 mm, total spring force 600 N and coefficient of friction 0.3? Assume the uniform pressure condition and both sides of the plate are used.

(6)   A conical clutch has a mean diameter of friction surface of 300 mm and a width of conical surface of 70 mm. The cone angle is 15° and the coefficient of friction between the friction surfaces 0.3. What will be the greatest power that can be transmitted without slipping at a speed of 20 rev/s if the maximum normal pressure is to be 60 kPa? Assume the constant wear condition.

(7)   A conical clutch has a mean diameter of friction surface of 150 mm and a cone angle of 15°. The coefficient of friction is 0.3. What is the torque required to produce slipping of the clutch if the axial force applied is 160 N? Assume the constant wear condition.

(8)   A conical clutch has a mean diameter of friction surface of 100 mm and a cone angle of 15°. The coefficient of friction is 0.3. The axial force applied is 180 N. The clutch connects a motor running at 900 rev/min with a shaft which is initially stationary. This shaft has a moment of inertia of 0.30 kg m^2. What is the time required for this shaft to attain full speed? Assume the constant wear condition.

(9)   A flat collar bearing has an internal diameter of 60 mm and an external diameter of 100 mm. The pressure exerted on the bearing surfaces may be assumed to be constant at 150 kPa. The coefficient of friction between the surfaces is 0.1. What is the power loss at 600 rev/min in overcoming friction at the bearing?

(10)   A vertical shaft of diameter 50 mm runs in a footstep bearing at 900 rev/min. The pressure between the bearing surfaces may be assumed to be constant at 100 kPa. The coefficient of friction between the surfaces is 0.06. What is the power loss due to friction?

(11)   How many collars will be needed for a multi-collar bearing if the collars have internal diameters of 60 mm and external diameters of

120 mm? The pressure between the bearing surfaces may be assumed to be constant at 160 kPa and the maximum axial load is 6.5 kN.

(12) An axial thrust of 20 kN along a shaft is taken by a conical pivot bearing. The bearing surface has an outer diameter of 160 mm and an inner diameter of 80 mm with a cone semi-angle of 60°. The coefficient of friction between the bearing surfaces is 0.04. If the pressure on the bearing surfaces can be assumed to be constant, what is the power loss due to friction at 240 rev/min?

(13) A collar bearing has an inner diameter of 100 mm and an outer diameter of 160 mm and runs in its housing on a film of lubricant of thickness 0.4 mm. If the viscosity of the lubricant is 0.1 N s/m², what is the power loss at the bearing when the shaft is rotating at 1200 rev/min?

(14) A vertical shaft is supported in a footstep bearing of diameter 60 mm, the base of the shaft being separated from the housing by a lubricant of thickness 0.2 mm. If the lubricant has a viscosity of 0.1 N s/m², what is the power loss when the shaft is rotating at 900 rev/min?

(15) A bearing consists of a shaft of diameter 140 mm rotating in a coaxial cylindrical housing. The housing and shaft are separated by a lubricant layer of thickness 0.3 mm. If the axial length of the surfaces is 200 mm, what is the power loss when the shaft is rotating at 300 rev/min? The lubricant has a viscosity of 0.1 N s/m².

(16) A journal bearing has a diameter of 50 mm and a length of 45 mm. It carries a radial load of 2 kN and the shaft runs at 900 rev/min. What is the power loss due to (a) friction when the bearing is unlubricated and the coefficient of friction is 0.05 and (b) viscous resistance when the bearing is lubricated with oil of viscosity 0.1 N s/m² and there is a constant radial clearance of 0.3 mm?

(17) Two pulleys, one of diameter 250 mm and the other 150 mm, are connected by a flat belt. The distance between the pulley centres is 500 mm. What are the angles of lap on each pulley? If the larger pulley rotates at 600 rev/min and the maximum permissible belt tension is 900 N, what will be the maximum power transmitted? The coefficient of friction is 0.25.

(18) Two pulleys, one of diameter 500 mm and the other 250 mm, are connected by a flat belt of mass 0.4 kg/m. The distance between the pulley centres is 1500 mm. The coefficient of friction is 0.3 and the maximum permissible belt tension is 900 N. What will be the power transmitted when the larger pulley rotates at 360 rev/min?

(19) A flat belt is installed with an initial tension of 600 N. The coefficient of friction between the belt and pulley is 0.3 and the angle of lap on the smaller pulley is 170°. The smaller pulley has a diameter of 100 mm and rotates at 900 rev/min. Determine the tension in the

slack and tight sections of the belt and the maximum power which the belt can transmit if it is assumed to have negligible mass.

(20)   A pulley is driven by a flat belt of width 100 mm, thickness 6 mm and density 1000 kg/m³. The angle of lap is 120° and the coefficient of friction between the belt and pulley is 0.3. If the maximum stress in the belt is not to exceed 8.0 MPa, what is the maximum power the belt can transmit and the maximum belt speed?

(21)   A pulley is driven by a V-belt of width 80 mm, thickness 4 mm and density 1200 kg/m³. The pulley has an angle of lap of 145°, rotates at 1200 rev/min and has a V-groove of angle 45°. If the stress in the belt is not to exceed 2.0 MPa, what is the maximum belt speed and power the belt can transmit?

(22)   Two pulleys, of diameters 200 mm and 100 mm, are connected by a V-belt of mass 0.2 kg/m. The distance between the pulley centres is 1000 mm. The V-groove has an angle of 45° and the coefficient of friction is 0.3. What will be the maximum belt speed if the maximum belt tension is 400 N?

(23)   A multiple V-belt drive is required to transmit 20 kW. The smaller pulley has a diameter of 150 mm, an angle of lap of 160°, and rotates at 1200 rev/min. The V-grooves have angles of 40° and the coefficient of friction between the belt and pulleys is 0.3. The belt material has a mass of 0.5 kg/m. If the tension in a belt is restricted to 1000 N, how many belts will be required?

(24)   A rope drive is required to transmit 20 kW. The smaller pulley has a diameter of 1.0 m, an angle of lap of 190°, and rotates at 180 rev/min. The pulley has V-grooves of angle 45° and the coefficient of friction between the belt and pulleys is 0.3. If the tension in a rope is restricted to 600 N, how many ropes are required?

Chapter 15

Gyroscopic motion

15.1 Vectors

Angular displacement is a vector quantity since it has both a magnitude and a direction. In order to completely specify an angular displacement by a vector, the vector must give the direction of the axis of the angular rotation, whether the angular displacement is clockwise or anticlockwise and the magnitude of the displacement. This is achieved by the convention of drawing a vector as a line with a length related to the displacement and an arrowhead in the direction a right-handed corkscrew would move to give by its rotation the angular displacement. Thus for the angular displacement shown in Figure 15.1, the vector would be a line at right angles to the plane of the paper and pointing upwards out of the paper.

Fig. 15.1 Angular displacement.

Angular velocity is the rate of change of angular displacement with time. It has both magnitude and direction and so is a vector quantity. As with angular displacement, angular velocity can be represented by a vector drawn as a line with a length related to the magnitude of the velocity and an arrowhead in the direction a right-handed corkscrew would move to give by its rotation the direction of the velocity.

Angular acceleration is the rate of change of angular velocity with time and likewise is a vector quantity. Angular acceleration can be represented by an arrowheaded line in the same way as angular velocity. There can be an angular acceleration when the magnitude of an angular velocity is changing and/or when the direction of the velocity is changing. Thus we would have a disc rotating with a constant angular velocity about some axis and then the axis changes without the size of the angular velocity changing. Because the direction is changing there is an angular acceleration. In section 11.2 a comparable situation was discussed for linear acceleration, an object moving with constant speed in a circular path has a changing velocity and hence an acceleration, the centripetal acceleration.

To produce an angular acceleration a torque is required. *Torque* is a vector quantity and can be represented by an arrowheaded line drawn using the right-handed corkscrew rule in the same way as for angular displacement.

The *angular momentum* of a body about some axis is the product of its moment of inertia I and its angular velocity ω about that axis. Angular momentum is thus a vector quantity which is represented by a line drawn with a length related to the magnitude of the momentum and in the direction given by the right-handed corkscrew for the angular velocity. The angular equivalent of Newton's second law of motion can be expressed as: the rate of change of vector angular momentum of a body is proportional to the applied torque. Thus if the total external torque acting on a system is zero then the total vector angular momentum of the system remains unchanged. This is the law of *conservation of angular momentum*. In the above I have included the word vector in front of the term angular momentum to emphasise that angular momentum can change if either its size or its direction changes.

15.2 Gyroscopic couple

Consider figure 15.2 which represents a rotor of moment of inertia I rotating with an angular velocity ω about the axis OX, this being termed the *spin axis*. We can represent the angular momentum $I\omega$, using the right-hand corkscrew rule, by the vector oa. Now suppose the spin axis rotates, the term *precessed* being used for such a rotation. We will consider the spin axis to be rotating about the perpendicular axis OY with an angular velocity ω_p,

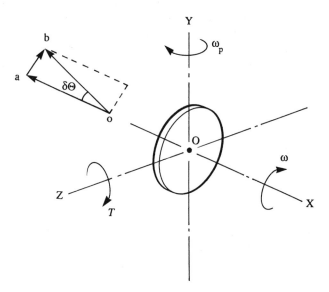

Fig. 15.2 Gyroscopic effect.

i.e. the precessional angular velocity. If this change results in the axis OX rotating through an angle $\delta\theta$ in a time δt then the direction of the angular momentum vector must have changed $\delta\theta$ and thus be represented by vector ob. The change in angular momentum necessary for this change is represented by the vector ab. For small angular changes ab $\approx$ oa x $\delta\theta$. But oa $= I\omega$ and so

change in angular momentum $= I\omega\delta\theta$

The rate of change of angular momentum is

$$\text{rate of change of angular momentum} = I\omega\frac{\delta\theta}{\delta t}$$

But a change in angular momentum can only be produced by the application of a torque T. Since

$T =$ rate of change of angular momentum

then, in the limit as $\delta\theta$ tends to zero,

$$T = I\omega\,\frac{d\theta}{dt}$$

But $\omega_p = d\theta/dt$, and so

$$T = I\omega\omega_p \qquad [1]$$

The torque, or couple, required to produce the change in angular momentum vector ab will be represented by a vector in the same direction as ab. This means that the torque is clockwise looking along the axis OZ, this axis being mutually perpendicular to the axes OX and OY. This torque is generally referred to as the *gyroscopic couple*.

Whenever the axis of spin of a body changes its direction, a gyroscopic couple must have been applied. This couple is usually applied through the bearings supporting the shaft and rotor. The reaction of the shaft on the bearings is equal and opposite to the action of the bearing on the shaft. There is thus a *reaction couple* which is equal in size but in the opposite direction to the gyroscopic couple. Figure 15.3 illustrates this, the reaction couple being FL. Thus

$$FL = I\omega\omega_p \qquad [2]$$

Example

The rotor of a ship's engine is effectively a mass of 60 kg with a radius of gyration 140 mm and rotates with an angular speed of 3000 rev/min. The rotor is centrally placed between horizontal bearings 2.0 m apart, as in figure 15.4. What are the reactions at each bearing as a result of the ship turning at 0.4 rad/s?

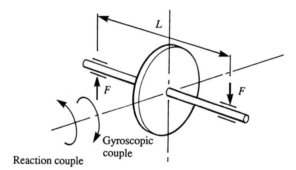

Fig. 15.3 Reaction couple.

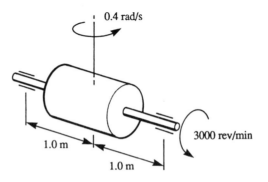

Fig. 15.4 Example.

The gyroscopic couple resulting from the ship turning is $I\omega\omega_p$ and the reaction couple at the bearings is FL. Thus, as in equation [2],

$$FL = I\omega\omega_p$$
$$F = \frac{60 \times 0.140^2 \times 2\pi \times 50 \times 0.4}{2.0} = 73.9\,\text{N}$$

15.2.1 Examples of gyroscopic motion

Figure 15.5 shows the basic form of what is termed a *gyroscope*. This figure shows a rotor which can rotate about its axis and is also free to rotate about two perpendicular axes. The term *gimbals* is used for the frames and their bearings. The rotor can rotate about its own axis YY and is mounted in an inner frame which is free to rotate about the XX-axis. This frame in turn is mounted in an outer frame which is free to rotate about the ZZ-axis. The mounting allows the rotor to be unaffected by any external torques, assuming friction at the bearings for each frame is negligible. Thus its angular momentum will never precess but maintain its fixed orientation of rate of revolution

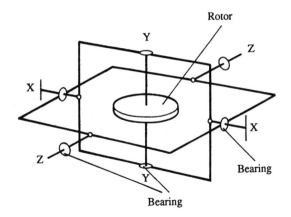

Fig. 15.5 Basic form of gyroscope.

whatever motions occur to the base of the instrument. Such an instrument is the basis of the gyrocompass. Inertial navigational instruments for aircraft use the same principle, the rotor being spun by means of a synchronous motor whose rotor is mounted on the axis of the spinning rotor. The outer gimbal is free to rotate in a bearing located in the aircraft frame. The movement of the outer gimbal when seen relative to the plane of the rotor can thus be used to indicate whether the aircraft is climbing or diving, banking to the left or right.

Figure 15.6 shows the *gyroscopic top*. If this top is not spinning and placed with its tip on a pivot then the weight W of the top will result in a torque which causes it to rotate clockwise in a vertical circle about the pivot and fall off the pivot. If, however, the top is spinning when placed on the pivot then the torque due to W causes the top to have a precessional angular velocity and move in a horizontal plane without falling.

Gyroscopic couples occur with any spinning rotor when an external torque is applied about an axis perpendicular to the axis of spin. The precession occurs about an axis which is at right angles to both the axis of spin and the

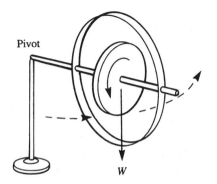

Fig. 15.6 Precessing top.

axis of the external torque (see figure 15.2). To illustrate this, consider a vehicle such as a train rounding a curve (figure 15.7). A pair of wheels and their axle will have a spin velocity ω of v/r, where v is the linear velocity at the rim of a wheel and r the radius of the wheel. The precessional velocity is that of the train round the curve. Thus for a curve of radius R the precessional velocity is v/R. The gyroscopic couple resulting from this is thus given by equation [1] as $I(v/r)(v/R)$. The angular momentum vector oa turns to ob and consequently the change in momentum vector required to give the precessional velocity is ab. This vector ab means a clockwise couple looking backwards which tends to overturn the vehicle outwards, so increasing the load on the outer rail and reducing it on the inner rail.

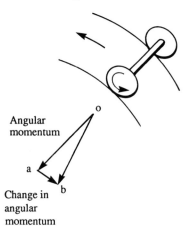

Fig. 15.7 Train rounding a curve.

An example where the gyroscopic couple can be used beneficially is in ship stabilisation. The rolling of a ship is uncomfortable to the passengers and stabilisers can be used to reduce the rolling. The waves apply a rolling moment to the ship and thus to reduce the amount of roll an opposing moment is required. Figure 15.8 shows the basic principle of the gyroscopic stabiliser devised by E.A. Sperry in 1920. The spinning rotor is mounted with its axis YY vertical. A motor is used to produce a precessional velocity for the frame in which the rotor is mounted. A consequence of this is that a reaction couple is produced which opposes the roll of the ship.

Example

A railway carriage runs on rails 1.5 m apart. A pair of wheels and axle have a moment of inertia of 100 kg m^2, the wheels having a rolling radius of 650 mm. An imperfection in the mounting of the rails has resulted in one of the rails raising the wheel on one side by 2 mm over a distance of 200 mm. What is the gyroscopic couple produced by this when the train is travelling at 30 m/s?

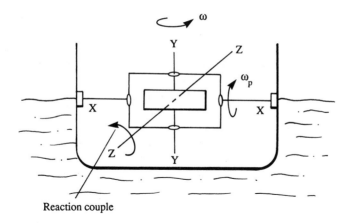

Fig. 15.8 Basic principles of the ship stabiliser.

The spin velocity of the wheels is $\omega = v/r = 30/0.650 = 46.15\,\text{rad/s}$. The imperfection leads to a precessional angular velocity of V/R, where V is the vertical wheel velocity and R the rail spacing. At 30 m/s the train covers a distance of 200 mm in $0.200/30 = 6.67 \times 10^{-3}\,\text{s}$. In this time the wheel moves a vertical distance of 2 mm. Thus the vertical velocity $V = 0.002/(6.67 \times 10^{-3}) = 0.300\,\text{m/s}$. Thus $\omega_r = 0.300/1.5 = 0.200\,\text{rad/s}$. Hence the gyroscopic couple is given by equation [1] as

$$T = I\omega\omega_p = 100 \times 46.15 \times 0.200 = 923\,\text{N m}$$

Problems

(1) A generator on a ship has a rotor of mass 1200 kg and radius of gyration 400 mm and rotates at 600 rev/min. What will be the gyroscopic couple exerted on the ship when the ship steams at 10 m/s in a curve of radius 200 m.

(2) The turbine rotor of a ship has a moment of inertia of $10\,000\,\text{kg m}^2$ and rotates at 2400 rev/min. If the ship pitches with an angular velocity of 0.05 rad/s what will be the gyroscopic couple on the holding-down bolts of the turbine?

(3) A generator on a ship has its rotor with its axis parallel to the central axis of the ship. The rotor has a moment of inertia of $200\,\text{kg m}^2$ and revolves at 360 rev/min. What will be the gyroscopic couple transmitted to the ship when it steams at 10 m/s round a curve of radius 200 m?

(4) A locomotive runs on rails 1.5 m apart. A pair of wheels and axle have a moment of inertia of $400\,\text{kg m}^2$, the wheels having a rolling radius of 2.0 m. An imperfection in the ballasting of the rails has resulted in one of the wheels on one side falling by 10 mm over a

distance of 400 mm. What is the gyroscopic couple produced by this when the locomotive is travelling at 30 m/s?

(5)   A car travels on a horizontal road round a left-handed curve of radius 50 m at a speed of 15 m/s. The rotating parts of the engine and transmission have a moment of inertia of 2.0 kg m² and an angular speed of 2000 rev/min in a clockwise direction when viewed from the front of the car. What is the gyroscopic couple?

(6)   The moment of inertia of the propeller on a single-engined aeroplane is 1.4 kg m² and the direction of rotation is clockwise when viewed from the front of the machine. The speed of rotation of the propeller is 1600 rev/min. What is the gyroscopic couple acting on the aeroplane and its effect when it is flying horizontally at 200 km/h in a right-handed turn of radius 200 m?

(7)   A flywheel of mass 40 kg and radius of gyration 300 mm rotates about its horizontal axis with an angular velocity of 100 rev/min. The horizontal shaft of the flywheel is supported on bearings 1.0 m apart. The flywheel axis is then made to rotate about a vertical axis through the centre of the wheel at an angular velocity of 50 rev/min. What is the load on the bearings of the wheel?

(8)   A car travels at 25 m/s round a horizontal curve of radius 80 m. Each of the wheels has a mass of 15 kg, a radius of gyration 300 mm about the axle and an effective diameter of 400 mm. The distance between the wheels on an axle is 1.3 m. What is the difference in the normal force acting on the wheels on the inside and outside of the curve as a result of the gyroscopic couple?

(9)   Rotating parts of a motor car engine and transmission have a moment of inertia of 2.5 kg m² and an angular velocity of 1000 rev/min in a clockwise direction when viewed from the front. The car travels round a left-hand curve of radius 20 m at a speed of 20 m/s. What is the size and effect of the gyroscopic couple on the axle loads? The engine shaft is along the longitudinal axis of the car.

Chapter 16

Free vibrations

16.1 Simple harmonic motion

An oscillation is said to be *simple harmonic motion* when the acceleration of the body is always directed towards a fixed point in its path and is proportional to its displacement from that point. Since the internal restoring force of the oscillating system is proportional to the acceleration ($F = ma$) then the restoring force is always directed towards a fixed point and is proportional to the displacement from that point.

Simple harmonic motion can be represented by a model, the motion of a point in a circular path with a constant angular velocity (figure 16.1). The length of the projection of the radius to the point on to the diameter of the circle represents the displacement relative to the circle centre. The displacement x at some instant is thus

$$x = A \cos \theta$$

where A, which is the amplitude of the oscillation, is the radius of the circle. If t is the time taken for OP to rotate through the angle θ then the angular velocity of the point round the circle ω is θ/t, hence

$$x = A \cos \omega t \qquad [1]$$

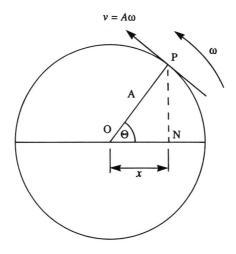

Fig. 16.1 Simple harmonic motion.

The linear velocity of the particle tangential to the circle $v = r\omega = A\omega$. The component of this velocity along the diameter of the circle represents the velocity of the oscillating body and is

$$v = -A\omega \sin \theta = -A\omega \sin \omega t \qquad [2]$$

The minus sign is included to show that the direction of the velocity is towards the circle centre O and in the opposite direction to which you would go if the displacement x was to be increased. The equation could have been obtained by just differentiating equation [1], i.e.

$$v = \frac{dx}{dt} = \frac{d(A \cos \omega t)}{dt} = -A\omega \sin \omega t$$

Since $\sin^2 \omega t + \cos^2 \omega = 1$, then

$$v^2 = A^2\omega^2 \sin^2 \omega t = A^2\omega^2(1 - \cos^2 \omega t) = \omega^2(A^2 - A^2 \sin^2 \omega t)$$
$$v = \omega\sqrt{(A^2 - x^2)} \qquad [3]$$

The maximum value of the velocity occurs when $x = 0$, i.e. maximum velocity $= \pm\omega A$. When $x = A$ the velocity is zero. Thus when $x = A$ there is no kinetic energy and when $x = 0$ the kinetic energy is a maximum, being $\frac{1}{2}mv^2 = \frac{1}{2}m\omega^2 A^2$.

The centripetal acceleration of point P is $\omega^2 r = \omega^2 A$. The component of this acceleration along the diameter represents the acceleration of the oscillating body and is

$$a = -\omega^2 A \cos \theta$$

The minus sign is because the direction of the acceleration is towards the centre O and in the opposite direction to which you would go if the displacement was to be increased. This equation could have been obtained by differentiating equation [2], i.e.

$$a = \frac{dv}{dt} = \frac{d(-A\omega \sin \omega t)}{dt} = -A\omega^2 \cos \omega t$$

Since $x = A \cos \theta$, then

$$a = -\omega^2 x \qquad [4]$$

The acceleration is thus proportional to the displacement. The acceleration is a maximum when $x = A$, i.e. maximum acceleration $= -\omega^2 A$, and zero when $x = 0$.

The restoring force F acting on the mass m which is oscillating with this simple harmonic motion is ma and so, since $a = -\omega^2 x$, is

$$F = -m\omega^2 x \qquad [5]$$

and is always directed towards the point O.

The *periodic time* T is the time taken to complete one oscillation, i.e. for point P to move once round the circle. Thus, since one complete path is a motion through 2π radians, the angular velocity $\omega = 2\pi/T$. The *frequency f* is the number of oscillations per second, i.e. the number of complete revolutions of P per second. Thus $T = 1/f$ and $\omega = 2\pi f$, hence since ω is just the frequency multiplied by 2π it is often referred to as the *angular frequency*. So, since $a = -\omega^2 x$, then

$$T = \frac{1}{f} = 2\pi\sqrt{\frac{x}{a}} = 2\pi\sqrt{\frac{\text{displacement}}{\text{acceleration}}} \qquad [6]$$

Example

An object moving with simple harmonic motion has a frequency of 5 Hz with an amplitude of 100 mm. What is (a) the maximum velocity, (b) the velocity 50 mm from the equilibrium position, (c) the maximum acceleration, (d) the acceleration 50 mm from the equilibrium position, (e) the time taken to move from the extremity of the oscillation to a distance of 50 mm from the equilibrium position?

(a) Since $\omega = 2\pi f = 10\pi\,\text{s}^{-1}$, the maximum velocity is $\pm A\omega = \pm 0.100 \times 10\pi = \pm 3.1\,\text{m/s}$.

(b) Using equation [3],

$$v = \omega\sqrt{(A^2 - x^2)} = 10\pi\sqrt{(0.100^2 - 0.050^2)} = \pm 2.7\,\text{m/s}$$

(c) The maximum acceleration is $-\omega^2 A = -(10\pi)^2 \times 0.100 = -98.7\,\text{m/s}^2$.

(d) Using equation [4], $a = -\omega^2 x = -(10\pi)^2 \times 0.050 = -49.3\,\text{m/s}^2$.

(e) The displacement at a particular time is given by $x = A\cos\omega t$, hence $\cos 10\pi t = x/A = 50/100$. Thus $10\pi t$ is an angle of $60°$ or $\pi/3$ radians. Hence $10\pi t = \pi/3$ and $t = 0.033\,\text{s}$.

16.1.1 Angular simple harmonic motion

Simple harmonic motion can also be defined in terms of angular displacements, the motion being simple harmonic if the angular acceleration α is proportional to the angular displacement θ from the equilibrium position and directed towards the equilibrium position. Similar equations can be developed to those used for linear simple harmonic motion. Thus

$$\theta = \theta_{max}\cos\omega t \qquad [7]$$

where θ_{max} is the maximum angular displacement, i.e. the angular amplitude. The angular velocity is $d\theta/dt$ and is thus

$$\text{angular velocity} = -\omega\theta_{max}\sin\omega t$$
$$(\text{angular velocity})^2 = \omega^2\theta_{max}^2\sin^2\omega t$$
$$= \omega^2\theta_{max}^2(1 - \cos^2\omega t)$$
$$= \omega^2\theta_{max}^2(1 - \theta^2/\theta_{max}^2)$$

Hence

$$\text{angular velocity} = \omega\sqrt{(\theta_{max}^2 - \theta^2)} \qquad [8]$$

Note that ω is not the angular velocity of the oscillating particle but $\omega = 2\pi f = 2\pi/T$, often being called the *angular frequency*. The angular acceleration is d(angular velocity)/dt and is thus

$$\alpha = \frac{d(-\omega\theta_{max}\,\sin\omega t)}{dt} = -\omega^2\theta_{max}\cos\omega t$$

$$\alpha = -\omega^2\theta \qquad\qquad [9]$$

Since the periodic time is $2\pi/\omega$ then equation [9] gives

$$T = \frac{1}{f} = 2\pi\sqrt{\frac{\theta}{\alpha}} \qquad\qquad [10]$$

16.2   Linear vibrations of an elastic system

An elastic system can be represented by a mass m suspended from a vertical spring (figure 16.2). If the mass is given a displacement x from its equilibrium position then the spring exerts a restoring force F of kx on it, assuming the spring obeys Hooke's law. k is a constant called the spring stiffness (i.e. the restoring force per unit displacement). When released, this force gives the mass an acceleration a.

Thus we have

$$F = kx = ma$$

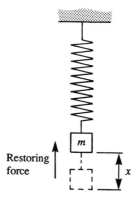

Fig. 16.2 Mass on spring.

Thus $a = kx/m$ and so the acceleration is proportional to the displacement x. Since the force, and hence the acceleration, is always directed towards the equilibrium position then the motion of the mass is simple harmonic motion. Hence, using equation [6]

$$T = \frac{1}{f} = 2\pi\sqrt{\frac{x}{a}} = 2\pi\sqrt{\frac{m}{k}} \qquad\qquad [11]$$

The mass m on the end of the spring will produce a static deflection x_{st} due to its weight mg. Thus $mg = kx_{st}$, and so equation [11] can be written as

$$T = 2\pi \sqrt{\frac{x_{st}}{g}} \qquad \qquad [12]$$

Since $\sqrt{g}$ is approximately equal to π then

$$T \approx 2\sqrt{x_{st}} \qquad \qquad [13]$$

Example

A horizontal platform rests on four vertical springs, each of stiffness 15 kN/m (figure 16.3). What will be the frequency of oscillation of the table if it has a mass of 2.0 kg?

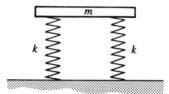

Fig. 16.3 Example.

The total restoring force is the sum of the forces acting on each spring, i.e. $F = kx + kx + kx + kx = 4kx$. We can consider there to be effectively a single spring with a force constant of $4k$. But $F = ma$, hence $4kx = ma$ and thus using equation [6]

$$f = \frac{1}{2\pi} \sqrt{\frac{a}{x}} = \frac{1}{2\pi} \sqrt{\frac{4k}{m}} = \frac{1}{2\pi} \sqrt{\left(\frac{4 \times 15 \times 10^{3}}{2.0}\right)} = 27.6 \, \text{Hz}$$

Example

An object of mass 1.0 kg is suspended from a spring of stiffness 500 N/m which in turn is suspended from a spring of stiffness 1000 N/m (figure 16.4). What is the periodic time of the oscillation?

The force F acting on the first spring will be the same as that acting on the second spring. Thus for the extension x_1 of the first spring, stiffness k_1 will be F/k_1 and for that of the second spring x_2, stiffness k_2 will be F/k_2. The total extension x is thus

$$x = x_1 + x_2 = F/k_1 + F/k_2$$

But $F = ma$ and so

$$x = ma \, (1/k_1 + 1/k_2)$$

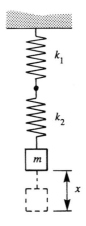

Fig. 16.4 Example.

Hence, using equation [6]

$$T = \frac{1}{f} = 2\pi \sqrt{\frac{x}{a}} = 2\pi \sqrt{\left[\frac{m(k_1 + k_2)}{k_1 k_2}\right]}$$

$$= 2\pi \sqrt{\left[\frac{1.0(500 + 1000)}{500 \times 1000}\right]} = 0.34\,s$$

Example

A uniform bar of mass m and length L is pivoted at one end and attached by a vertical spring of stiffness k at the other end to a fixed support. The bar is horizontal when in equilibrium. What is the periodic time of the oscillations of the bar?

Figure 16.5 shows the arrangement when the bar is given an initial small angular deflection θ. The extension x of the spring is thus approximately $L\theta$. The restoring force $F = kx = kL\theta$. Hence the restoring torque is $T = FL = kL^2\theta$. Since $T = I\alpha$ then $I\alpha = kL^2\theta$. The moment of inertia of the bar about the axis through the hinge is $mL^2/3$. Thus $mL^2\alpha/3 = kL^2\theta$ and so $\theta/\alpha = m/3k$. Hence, using equation [10]

$$T = \frac{1}{f} = 2\pi \sqrt{\frac{\theta}{\alpha}} = 2\pi \sqrt{\frac{m}{3k}}$$

16.2.1 The pendulum

A *simple pendulum* consists of a concentrated mass m which swings on the end of a mass-less string of length L (figure 16.6). If the pendulum bob is pulled to one side to give an angular displacement θ of the string then the restoring moment is $mgL \sin \theta$. Hence, when released,

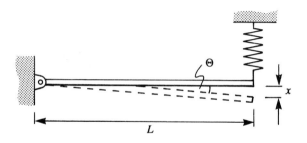

Fig. 16.5 Example.

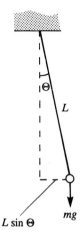

Fig. 16.6 Simple pendulum.

$$mgL \sin \theta = I\alpha$$

where I is the moment of inertia of the bob about the axis of suspension of the pendulum. Thus $I = mL^2$ and if θ is small enough we can write $\sin \theta \approx \theta$ and so

$$mg\theta \approx mL^2\alpha$$

Thus $\theta/\alpha = L/g$ and so equation [10] gives

$$\text{periodic time} = \frac{1}{f} = 2\pi \sqrt{\frac{\theta}{\alpha}} = 2\pi \sqrt{\frac{L}{g}} \qquad [14]$$

If the mass of the pendulum is not concentrated at a point the term simple is not used to describe the pendulum but the pendulum is termed *compound* and can be considered in general to be of the form shown in figure 16.7. The weight mg of the pendulum can be considered to act at the centre of gravity G, a distance h from the point of suspension O. When the pendulum is displaced through an angle θ then the restoring moment is $mgh \sin \theta$. For

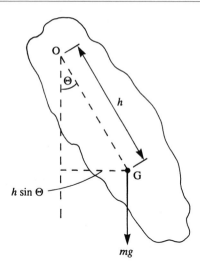

Fig. 16.7 Compound pendulum.

small angles this becomes approximately $mgh\theta$. When released, an angular acceleration α is produced where

$$mgh\theta \approx I_O\alpha$$

with I_O being the moment of inertia of the pendulum about an axis through O. If I_G is the moment of inertia about the centre of gravity G then by the parallel axis theorem $I_O = I_G + mh^2$. The moment of inertia I_G can be represented by mk_g^2, where k_g is the radius of gyration of the pendulum about the centre of gravity. Thus

$$mgh\theta \approx (mk_g^2 + mh^2)\alpha$$

Thus $\theta/\alpha = (k_g^2 + h^2)/gh$ and so equation [10] gives

$$\text{periodic time} = \frac{1}{f} = 2\pi \sqrt{\frac{\theta}{\alpha}} = 2\pi \sqrt{\left(\frac{k_g^2 + h^2}{gh}\right)} \qquad [15]$$

Example

A connecting rod has a mass 5.0 kg and radius of gyration about an axis through its centre of gravity of 170 mm. The connecting rod is supported on a knife edge inside one of the bearings so that it can oscillate like a pendulum about the knife edge, this being 200 mm above the centre of gravity. What is the periodic time of the oscillations?

Using equation [15],

$$\text{periodic time} = 2\pi \sqrt{\left(\frac{k_g^2 + h^2}{gh}\right)} = 2\pi \sqrt{\left(\frac{0.170^2 + 0.200^2}{9.8 \times 0.200}\right)}$$

$$= 1.2\,\text{s}$$

16.3   Torsional vibrations of an elastic system

Consider an elastic system consisting of a rod attached at one end to a rigid support and at the free end to a rotor of moment of inertia I (figure 16.8). In the same way as with a spring being stretched we can write $F = kx$, for the relationship between the force F and the resulting extension x, then we can write for the twisting of the rod $T = q\theta$ for the relationship between the torque T and the angle of twist θ. With the spring k is referred to as the stiffness, with the twisted rod q is the torsional stiffness. Thus when the free end of the rod in figure 16.8 is twisted through angle θ then the restoring torque $T = q\theta$. When released this torque gives the rotor an angular acceleration α where $T = I\alpha$. Thus $I\alpha = q\theta$. The angular acceleration is proportional to the angular displacement and since the torque, and hence acceleration, is always directed towards the equilibrium position the motion is simple harmonic. Thus, since $\theta/\alpha = I/q$, equation [10] gives

$$\text{periodic time} = \frac{1}{f} = 2\pi \sqrt{\frac{\theta}{\alpha}} = 2\pi \sqrt{\frac{I}{q}} \qquad [16]$$

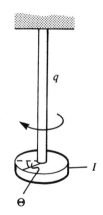

Fig. 16.8 Angular motion of an elastic system.

If the rod has a length L, a second moment of area J and modulus of rigidity G then since $T = GJ\theta/L$ (see chapter 6) and $T = q\theta$, we have $q = GJ/L$ and equation [16] can be written as

$$\text{periodic time} = \frac{1}{f} = 2\pi \sqrt{\left(\frac{IL}{GJ}\right)} \qquad [17]$$

Example

A vertical steel wire of length 1.0 m and diameter 2.0 mm is fixed to a rigid support at its upper end and has a horizontal uniform disc of moment of inertia 0.040 kg m^2 attached to the free end. What will be the periodic time of the torsional oscillations of the disc if the wire has a modulus of rigidity of 80 GPa?

The second moment of area of the wire is $J = \pi d^4/32 = \pi \times 0.0020^4/32 = 1.57 \times 10^{-12}\,\text{m}^4$. Hence, using equation [17],

$$\text{periodic time} = 2\pi \sqrt{\left(\frac{IL}{GJ}\right)} = 2\pi \sqrt{\left(\frac{0.040 \times 1.0}{80 \times 10^9 \times 1.57 \times 10^{-12}}\right)}$$

$$= 3.55\,\text{s}$$

16.3.1 Rotor systems

Consider a single rotor system which has a disc of moment of inertia I attached to one end of a shaft, as in figure 16.9. When the rotor is displaced through an angle θ, putting the shaft in torsion, then the restoring torque is $GJ\theta/L$ (see chapter 6). Thus $GJ\theta/L = I\alpha$, where α is the angular acceleration of the disc and I the moment of inertia of the disc about its axis if the moment of inertia of the shaft can be ignored. Hence $\alpha/\theta = GJ/IL$ and so equation [16] gives for the periodic time

$$\text{periodic time} = \frac{1}{f} = 2\pi \sqrt{\frac{\theta}{\alpha}} = 2\pi \sqrt{\frac{IL}{GJ}} \qquad [18]$$

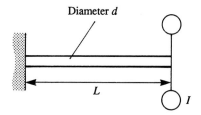

Fig. 16.9 Single rotor system.

Now consider a single rotor system where the rotor is attached to the shaft between two fixed ends, as in figure 16.10. When the rotor is displaced through an angle θ then the two parts of the shaft are put in torsion. The restoring torque due to twisting the shaft to the left of the rotor is $GJ_1\theta/L_1$, where J_1 is the second moment of area of that part of the shaft. The restoring torque due to twisting the shaft to the right of the rotor is $GJ_2\theta/L_2$, where J_2 is its second moment of area. It has been assumed that both shafts have the same value of G. The total restoring torque is thus $(GJ_1\theta/L_1) + (GJ_2\theta/L_2)$. Hence, if I is the moment of inertia of the rotor, that of the shafts being ignored,

$$G\theta \left(\frac{J_1}{L_1} + \frac{J_2}{L_2}\right) = I\alpha$$

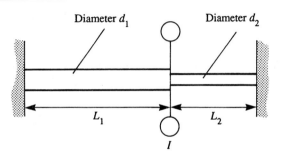

Fig. 16.10 Single rotor system.

Thus, using equation [16]

$$\text{periodic time} = \frac{1}{f} = 2\pi \sqrt{\frac{\theta}{\alpha}} = 2\pi \sqrt{\left[\frac{I}{G}\left(\frac{1}{J_1/L_1 + J_2/L_2}\right)\right]} \qquad [19]$$

Now consider the case of a shaft supported on bearings and carrying rotors at each end, as in figure 16.11. When set into torsional oscillations, the rotors at any instant will be moving in opposite directions and there will be a point along the shaft, called the node, at which there will be no motion. Because of this we can consider the system to be effectively two separate single rotor systems of the form shown earlier in figure 16.9. For the system to the left of the node, equation [18] can be written as

$$\text{periodic time} = \frac{1}{f} = 2\pi \sqrt{\frac{I_1 L_2}{GJ}} \qquad [20]$$

and for the part to the right

$$\text{periodic time} = \frac{1}{f} = 2\pi \sqrt{\frac{I_2 L_2}{GJ}} \qquad [21]$$

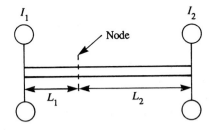

Fig. 16.11 Two rotor system.

Since the two parts of the shaft must have the same periodic time, the same frequency, then for the above two equations [20] and [21] to be identical we must have

$$I_1 L_1 = I_2 L_2 \qquad [22]$$

With known values of the moments of inertia for the rotors and a shaft of total length L, i.e. $L = L_1 + L_2$, then equation [20] enables L_1 and L_2 to be obtained and hence the periodic time of the oscillation obtained using either equation [20] or [21].

Example

A rotor with a moment of inertia of $100\,\text{kg m}^2$ is fixed between the ends of two shafts, one of which has a length of $1.0\,\text{m}$ and a diameter of $75\,\text{mm}$ and the other a length of $0.5\,\text{m}$ and diameter $60\,\text{mm}$ (as in figure 16.10). What is the frequency of torsional oscillations of the rotor? For the shaft $G = 80\,\text{GPa}$.

Since $J = \pi d^4/32$ then equation [21] can be written as

$$f = \frac{1}{2\pi}\sqrt{\left[\frac{G}{I}\left(\frac{J_1}{L_1} + \frac{J_2}{L_2}\right)\right]} = \frac{1}{2\pi}\sqrt{\left[\frac{\pi G}{32I}\left(\frac{d_1^4}{L_1} + \frac{d_2^4}{L_2}\right)\right]}$$

$$= \frac{1}{2\pi}\sqrt{\left[\frac{\pi \times 80 \times 10^9}{32 \times 100}\left(\frac{75^4}{1.0} + \frac{60^4}{0.5}\right) \times 10^{-12}\right]} = 10.7\,\text{Hz}$$

Example

Two wheels, with moments of inertia of $1.5\,\text{kg m}^2$ and $2.0\,\text{kg m}^2$, are mounted $1.0\,\text{m}$ apart on a shaft of diameter $40\,\text{mm}$, as in figure 16.11. What is the position of the node and the frequency of torsional oscillations? For the shaft $G = 80\,\text{GPa}$.

If L_1 is the distance of the node from the $1.5\,\text{kg m}^2$ wheel and L_2 that from the $2.0\,\text{kg m}^2$ wheel, then we must have $L_1 + L_2 = 1.0$ and, using equation [22], $1.5L_1 = 2.0L_2$. Thus $1.5L_1 = 2.0(1.0 - L_1)$ and so $L_1 = 0.57\,\text{m}$. The frequency of torsional oscillations is given by equation [20] as

$$f = \frac{1}{2\pi}\sqrt{\left(\frac{GJ}{I_1 L_1}\right)} = \frac{1}{2\pi}\sqrt{\left(\frac{80 \times 10^9 \times 0.040^4 \times \pi/32}{1.5 \times 0.57}\right)} = 24.4\,\text{Hz}$$

16.4 Transverse vibrations of beams

Consider a beam of length L resting on two supports at its ends and subject to a single concentrated load of mass m at its midpoint (figure 16.12). This produces a static deflection of the beam of y_{st} at the point of application of the load. The beam is then, at this point, pulled down a further amount y. The force required to give a deflection y for such a beam is $F = 48EIy/L^3$ (see chapter 4). This is thus the restoring force due to the stiffness of the beam. When released the beam has an acceleration a given by $F = ma$. Thus

$$ma = \frac{48EIy}{L^3}$$

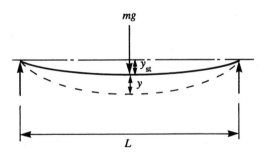

Fig. 16.12 Light beam with single concentrated load.

The acceleration a is thus proportional to the displacement y from the equilibrium position, and since the restoring force, and hence acceleration, is always directed towards the equilibrium position the resulting motion is simple harmonic oscillations. Thus, using equation [6],

$$T = \frac{1}{f} = 2\pi \sqrt{\frac{y}{a}} = 2\pi \sqrt{\frac{mL^3}{48EI}} \qquad [23]$$

The load mg on the beam produces a static deflection y_{st}, where $y_{st} = mgL^3/48EI$ (see chapter 4). Thus equation [23] can be written as

$$T = \frac{1}{f} = 2\pi \sqrt{\frac{y_{st}}{g}} \qquad [24]$$

In general for all light beams subject to a single concentrated load, whatever the form and position of the supports or the position at which the load is applied, the restoring force F is of the form $kEIy/L^3$, where k is some constant (in the above example). Then

$$T = \frac{1}{f} = 2\pi \sqrt{\frac{y}{a}} = 2\pi \sqrt{\frac{mL^3}{kEI}} \qquad [25]$$

The load mg on the beam produces a static deflection y_{st}, where $y_{st} = mgL^3/kEI$. Thus equation [25] can be written as

$$T = \frac{1}{f} = 2\pi \sqrt{\frac{y_{st}}{g}} \qquad [26]$$

which then becomes a general equation for transverse oscillations of beams. Since $\sqrt{g}$ is approximately π, equation [26] is often approximated to

$$T \approx 2\sqrt{y_{st}} \qquad [27]$$

Example

A steel strip of length 200 mm, width 25 mm and 5 mm thick, is used as a cantilever, being clamped at one end and carrying a mass 10 kg at the free end. Neglecting the mass of the cantilever, what is the frequency of the transverse oscillations if the tensile modulus of the strip is 200 GPa?

For a cantilever the static deflection y_{st} of the free end when subject to a force of mg, with $I = bt^3/112$, is (see chapter 4)

$$y_{st} = \frac{mgL^3}{3EI} = \frac{10 \times 9.8 \times 0.200^3}{3 \times 200 \times 10^9 \times 0.025 \times 0.005^3/12} = 0.0050\,\text{m}$$

Thus, using equation [26],

$$f = \frac{1}{2\pi}\sqrt{\frac{g}{y_{st}}} = \frac{1}{2\pi}\sqrt{\frac{9.8}{0.0050}} = 7.0\,\text{Hz}$$

16.4.1   Beams with uniformly distributed loads

Consider a uniform beam of length L between two supports and with a mass per unit length of m (figure 16.13). This loading will produce static deflections at each point along the beam length. Thus at some distance x from one support the static deflection is y_{st}. If the beam is now pulled down a further amount, say y at distance x, and released then transverse oscillations will occur. A reasonable approximation for the transverse oscillation frequency can be obtained if it is assumed that at any value of x the deflection y is proportional to y_{st}, i.e. $y = cy_{st}$ with c being a constant.

Consider a segment of the beam of length δx a distance x from one support. The weight of this segment is $mg\delta x$. This gives a static deflection y_{st}. The force F required to deflect the beam a further distance y is $cmg\delta x$. Hence the work done by this force moving through a distance y is $\frac{1}{2}Fy$, i.e.

$$\text{work done} = \tfrac{1}{2}(cmg\delta x)\,y = \tfrac{1}{2}(cmg\delta x)\,cy_{st} = \tfrac{1}{2}c^2 mgy_{st}\delta x$$

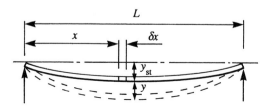

Fig. 16.13 Beam with uniformly distributed load.

The work done for the entire beam is thus

$$\text{work done} = \tfrac{1}{2}c^2 mg \int_0^L y_{st}\,dx \qquad\qquad [28]$$

This work is the energy transferred to the beam as a result of it being pulled down the extra distance y and is thus the potential energy of the beam at the maximum displacement of its transverse oscillation. This potential energy is all transformed to kinetic energy when the oscillating beam passes through its equilibrium position. But for simple harmonic motion the maximum velocity, i.e. the velocity when at the equilibrium position, is ωy (see earlier this chapter). Thus the kinetic energy for the element δx at the equilibrium position is

$$\text{kinetic energy} = \tfrac{1}{2}m\delta x(\omega y)^2 = \tfrac{1}{2}m\delta x\omega^2 c^2 y_{st}^2$$

The total kinetic energy of the beam is thus

$$\text{kinetic energy} = \tfrac{1}{2}m\omega^2 c^2 \int_0^L y_{st}^2 \, dx \qquad [29]$$

Thus equating equations [28] and [29],

$$\tfrac{1}{2}m\omega^2 c^2 \int_0^L y_{st}^2 \, dx = \tfrac{1}{2}c^2 mg \int_0^L y_{st} \, dx$$

$$\omega^2 = g \frac{\displaystyle\int_0^L y_{st} \, dx}{\displaystyle\int_0^L y_{st}^2 \, dx}$$

Since $\omega = 2\pi f = 2\pi/T$, where f is the frequency and T the periodic time, then

$$f = \frac{1}{T} = \frac{1}{2\pi} \sqrt{g \frac{\displaystyle\int_0^L y_{st} \, dx}{\displaystyle\int_0^L y_{st}^2 \, dx}} \qquad [30]$$

For a uniform beam of length L supported at its ends

$$y_{st} = \frac{mg}{24EI}(L^3 x - 2Lx^3 + x^4) \qquad [31]$$

Carrying out the integrations required for equation [30] results in

$$f = \frac{1}{T} = 4.935 \sqrt{\frac{EI}{mgL^4}} \qquad [32]$$

But the maximum static deflection y_{max} of the beam occurs at its midpoint, i.e. when $x = \tfrac{1}{2}L$, and is given by equation [31] as

$$y_{max} = \frac{mg}{24EI}\left(\frac{L^4}{2} - \frac{2L^4}{8} + \frac{L^4}{16}\right) = \frac{5mgL^4}{16 \times 24EI}$$

Hence equation [32] can be written as

$$f = \frac{1}{T} = \frac{0.564}{\sqrt{y_{max}}} \qquad [33]$$

For a cantilever, with x measured from the fixed end,

$$y_{st} = \frac{mg}{24EI}(6L^2 x^2 - 4Lx^3 + x^4)$$

with y_{max} occurring at $x = L$. Equation [30] thus gives

$$f = \frac{1}{T} = 1.755 \sqrt{\frac{EI}{mgL^4}} = \frac{0.624}{\sqrt{y_{max}}} \qquad [34]$$

For a built-in uniform beam of length L

$$y_{st} = \frac{mg}{24EI} x^2 (L - x)^2$$

with y_{max} occurring at $x = \frac{1}{2}L$. Equation [30] thus gives

$$f = \frac{1}{T} = 11.2 \sqrt{\frac{EI}{mgL^4}} = \frac{0.572}{\sqrt{y_{max}}} \qquad [35]$$

Example

What is the frequency of transverse oscillation of an I-section beam resting on two supports 5.0 m apart and which carries a uniformly distributed load of 250 kg/m? The tensile modulus of the beam is 200 GPa and the second moment of area $85 \times 10^{-6} \, m^4$.

Using equation [32],

$$f = 4.935 \sqrt{\frac{EI}{mgL^4}} = 4.935 \sqrt{\frac{200 \times 10^9 \times 85 \times 10^{-6}}{250 \times 9.8 \times 5^4}} = 16.4 \, Hz$$

16.4.2 Dunkerley's emperical method

Consider the problem of determining the transverse frequency of oscillation for a beam subject to a number of loads. The problem can be tackled in the same way as for the distributed load problem above, i.e. equating the work done to the kinetic energy of the oscillating beam at the equilibrium position. However a more convenient, though approximate, way is to use *Dunkerley's empirical method*. This method relates the frequency f with all the loads acting to the frequencies given when each load acts alone. If f_1 is the frequency given when the concentrated load m_1g acts alone, f_2 the frequency for the concentrated load m_2g acting alone, f_3 the frequency for the concentrated load m_3g acting alone, etc., and f_0 the frequency for the uniformly distributed mass of the beam itself, then

$$\frac{1}{f^2} = \frac{1}{f_1^2} + \frac{1}{f_2^2} + \frac{1}{f_3^2} + \dots + \frac{1}{f_0^2} \qquad [36]$$

But equation [27] gives $f_1 = 1/2\sqrt{y_{st1}}$, $f_2 = 1/2\sqrt{y_{st2}}$, and $f_3 = 1/2\sqrt{y_{st3}}$, where y_{st1} is the static deflection for load m_1g alone, y_{st2} for load m_2g alone, and y_{st3} for load m_3g alone. If the beam is simple supported at its ends then equation [33] gives $f_0 = 0.564/\sqrt{y_{max}}$, where y_{max} is the maximum static deflection of the beam. Thus equation [36] can be written, for a simply supported beam, as

$$\frac{1}{f^2} = 4y_{st1} + 4y_{st2} + 4y_{st3} + \ldots + \frac{y_{max}}{0.564^2}$$

$$f^2 = \frac{1}{4(y_{st1} + y_{st2} + y_{st3} + \ldots y_{max}/1.27)} \tag{37}$$

Example

A beam of length 1.0 m is supported at its ends and carries loads of 10 kN at 1.0 m and 20 kN at 2.0 m from one end. If the beam has a tensile modulus of 200 GPa and a second moment of area of 100×10^{-6} m^4, what is the frequency of transverse oscillations of the beam? Neglect the mass of the beam.

For the load at 1.0 m acting alone, i.e. an unsymmetrical load, the static deflection y_{st1} is given by (see chapter 4, *a* being the distance to the load from one end and *b* the distance from the other)

$$y_{st1} = \frac{mga^2b^2}{3EIL} = \frac{10 \times 10^3 \times 1.0^2 \times 2.0^2}{3 \times 200 \times 10^9 \times 100 \times 10^{-6}} = 6.7 \times 10^{-4} \text{ m}$$

For the second load the static deflection y_{st2} is given by

$$y_{st1} = \frac{mga^2b^2}{3EIL} = \frac{20 \times 10^3 \times 2.0^2 \times 1.0^2}{3 \times 200 \times 10^9 \times 100 \times 10^{-6}}$$

$$= 13.3 \times 10^{-4} \text{ m}$$

Thus, using equation [37],

$$f^2 = \frac{1}{4(y_{st1} + y_{st2})} = \frac{1}{4(6.7 \times 10^{-4} + 13.3 \times 10^{-4})}$$

Hence $f = 11.2$ Hz.

16.5 Whirling of shafts

A rotor mounted on a shaft will generally have its centre of gravity displaced from the axis of rotation of the shaft. This may be due to bending of the shaft under the action of the rotor weight, a lack of straightness of the shaft, or eccentric mounting of the rotor. Consider a rotor of mass M (figure 16.14) with its centre of gravity displaced e from the shaft axis, e being termed the *eccentricity*. If the mass of the shaft is negligible then, when not rotating, the rotor will cause a static deflection y_{st} of the shaft. When the shaft rotates with an angular velocity ω then the rotor mass can be considered to be rather like a concentrated mass on the end of a string being whirled round in a vertical circle. With such a motion the string extends to provide the tension to keep the mass in its circular motion. With the rotating shaft, the shaft bends more to provide the force necessary for the circular motion. Suppose the shaft bends more so that the centre of gravity of the rotor is displaced by a further distance y. The force F keeping this mass rotating in

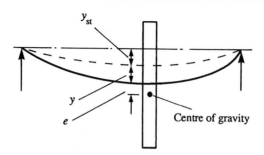

Fig. 16.14 Whirling.

its circular path is force needed to bend the shaft by this extra displacement y and is thus $F = (48EI/L^3)y$. This force provides the centripetal force for the circular motion. For circular motion we must have $F = mr\omega^2$ with r being the radius of the circle (see chapter 11), then since the radius of motion of the centre of gravity is $(y + e)$

$$m(y + e)\omega^2 = (48EI/L^3)y$$

$$y = \frac{e}{(48EI/m\omega^2 L^3) - 1} \qquad [38]$$

But for a shaft supported at its ends and with a concentrated load mg at its centre, the static deflection $y_{st} = mgL^3/48EI$ and so equation [38] can be written as

$$y = \frac{e}{(g/\omega^2 y_{st}) - 1} \qquad [39]$$

When the denominator of equation [39] is zero, i.e. $g/\omega^2 y_{st} = 1$, then y is infinite. The angular speed of rotation at which this occurs is called the *whirling* or *critical* speed ω_c. Thus

$$\omega_c = \sqrt{\frac{g}{y_{st}}} \qquad [40]$$

The frequency of transverse oscillations of such a shaft is given by equation [26] as

$$f_t = \frac{1}{2\pi} \sqrt{\frac{g}{y_{st}}}$$

Thus if we write for the angular frequency of the transverse oscillations $\omega_t = 2\pi f_t$ then we must have the whirling speed ω_c equal to the angular frequency of the transverse oscillations ω_t. Equation [39] can be written as

$$y = \frac{e}{(\omega_t^2/\omega^2) - 1} = \frac{e}{(\omega_c^2/\omega^2) - 1} \qquad [41]$$

Figure 16.15 shows how the deflection y depends on the angular speed of rotation ω. At the whirling speed the deflection y can be very large, limited

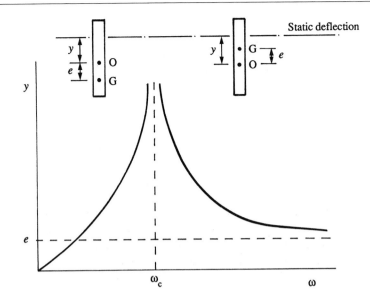

Fig. 16.15 Effect of ω on amplitude Y.

only by internal damping in the shaft from becoming infinite. For this reason speeds close to the whirling speed must be avoided in order to prevent damage to the shaft. When the speed is less than the whirling speed the centre of gravity G lies below the centre of the rotor O and e can be considered to be a positive quantity since it adds to y. When the speed is greater than the whirling speed the centre of gravity G lies above the centre of the rotor O and e can be considered to be negative since it subtracts from y. When ω is much greater than ω_c then y becomes equal to e and so the centre of gravity becomes equal to the static deflection.

Example

What is the whirling speed for a light shaft of diameter 25 mm, supported in flexible bearings 0.80 m apart, when carrying a rotor of mass 10 kg at its midpoint? If the rotor has an eccentricity of 0.5 mm from its geometrical axis, what will be the range of speeds for which the deflection will exceed 2 mm? The tensile modulus for the shaft is 200 GPa.

Equation [38],

$$y = \frac{e}{(48EI/m\omega^2 L^3) - 1}$$

gives the critical speed as being when the denominator of the equation is zero, i.e. when

$$\omega_c^2 = \frac{48EI}{mL^3} = \frac{48 \times 200 \times 10^9 \times \pi \times 0.025^4/64}{10 \times 0.80^3}$$

Hence $\omega_c = 190$ rad/s and so the frequency is $190/2\pi = 30.2$ Hz.

Using equation [41], and since e can be either positive or negative

$$y = \pm \frac{e}{(\omega_c^2/\omega^2) - 1} = 0.002 = \pm \frac{0.0005}{(190^2/\omega^2) - 1}$$

$$\pm 0.0005\omega^2 = 72.2 - 0.002\omega^2$$

Hence when e is positive $\omega = 170$ rad/s and so the frequency $= 170/2\pi = 27$ Hz, when e is negative then $\omega = 219$ rad/s and so the frequency is $219/2\pi = 35$ Hz.

Problems

(1) An object of mass 2 kg moves with simple harmonic motion with a frequency of 1.5 Hz and an amplitude of 200 mm. What is (a) the maximum velocity, (b) the velocity at 100 mm, (c) the maximum acceleration, (d) the acceleration at 100 mm, (e) the maximum restoring force, (f) the restoring force at 100 mm?

(2) An object moves with simple harmonic motion of amplitude 0.10 m and periodic time 0.50 s. What will be the time taken for the object to move from 0.20 m from its equilibrium position to 0.10 m?

(3) An object moves with simple harmonic motion of amplitude 200 mm and a frequency of 1.0 Hz. What will be the distance moved during the first 0.20 s from the extremity of the oscillation and the velocity at that point?

(4) An element of a machine has a reciprocating motion which is simple harmonic with an amplitude of 100 mm and a frequency of 4 oscillations per second. If the element has a mass of 2 kg, what is the restoring force acting on the element, and its velocity, when it is 50 mm from its mid-stroke position?

(5) An object is suspended from the end of a vertically suspended spring and oscillates with simple harmonic motion. At equilibrium the static deflection of the spring is 0.10 m. What is the frequency of the oscillation?

(6) A bar of mass 5 kg with a radius of gyration of 0.7 m is pivoted at one end and attached to a vertical spring, stiffness 200 N/m, a distance 0.50 m from the pivot (figure 16.16). At equilibrium the bar is horizontal. What is the frequency of oscillation of the bar?

(7) An object of mass 2.0 kg is supported by two springs, one of stiffness 20 kN/m and the other 30 kN/m. What will be the frequency of oscillation when the springs are connected (a) in series, (b) in parallel?

(8) Derive an equation for the frequency of oscillation of the bar, of insignificant mass, shown in figure 16.17 when there are point

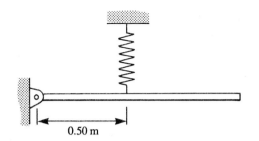

Fig. 16.16 Problem 6.

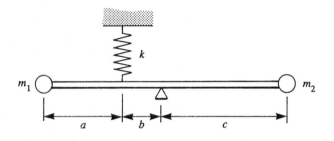

Fig. 16.17 Problem 8.

masses of m_1 and m_2 at the ends. The bar is pivoted a distance c from the mass m_2 and attached to a vertical spring, stiffness k, a distance a from mass m_1. At equilibrium the bar is horizontal.

(9)   A solid cylinder of diameter 400 mm and mass 50 kg is suspended with its axis vertical by a wire from a rigid support. If the wire has a stiffness of 20 N m/rad, what will be the frequency of the torsional oscillations?

(10)   A disc with a moment of inertia of $0.06 \, \mathrm{kg \, m^2}$ is suspended by a vertical wire from a rigid support. The wire has a length of 0.50 m, a diameter of 10 mm and a modulus of rigidity of 80 GPa. What is the frequency of the torsional oscillations of the disc?

(11)   A rotor with a moment of inertia of $20 \, \mathrm{kg \, m^2}$ is fixed between the ends of two shafts, one of which has a length of 1.0 m and a diameter of 40 mm and the other a length of 0.5 m and diameter 30 mm (as in Figure 16.10). What is the frequency of torsional oscillations of the rotor? For the shaft, $G = 80 \, \mathrm{GPa}$.

(12)   A rotor with a moment of inertia of $30 \, \mathrm{kg \, m^2}$ is fixed to the midpoint of a uniform shaft of length 0.5 m and diameter 40 mm, the ends of the shaft being rigidly fixed. What is the frequency of torsional oscillations of the rotor? For the shaft, $G = 80 \, \mathrm{GPa}$.

(13)   Two wheels are mounted 1.5 m apart on a shaft of diameter 50 mm.

If the wheels have moments of inertia of $15\,\text{kg}\,\text{m}^2$ and $25\,\text{kg}\,\text{m}^2$, what will be the position of the mode and the frequency of torsional oscillations? For the shaft, $G = 80\,\text{GPa}$.

(14)   A uniform rod of length 2.0 m oscillates as a pendulum about an axis perpendicular to its length and 0.6 m from one end. What is the periodic time of the oscillations? $(I_G = mL^2/12.)$

(15)   A thin circular hoop of diameter d hangs on a nail and swings back-and-forth. What is the periodic time of the oscillations? (Hint: consider it as a compound pendulum.)

(16)   A heavy uniform disc is suspended by three wires, each of length L, placed symmetrically round the disc rim, as in figure 16.18. The table is then rotated through a small angle and set into torsional oscillations. What is the periodic time of the oscillations?

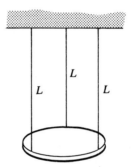

Fig. 16.18 Problem 16.

(17)   A connecting rod of mass 400 kg is suspended on a knife edge which is 0.90 m from its centre of gravity. For small oscillations as a pendulum the periodic time is found to be 2.5 s, what is the moment of inertia of the rod about an axis through its centre of gravity and perpendicular to the plane of the oscillation?

(18)   A steel strip of length 150 mm, width 10 mm and 1 mm thick, is used as a cantilever, being clamped at one end and carrying a mass of 200 g at the free end. Neglecting the mass of the cantilever, what is the frequency of the transverse oscillations if the tensile modulus of the strip is 200 GPa?

(19)   A steel strip, width 10 mm and thickness 1 mm, rests on knife edge supports 200 mm apart. What will be the frequency of the transverse oscillations of the strip when there is a mass of 150 g attached to the midpoint of the strip? The mass of the strip may be neglected and the tensile modulus taken as 200 GPa.

(20)   A horizontal shaft of diameter 25 mm is supported between two

bearings 2.0 m apart and carries a load of 1 kN at a point 0.5 m from one bearing. What is the frequency of transverse vibrations of the shaft if it has a tensile modulus of 200 GPa?

(21)   What is the transverse frequency of oscillation of an I-section cantilever of length 3.0 m when subject to a uniformly distributed load of 30 kg/m if it has a second moment of area of 40×10^{-6} m^4 and a tensile modulus of 200 GPa?

(22)   An I-section beam is supported at its ends and has a span of 3.0 m. The mass per metre of the beam is 150 kg/m and it has a second moment of area of 120×10^{-6}m^4. The tensile modulus is 200 GPa. What is the transverse frequency of oscillation of the beam?

(23)   A uniform beam of length L is supported at each end and carries loads of F at distances of $L/3$ and $2L/3$ from one end. What is the frequency of transverse oscillations of the beam? Neglect the mass of the beam.

(24)   A uniform beam of length 8.0 m is supported at each end and carries concentrated loads of 30 kN a distance of 2.0 m from one end, 20 kN a distance of 4.0 m from and 30 kN at 6.0 m from the same end. Neglecting the mass of the beam, what is the frequency of transverse oscillations? The beam has a tensile modulus of 200 GPa and a second moment of area of 100×10^{-6}m^4.

(25)   A cantilever of length 4.0 m carries concentrated loads of 20 kN a distance of 2.0 m from the fixed end and 10 kN at the free end. Neglecting the mass of the beam, what is the frequency of transverse oscillations? The cantilever has a tensile modulus of 200 GPa and a second moment of area of 100×10^{-6}m^4.

(26)   An I-section beam of mass 50 kg/m and second moment of area 120×10^{-6}m^4 has a span of 4.0 m and carries a concentrated load of 10 kN at its midpoint. What is the frequency of transverse oscillations of the beam? The beam has a tensile modulus of 200 GPa.

(27)   A rotor of mass 80 kg is fixed at the midpoint of a shaft of length 0.60 m and diameter 40 mm, the shaft being supported at its ends. If the shaft has a density of 7.8 Mg/m^3 and tensile modulus of 200 GPa, what is the transverse frequency of oscillation of the system?

(28)   An I-section beam has a mass of 52 kg/m, a tensile modulus of 200 GPa, and a second moment of area of 120×10^{-6} m^4. It has a length of 6.0 m and is supported at each end. What is the transverse frequency of oscillation of the beam when it is carrying a concentrated load of 1000 kg a distance of 2.4 m from one end?

(29)   A shaft of diameter 10 mm is supported on two flexible bearings 0.50 m apart and carries a rotor of mass 5.0 kg at the centre of its span. What is the whirling speed of the system and what will be the

deflection of the shaft when the angular speed is 10 rev/s? The eccentricity of the rotor relative to its centre is 0.05 mm.

(30) A shaft of diameter 12 mm is supported on two flexible bearings 300 mm apart and carries a rotor of mass 10 kg at a point 125 mm from one bearing. What is the whirling speed?

(31) What is the maximum eccentricity permissible for a rotor carried on a light shaft between two flexible bearings if when the shaft rotates at twice the whirling speed the deflection produced by the rotation is not to exceed 0.05 mm?

(32) What is the maximum deflection produced by a rotor mounted centrally on a light shaft mounted between two flexible bearings when the speed of rotation is 10% above the whirling speed if the eccentricity of the rotor is 0.10 mm?

(33) A rotor mounted on a light shaft between two flexible bearings has an eccentricity of 0.50 mm and a whirling speed of 10 rev/s. What is the speed range over which the deflection of the rotor from the static position will exceed 1.3 mm?

Chapter 17

Damped and forced oscillations

17.1 Free oscillations

Mechanical systems can often be considered to be capable of being re-presented by a spring suspended from some rigid support with a weight attached to the lower end of the spring. When the mass is pulled down and then released, oscillations occur. In the ideal situation these oscillations would continue indefinitely, the motion then being referred to as *free or natural oscillations*. In practice the oscillations die away with time and so we have to include with the model a means of providing damping.

Figure 17.1 shows a free oscillation system with no damping. The restoring force acting on the mass m when it is pulled down some distance x and then released is the force exerted by the stretched spring, i.e. $-kx$ where k is the stiffness and a minus sign is included because the restoring force is in the opposite direction to that in which x increases. This force causes an acceleration a of the mass. Thus $-kx = ma$. But acceleration is the rate of change of velocity with time dv/dt and velocity v is the rate of change of displacement with time x, i.e. dx/dt. Thus $a = d(dx/dt)/dt = d^2x/dt^2$. Hence

$$-kx = m\,\frac{d^2x}{dt^2}$$

or, when rearranged,

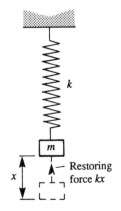

Fig. 17.1 Free oscillations.

$$m \frac{d^2x}{dt^2} + kx = 0 \qquad [1]$$

From the previous chapter we know that this equation must describe an object moving with simple harmonic motion and thus the solution to this differential equation is $x = A \cos \omega_n t$, where A is the amplitude and ω_n the natural angular frequency. This gives

$$\text{acceleration} = \frac{d^2x}{dt^2} = -A\omega_n^2 \cos \omega_n t = -\omega_n^2 x$$

Hence equation [1] gives

$$-m\omega_n^2 x + kx = 0$$

and so $k = m\omega_n^2$. Thus equation [1] can be written as

$$\frac{d^2x}{dt^2} + \omega_n^2 x = 0 \qquad [2]$$

Equations [1] and [2] are called second order differential equations.

Note that the solution to the above equation is, as given above, $x = A \cos \omega_n t$ but that in general it will be found to be quoted as

$$x = a \cos \omega_n t + b \sin \omega_n t$$

However, since for the situation we have used $x = A$ when $t = 0$ and $dx/dt = 0$ when $t = 0$ then $A = a$ and $b = 0$ and so $x = A \cos \omega_n t$.

17.1.1 Free oscillations with angular motion

Similar equations to [1] and [2] can be derived for angular motion. Thus the restoring force acting on a body with moment of inertia I when it is rotated through some angle θ, as when a shaft is twisted, is $-q\theta$, where q is the torsional stiffness. When released then $-q\theta = I\alpha$, where α is the angular acceleration. But $\alpha = d^2\theta/dt^2$ and so

$$-q\theta = I \frac{d^2\theta}{dt^2}$$

$$I \frac{d^2\theta}{dt^2} + q\theta = 0 \qquad [3]$$

Since this equation describes simple harmonic motion the solution is (see previous chapter) $\theta = \theta_{max} \cos \omega_n t$. This gives $d^2\theta/dt^2 = -\theta_{max}\omega_n^2 \cos \omega_n t = -\omega_n^2 \theta$. Hence equation [3] gives $-I\omega_n^2\theta + q\theta = 0$ and so $q = I\omega_n^2$. Equation [3] can thus be written as

$$\frac{d^2\theta}{dt^2} + \omega_n^2 \theta = 0 \qquad [4]$$

17.2 Damped oscillations

Consider a mass suspended from a vertical spring where there is *damping* (figure 17.2). Damping is often what is termed *viscous damping* and represented by the movement of a piston in a container, i.e. a dash pot. Such damping produces a resistive force which is proportional to velocity. When the mass is pulled down a distance x then the restoring force is $-kx$. When the mass is released the forces acting on the mass are the restoring force and the damping force. The force due to the damping is proportional to the rate at which the displacement of the piston is changing, i.e. proportional to dx/dt, and can thus be written as $-c\,dx/dt$, where c, which is the damping force per unit velocity, is a constant called the *damping coefficient*. The minus sign is because the damping force is in the opposite direction to the dx/dt. Hence the net force acting on the mass is

$$\text{net force} = -kx - c\frac{dx}{dt}$$

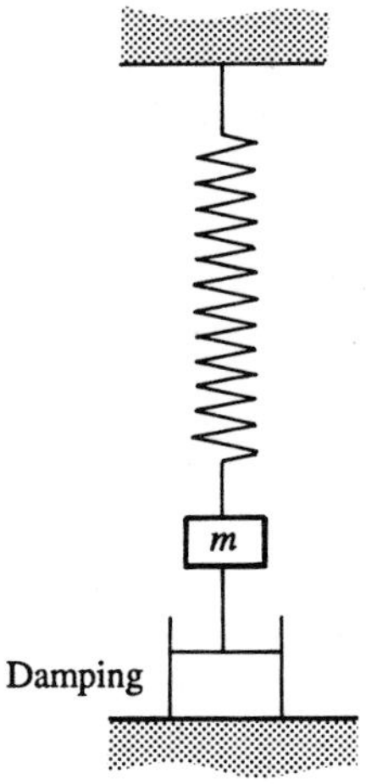

Fig. 17.2 Damped oscillations.

This net force causes an acceleration a of the mass, with $a = d^2x/dt$. Hence

$$-kx - c\frac{dx}{dt} = m\frac{d^2x}{dt^2}$$

or, when rearranged,

$$m\frac{d^2x}{dt^2} + c\frac{dx}{dt} + kx = 0 \qquad\qquad [5]$$

There are a number of alternative ways of writing this equation, one way being

$$\frac{d^2x}{dt^2} + 2\zeta\omega_n\frac{dx}{dt} + \omega_n^2 x = 0 \qquad\qquad [6]$$

where $2\zeta\omega_n = c/m$ and $\omega_n^2 = k/m$, with ω_n being the natural angular frequency for the free oscillations. The term ζ is called the *damping ratio* (see later for a discussion of the significance of this term). Sometimes instead of the damping ratio the substitution made is for $2\mu = c/m$ to give

$$\frac{d^2x}{dt^2} + 2\mu\frac{dx}{dt} + \omega_n^2 x = 0$$

If there was no damping and there was free oscillations then the $c\,dx/dt$ or $2\zeta\omega_n\,dx/dt$ or $2\mu\,dx/dt$ term is zero, i.e. $c = \zeta = \mu = 0$, and the equation becomes identical with equation [4].

The way in which x varies with time t depends on the value of the damping factor ζ. With $\zeta > 1$ (or $\mu > \omega_n$ or c greater than some critical value c_c) then there are no oscillations but when the mass, after being pulled down and given an initial displacement x, is released then x slowly decreases to zero (figure 17.3). Such a motion is said to be *over damped*. With $\zeta < 1$ (or $\mu < \omega_n$ or $c < c_c$) the mass, after being released, oscillates about the $x = 0$ value with oscillations of steadily decreasing amplitudes until eventually it comes to rest (figure 17.3). Such a motion is said to be *under damped*. When $\zeta = 1$ (or $\mu = \omega_n$ or $c = c_c$) the deflection of the mass decays to zero as quickly as is possible without any oscillations occurring (figure 17.3). Such a motion is said to be *critically damped*. Thus, since for a given system ζ is proportional to the damping coefficient, the damping ratio can be considered to be the ratio c/c_c, where c_c is the damping coefficient when there is critical damping.

The equations describing how x varies with time for different values of ζ are derived in the following section.

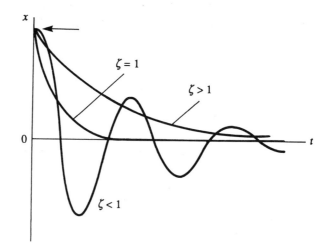

Fig. 17.3 The effect of the damping factor.

17.2.1 *Solving the damped differential equation*

To solve the differential equation [6] for the damped motion we can try a solution of the form

$$x = Ae^{st}$$

With such a solution

$$\frac{dx}{dt} = Ase^{st}$$

$$\frac{d^2x}{dt^2} = As^2e^{st}$$

Thus equation [6] becomes

$$As^2e^{st} + 2\zeta\omega_n Ase^{st} + \omega_n^2 Ae^{st} = 0$$

$$s^2 + 2\zeta\omega_n s + \omega_n^2 = 0 \tag{7}$$

Thus $x = Ae^{st}$ can only be a solution provided the above equation equals 0. Equation [7] is called the *auxiliary equation*. The roots of this quadratic equation can be obtained by using the formula for the roots of an equation of the form $ax^2 + bx + c = 0$, i.e.

$$x = \frac{-b \pm \sqrt{(b^2 - 4ac)}}{2a}$$

Thus, for equation [7],

$$s = \frac{-2\zeta\omega_n \pm \sqrt{(4\zeta^2\omega_n^2 - 4\omega_n^2)}}{2}$$

$$s = -\zeta\omega_n \pm \omega_n\sqrt{(\zeta^2 - 1)} \tag{8}$$

When ζ^2 is greater than 1 the square root term gives the square root of a positive number and there are two different real roots s_1 and s_2, i.e.

$$s_1 = -\zeta\omega_n + \omega_n\sqrt{(\zeta^2 - 1)}$$

$$s_2 = -\zeta\omega_n - \omega_n\sqrt{(\zeta^2 - 1)}$$

and so the general solution for x is

$$x = A\exp s_1 t + B\exp s_2 t \tag{9}$$

where A and B are constants. Thus when the damping factor is greater than 1 we have s_1 and s_2 as negative numbers and the variation of the displacement with time is the sum of two decaying exponentials. This means no oscillation but a displacement which slowly dies to zero with time. For such conditions the system is said to be *over damped*.

When $\zeta = 1$ there are two equal roots with $s_1 = s_2 = -\omega_n$. For this condition, which is called *critically damped*,

$$x = (At + B)\exp -\omega_n t \tag{10}$$

It may seem that the solution for this case should be $x = Ae^{st}$, but such a solution, with just one constant A, is not capable of satisfying the initial conditions for a second order system.

When ζ^2 is less than 1 we have the square root of a negative number. There are two complex roots since the roots both involve the square root of minus one. Thus equation [8] can be written as

$$s = -\zeta\omega_n \pm \omega_n\sqrt{(\zeta^2 - 1)} = -\zeta\omega_n \pm \omega_n\sqrt{(-1)}\,(1 - \zeta^2)$$

and so writing j for $\sqrt{(-1)}$,

$$s = -\zeta\omega_n \pm j\omega_n\sqrt{(1 - \zeta^2)}$$

If we let

$$\omega = \omega_n\sqrt{(1 - \zeta^2)} \qquad\qquad\qquad [11]$$

then

$$s = -\zeta\omega_n \pm j\omega$$

and so the two roots can be written as

$$s_1 = -\zeta\omega_n + j\omega$$
$$s_2 = -\zeta\omega_n - j\omega$$

The term ω is the angular frequency of the motion when it is damped (this is made clear by the result obtained in equation [12] when the oscillation ends up being described by an equation of the form $\cos \omega t$). The solution under these conditions is

$$x = A\exp(-\zeta\omega_n + j\omega)t + B\exp(-\zeta\omega_n + j\omega)t$$
$$x = \exp(-\zeta\omega_n t)[A\exp(j\omega t) + B\exp(-j\omega t)]$$

But

$$\exp j\omega t = \cos \omega t + j\sin \omega t$$

and

$$\exp -j\omega t = \cos \omega t - j\sin \omega t$$

Hence

$$x = \exp(-\zeta\omega_n t)[A\cos \omega t + jA\sin \omega t + B\cos \omega t - jB\sin \omega t]$$
$$x = \exp(-\zeta\omega_n t)[(A + B)\cos \omega t + j(A - B)\sin \omega t]$$

If we substitute constants P and Q for $(A + B)$ and $j(A - B)$, then

$$x = \exp(-\zeta\omega_n t)(P\cos \omega t + Q\sin \omega t) \qquad\qquad [12]$$

This equation describes an oscillation which has an amplitude which decays exponentially with time. For such conditions the system is said to be *under damped*.

Example

A mass of 1.0 kg is suspended from a vertical spring of stiffness 100 N/m and is provided with dash pot damping which provides a damping force of 2.5 N when the velocity of the mass is 0.25 m/s. What is the frequency of the undamped (free) motion, the damping coefficient, the damping factor and the frequency of the damped motion?

The frequency f_n of free motion is given by equation [11] of the previous chapter as

$$f_n = \frac{1}{2\pi} \sqrt{\frac{k}{m}} = \frac{1}{2\pi} \sqrt{\frac{100}{1.0}} = 1.59 \text{ Hz}$$

The damping coefficient c is the damping force per unit velocity and so is $2.5/0.25 = 10$ N s/m. Since $2\zeta\omega_n = c/m$ then the damping factor $\zeta = 10/(2 \times 1.59 \times 2\pi \times 1.0) = 0.50$. Equation [11] then gives

$$\omega = \omega_n\sqrt{(1 - \zeta^2)} = 1.59 \times 2\pi \sqrt{(1 - 0.50^2)}$$

and so $f = \omega/2\pi = 1.38$ Hz.

17.2.2 *Logarithmic decrement*

For a damped oscillation described by equation [12], i.e.

$$x = \exp(-\zeta\omega_n t)(P\cos\omega t + Q\sin\omega t)$$

the angular frequency is ω and so the periodic time T is $2\pi/\omega$. But equation [11] gives

$$\omega = \omega_n\sqrt{(1 - \zeta^2)}$$

and so the periodic time, and frequency f, is

$$T = \frac{1}{f} = \frac{2\pi}{\omega} = \frac{2\pi}{\omega_n\sqrt{(1 - \zeta^2)}} \qquad [13]$$

Suppose initially the mass is pulled down a distance x_1 and then released. Then $x = x_1$ when $t = 0$. Equation [12] thus becomes

$$x_1 = \exp(-0)(P\cos 0 + Q\sin 0) = P$$

Suppose now that after one complete oscillation, i.e. the periodic time T, that the deflection is x_2 (figure 17.4). Then $x = x_2$ at $t = T = 2\pi/\omega$. Equation [12] then becomes

$$x_2 = \exp(-\zeta\omega_n T) (P\cos 2\pi + Q\sin 2\pi) = P\exp(-\zeta\omega_n T)$$

Hence

$$\frac{x_1}{x_2} = \frac{P}{P\exp(-\zeta\omega_n T)} = \exp(\zeta\omega_n T)$$

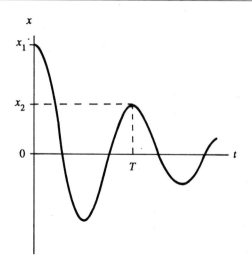

Fig. 17.4 Damped oscillations.

$$\ln\left(\frac{x_1}{x_2}\right) = \zeta\omega_n T = \frac{2\pi\zeta\omega_n}{\omega} \qquad [14]$$

The term $\ln(x_1/x_2)$ is known as the *logarithmic decrement*. Since equation [11] gives $\omega = \omega_n\sqrt{(1 - \zeta^2)}$, equation [14] can be written as

$$\ln\left(\frac{x_1}{x_2}\right) = \frac{2\pi\zeta}{\sqrt{(1 - \zeta^2)}} \qquad [15]$$

Example

A mass suspended from a spring is subject to viscous damping and oscillates with a damped frequency of 2.0 Hz and an amplitude which decreases by 20% in one complete oscillation. What is the damping ratio and the frequency of the undamped oscillations?

Using equation [15], with x_2 being 20% of x_1,

$$\ln\left(\frac{x_1}{x_2}\right) = \frac{2\pi\zeta}{\sqrt{(1 - \zeta^2)}} = \ln 5 = 1.6$$
$$(2\pi\zeta/1.6)^2 = 1 - \zeta^2$$

Hence $\zeta = 0.25$. Using equation [11]

$$\omega = \omega_n\sqrt{(1 - \zeta^2)}$$
$$2\pi \times 2.0 = \omega_n\sqrt{(1 - 0.25^2)}$$

Hence $\omega_n = 13.0\,\text{rad/s}$ and $f_n = \omega_n/2\pi = 2.07\,\text{Hz}$.

Example

A mass of 50 kg is suspended from a vertical spring of stiffness 10 kN/m. When pulled down and released it performs damped oscillations for which the amplitude is reduced to one tenth in 5 complete oscillations. If the damping is viscous, what is the damping ratio and the frequency of the damped oscillations?

With $x = x_1$ at $t = 0$ equation [12] becomes

$$x_1 = \exp(-0)(P\cos 0 + Q\sin 0) = P$$

With $x = x_5$ after 5 complete oscillations, i.e. a time $5T$ or $5(2\pi/\omega)$, where T is the periodic time and ω the damped angular frequency, equation [12] then becomes

$$x_5 = \exp(-\zeta\omega_n 5T)(P\cos 10\pi + Q\sin 10\pi)$$
$$= P\exp(-\zeta\omega_n 5T)$$

Hence

$$\frac{x_1}{x_5} = \frac{P}{P\exp(-\zeta\omega_n 5T)} = \exp(\zeta\omega_n 5T)$$

$$\ln\left(\frac{x_1}{x_5}\right) = \zeta\omega_n 5T = \frac{10\pi\zeta\omega_n}{\omega}$$

Since equation [11] gives $\omega = \omega_n\sqrt{(1 - \zeta^2)}$, the above equation can be written as

$$\ln\left(\frac{x_1}{x_5}\right) = \frac{10\pi\zeta}{\sqrt{(1 - \zeta^2)}}$$

Hence

$$\ln\left(\frac{1}{0.1}\right) = 2.3 = \frac{10\pi\zeta}{\sqrt{(1 - \zeta^2)}}$$

and $\zeta = 0.073$. The frequency f_n of free motion is given by equation [11] of the previous chapter as

$$f_n = \frac{1}{2\pi}\sqrt{\frac{k}{m}} = \frac{1}{2\pi}\sqrt{\frac{10 \times 10^3}{50}} = 2.25\,\text{Hz}$$

Using equation [11]

$$\omega = \omega_n\sqrt{(1 - \zeta^2)} = 2\pi \times 2.25\sqrt{(1 - 0.073^2)}$$

and $f = \omega/2\pi = 2.24\,\text{Hz}$.

17.2.3   Angular damped oscillations

Consider a rotor of moment of inertia I on a shaft of torsional stiffness q

and subject to viscous damping with a damping torque of c per unit angular velocity. If the shaft is given an angular displacement θ then the restoring force resulting from the twisted shaft is $q\theta$. When the shaft is released and motion occurs then the damping force is $c\,d\theta/dt$. Hence, as with the discussion for the oscillations of the damped spring earlier in this chapter,

$$I\alpha = I\frac{d^2\theta}{dt^2} = -q\theta - c\frac{d\theta}{dt}$$

$$I\frac{d^2\theta}{dt^2} + c\frac{d\theta}{dt} + q\theta = 0 \qquad [16]$$

With $\omega_n^2 = q/I$ and $2\zeta\omega_n = c/I$ then equation [16] can be written as

$$\frac{d^2\theta}{dt^2} + 2\zeta\omega_n\frac{d\theta}{dt} + \omega_n^2\theta = 0 \qquad [17]$$

Equations [16] and [17] are in the same form as those for the motion of a mass on the end of a spring, i.e. equations [5] and [6], and so have similar solutions.

Example

A rotor has a moment of inertia of $0.10\,\text{kg m}^3$ and is immersed in a viscous fluid (figure 17.5). The shaft to which the rotor is attached has a diameter of 10 mm, a length of 400 mm and a modulus of rigidity of 40 GPa. When the rotor is given an angular displacement and then released, it performs damped torsional oscillations. If the angular deflections for successive amplitudes are $5°$ and $3°$, what is the logarithmic decrement, the damping ratio and the frequency of the damped oscillations?

As with the oscillations of a mass on a spring, the logarithmic decrement is $\ln(\theta_1/\theta_2) = \ln(5/3) = 0.51$. Using the angular equivalent of equation [15],

$$\ln\left(\frac{\theta_1}{\theta_2}\right) = \frac{2\pi\zeta}{\sqrt{(1 - \zeta^2)}} = 0.51$$

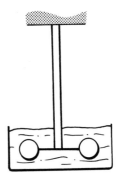

Fig. 17.5 Example.

Hence $\zeta = 0.081$. The natural frequency f_n is given by

$$f_n = \frac{1}{2\pi} \sqrt{\frac{GJ}{IL}} = \frac{1}{2\pi} \sqrt{\frac{40 \times 10^9 \times 0.01^4 \times \pi/32}{0.10 \times 0.400}} = 4.99 \, \text{Hz}$$

Using equation [11]

$$\omega = \omega_n \sqrt{(1 - \zeta^2)} = 2\pi \times 4.99 \sqrt{(1 - 0.081^2)}$$

and $f = \omega/2\pi = 4.97 \, \text{Hz}$.

17.3   Undamped forced oscillations

Consider a mass suspended from a spring, with no damping present, but the mass acted on by a force which varies sinusoidally with time, i.e. disturbing force $= F \cos \omega_f t$ with ω_f being the angular frequency of this force and F its maximum value (figure 17.6). The resulting oscillations of the mass are said to be *forced*, with ω_f being the *forcing frequency*. At some displacement x from the equilibrium position there will be restoring force $-kx$, where k is the spring stiffness. Thus the net force acting on the object is $(F \cos \omega_f t - kx)$. Hence

$$F \cos \omega_f t - kx = ma = m \frac{d^2x}{dt^2}$$

$$m \frac{d^2x}{dt^2} + kx = F \cos \omega_f t \qquad\qquad [18]$$

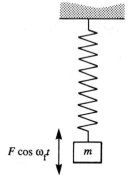

$F \cos \omega_f t$ m

Fig. 17.6 Forced oscillations.

As before we can write for the free oscillation frequency of the mass $\omega_n^2 = k/m$ and so equation [18] becomes

$$\frac{d^2x}{dt^2} + \omega_n^2 x = \frac{F}{m} \cos \omega_f t \qquad\qquad [19]$$

This second order differential equation can be solved by substituting for x

$$x = u + v \tag{20}$$

Then equation [19] becomes

$$\frac{d^2(u + v)}{dt^2} + \omega_n^2(u + v) = 0 + \frac{F}{m}\cos\omega_f t$$

This equation can be split into two parts,

$$\frac{d^2 u}{dt^2} + \omega_n^2 u = 0 \tag{21}$$

$$\frac{d^2 v}{dt^2} + \omega_n^2 v = \frac{F}{m}\cos\omega_f t \tag{22}$$

Equation [21] is just the equation for the free oscillations of the undamped system solved earlier this chapter, i.e. the solution of equation [2] as just simple harmonic motion. Thus, in general,

$$u = A\sin\omega_n t + B\cos\omega_n t \tag{23}$$

Equation [22] can be solved by assuming a solution of the form $v = C\cos\omega_f t$, and so since $dv/dt = C\omega_f\sin\omega_f t$ and $d^2v/dt^2 = -C\omega_f^2\cos\omega_f t$ equation [23] becomes

$$-C\omega_f^2\cos\omega_f t + \omega_n^2 C\cos\omega_f t = \frac{F}{m}\cos\omega_f t$$

and hence, cancelling out the cosine terms,

$$C = \frac{F/m}{\omega_n^2 - \omega_f^2}$$

Thus the solution for equation [22] is

$$v = \frac{F/m}{\omega_n^2 - \omega_f^2}\cos\omega_f t \tag{24}$$

Hence the complete solution for x is, when equations [23] and [24] are substituted in equation [20],

$$x = A\sin\omega_n t + B\cos\omega_n t + \frac{F/m}{\omega_n^2 - \omega_f^2}\cos\omega_f t \tag{25}$$

The solution thus consists of two parts, one due to the natural oscillations of the mass, i.e. the u part of the solution, and the other due to the forcing oscillation, i.e. the v part of the solution. Since all oscillating systems are subject to friction and damping, the natural oscillations of the mass will eventually die away. For this reason they are referred to as *transient*. Thus eventually all that will be left is the oscillation due to the forcing oscillation and so this is referred to as *steady state*. Thus the steady state oscillation of the mass that will occur after all transient have died away is

$$x_s = \frac{F/m}{\omega_n^2 - \omega_f^2}\cos\omega_f t \tag{26}$$

This is an oscillation of angular frequency ω_f and amplitude

$$\text{amplitude} = \frac{F/m}{\omega_n^2 - \omega_f^2} \qquad [27]$$

When $\omega_f = \omega_n$ then the amplitude is infinite (figure 17.7). This condition is called *resonance*. With $\omega_f < \omega_n$ then the amplitude term is positive and so the mass oscillates in phase with the forcing oscillation. With $\omega_f > \omega_n$ then the amplitude term is negative and so $x = -A\cos\omega_f t = A\cos(\omega_f t - 180°)$ and the mass oscillates 180° out of phase with the forcing oscillation. Thus resonance marks the transition from the mass oscillating in phase to 180° out of phase with the forcing oscillation.

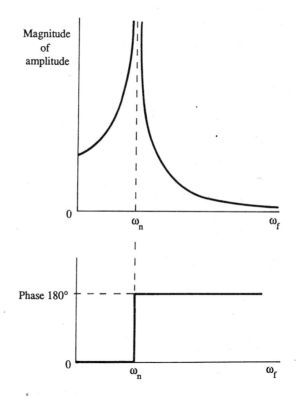

Fig. 17.7 Steady state amplitude and phase.

The force acting on the support for the spring is the algebraic sum of the force mg due to the weight of the mass stretching the spring plus the force resulting from the stretching of the spring as a result of the oscillations of the mass, i.e. kx. The maximum value of the force will thus be when the oscillation displacement is a maximum, i.e. equal to the amplitude A, and the resulting force adds to weight to give $mg + kA$. The minimum value of the force would be when it subtracts from the weight, i.e. $mg - kA$. Thus the maximum force on the support is, using equation [27],

$$\text{max. force} = mg + \frac{F/m}{\omega_n^2 - \omega_f^2} \qquad [28]$$

Similar equations to those obtained above can be obtained for a rotor of moment of inertia I supported by a shaft of torsional stiffness q and subject to a sinusoidal torque of $T\cos\omega_f t$, i.e.

$$\frac{d^2\theta}{dt^2} + \omega_n^2\theta = \frac{T}{I}\cos\omega_f t \qquad [29]$$

with $\omega_n^2 = q/I$. The solution of this equation is similar to that for the mass on the end of the spring.

Example

A mass of 30 kg is supported by a vertical spring of stiffness 25 kN/m and is acted on by a periodic force of amplitude 50 N and frequency 5.0 Hz. What is the steady state amplitude of the forced vibration and the maximum force acting on the support point for the spring?

The natural angular frequency ω_n is given by

$$\omega_n = \sqrt{\frac{k}{m}} = \sqrt{\frac{25 \times 10^3}{30}} = 28.9\,\text{rad/s}$$

Thus, since $\omega_f = 2\pi \times 5.0 = 31.4\,\text{rad/s}$, then the amplitude is given by equation [27] as

$$\text{amplitude} = \frac{F/m}{\omega_n^2 - \omega_f^2} = \frac{50/30}{28.9^2 - 31.4^2} = -0.011\,\text{m}$$

The minus sign is because the oscillation is 180° out of phase with the forcing oscillation. The force acting on the support is

$$\text{force} = mg + kx$$

and this is a maximum when x is equal to the amplitude. Thus

$$\text{max. force} = 30 \times 9.8 + 25 \times 10^3 \times 0.011 = 569\,\text{N}$$

17.3.1 Periodic support movement

In the above discussion it was assumed that a periodic force was acting on the mass suspended by a spring from a rigid support and directly resulting in a periodic movement of the mass. In some situations, however, the periodic force may act on the support and result in periodic movement of it (figure 17.8). Consider the support displacement y with time t to be given by

$$y = Y\cos\omega_f t$$

where Y is its maximum value and ω_f its angular frequency. At some instant

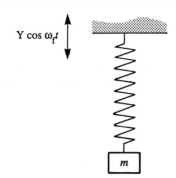

$Y \cos \omega_f t$

Fig. 17.8 Periodic support movement.

when the displacement of the mass from its equilibrium value is x then the spring will be of length $(x - y)$. Therefore the restoring force acting on the mass is $k(x - y)$. Thus, in the absence of damping,

$$-k(x - y) = ma = m \frac{d^2x}{dt^2}$$

$$m \frac{d^2x}{dt^2} + kx = ky = kY \cos \omega_f t$$

$$\frac{d^2x}{dt^2} + \omega_n^2 x = \omega_n^2 Y \cos \omega_f t \qquad [30]$$

where $\omega_n^2 = k/m$. This equation is of the same form as equation [22] and has a similar solution, namely

$$x_s = \frac{\omega_n^2 Y}{\omega_n^2 - \omega_f^2} \cos \omega_f t \qquad [31]$$

This is an oscillation of angular frequency ω_f and amplitude

$$\text{amplitude} = \frac{\omega_n^2 Y}{\omega_n^2 - \omega_f^2} \qquad [32]$$

Example

An instrument is rigidly attached to a platform which is supported by four springs, each having a stiffness 1.0 kN/m (figure 17.9). The floor to which the springs are attached is subject to a periodic displacement of amplitude 12 mm and angular frequency 10 rad/s. The total mass of the instrument plus platform is 20 kg and the instrument is constrained to move only vertically. What is the maximum vertical steady state displacement of the platform from the equilibrium position?

The natural angular frequency ω_n is given by

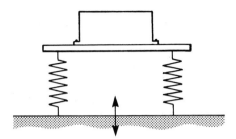

Fig. 17.9 Example.

$$\omega_n = \sqrt{\frac{k}{m}} = \sqrt{\frac{4 \times 1.0 \times 10^3}{20}} = 14.1\,\text{rad/s}$$

The maximum displacement is the amplitude and so given by equation [32] as

$$\text{amplitude} = \frac{\omega_n^2 Y}{\omega_n^2 - \omega_f^2} = \frac{14.1^2 \times 0.012}{14.1^2 - 10^2} = 0.024\,\text{m}$$

17.4 Damped forced oscillations

Consider a mass m suspended by a spring, stiffness k, from a rigid support and with viscous damping (figure 17.10). A sinusoidal disturbing force of $F\cos\omega_f t$ is then applied to the mass. At some displacement x from the equilibrium position there will be a restoring force kx and when it is moving with a velocity of dx/dt a damping force of $c\,dx/dt$, where c is the damping coefficient. Thus the net force acting on the object is $(F\cos\omega_f t - kx - c\,dx/dt)$. Hence

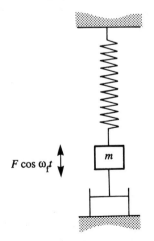

$F\cos\omega_f t$

Fig. 17.10 Damped forced oscillations.

$$F \cos \omega_f t - kx - c \frac{dx}{dt} = ma = m \frac{d^2x}{dt^2}$$

$$m \frac{d^2x}{dt^2} + c \frac{dx}{dt} + kx = F \cos \omega_f t \qquad [33]$$

As before we can write for the free oscillation frequency of the mass $\omega_n^2 = k/m$ and for the damping $2\zeta\omega_n = c/m$ (or as sometimes written $2\mu = c/m$), with ζ being the damping factor, so equation [33] becomes

$$\frac{d^2x}{dt^2} + 2\zeta\omega_n \frac{dx}{dt} + \omega_n^2 x = \frac{F}{m} \cos \omega_f t \qquad [34]$$

This second order differential equation can be solved by using $x = u + v$ and so obtaining,

$$\frac{d^2(u + v)}{dt^2} + 2\zeta\omega_n \frac{d(u + v)}{dt} + \omega_n^2(u + v) = 0 + \frac{F}{m} \cos \omega_f t$$

This equation can be split into two parts,

$$\frac{d^2u}{dt^2} + 2\zeta\omega_n \frac{du}{dt} + \omega_n^2 u = 0 \qquad [35]$$

$$\frac{d^2v}{dt^2} + 2\zeta\omega_n \frac{dv}{dt} + \omega_n^2 v = \frac{F}{m} \cos \omega_f t \qquad [36]$$

Equation [35] is just the equation for the damped oscillations with no disturbing force of the system solved earlier this chapter, i.e. the solution of equation [6]. This motion will however die away with time, being transient. Thus if we are concerned with just the steady state motion then this is just the solution of equation [36] that is required. Equation [36] can be solved by assuming a solution of the form $v = C \sin \omega_f t + D \cos \omega_f t$, substituting in equation [36] and then equating the coefficients of all the sine terms and equating all the coefficients of the cosine terms. Solving the two resulting simultaneous equations leads to values for C and D. Hence v and so the steady state value of x is given by

$$x_s = \frac{F/m}{\sqrt{[4\zeta^2\omega_n^2\omega_f^2 + (\omega_n^2 - \omega_f^2)^2]}} \cos(\omega_f t - \theta) \qquad [37]$$

where

$$\theta = \tan^{-1}\left(\frac{2\zeta\omega_n\omega_f}{\omega_n^2 - \omega_f^2}\right) \qquad [38]$$

The amplitude of the resulting oscillation is given by the first part of equation [37] and the phase angle by which the oscillation lags behind the forcing oscillation by θ. The amplitude and the phase angle depend not only on the values of ω_n and ω_f but also the damping factor ζ. With $\zeta = 0$, i.e. no damping, equation [37] becomes the same as equation [26] for undamped forced oscillation. Figure 17.11 shows graphs of how the magnitude of the

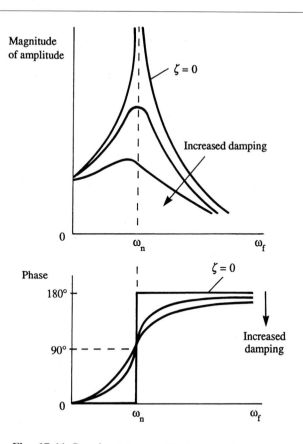

Fig. 17.11 Steady state amplitude and phase.

amplitude and the phase angle vary with forcing frequency for a number of different damping factors.

There is an alternative way of solving this problem in order to obtain the steady state motion. Acting on the object are a number of forces, each of which varies sinusoidally with time. But a quantity which varies sinusoidally with time can be represented by the rotation of a radius line of length equal to the amplitude of the force and with an angular velocity which equals the angular frequency of the force (see the previous chapter). Such a rotating radius line is often referred to as a vector or phasor. If two such forces are oscillating with a 90° phase difference then the vectors will always be drawn with a 90° angle between them. Consider the effect of all the forces acting on the system to be a displacement of the form $x = A \cos(\omega_f t - \theta)$, then the spring force $-kx$ is

$$-kx = -kA \cos(\omega_f t - \phi) = kA \cos(\omega_f t - \phi - 180°)$$

and so is a vector of length kA and lagging the line drawn to represent the direction of x by 180°. The damping force $-c \, dx/dt$ is

$$-c\frac{dx}{dt} = -c\omega_f A\sin(\omega_f t - \phi) = c\omega_f A\cos(\omega_f t - \phi - 90°)$$

and so is a vector of length $c\omega_f A$ and lagging the line drawn to represent the direction of x by 90°. The disturbing force is $F\cos\omega_f t$

$$F\cos\omega_f t = F\cos(\omega_f t - \phi + \phi)$$

and so is a vector of length F and leading the line drawn to represent the direction of x by ϕ. The resultant of these three forces is

$$-m\frac{d^2x}{dt^2} = -m\frac{d^2[A\cos(\omega_f t - \phi)]}{dt^2} = m\omega_f^2 A\cos(\omega_f t - \phi)$$

and so is a vector of length $m\omega_f^2 A$ and in the same direction as the line representing x. Figure 17.12 shows these vectors, the diagram being a vector representation of equation [33].

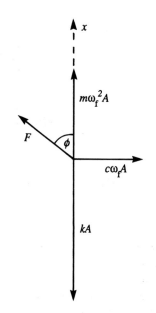

Fig. 17.12 Vectors for damped forced oscillations.

Since the $m\omega_f^2 A$ force is the resultant of the other three forces, then resolving the vectors horizontally we must have

$$F\sin\phi = c\omega_f A \tag{39}$$

and resolving vertically

$$m\omega_f^2 A + F\cos\phi = kA \tag{40}$$

Since $\sin^2\phi + \cos^2\phi = 1$ then equations [39] and [40] give

$$F^2 = (c\omega_f A)^2 + (kA - m\omega_f^2 A)^2 = A^2[c^2\omega_f^2 + (k - m\omega_f^2)^2]$$

Thus the amplitude A of the resulting displacement is

$$A = \frac{F}{\sqrt{[c^2\omega_f^2 + (k - m\omega_f^2)^2]}} \tag{41}$$

If we substitute using $\omega_n^2 = k/m$ and $2\zeta\omega_n = c/m$ then equation [41] becomes the same as the amplitude term in equation [37]. Equations [39] and [40] can be rearranged to give $\tan\phi$, i.e.

$$\tan\phi = \frac{F\sin\phi}{F\cos\phi} = \frac{c\omega_f A}{kA - m\omega_f^2 A} = \frac{c\omega_f}{k - m\omega_f^2} \tag{42}$$

This equation is the same as equation [38].

We can use figure 17.12 to obtain the force transmitted by the oscillating mass, as a consequence of the oscillations, to its supports. The force is that transmitted by the spring, i.e. kx, and that by the damping, i.e. $c\,dx/dt$. These two forces are 90° out of phase (see figure 17.12) and thus the transmitted force is the vector sum. This has a maximum value when the displacement equals the amplitude A, and so

$$\text{max. force} = \sqrt{[(kA)^2 + (c\omega_f A)^2]} = A\sqrt{[k^2 + c^2\omega_f^2]} \tag{43}$$

with A being given by equation [41]. The total force transmitted to the supports will be the sum of the above force due to the oscillations and the static force due to the weight of the supported mass.

Similar equations can be derived for forced damped angular oscillations when the forcing agent for a rotor of moment of inertia I on a shaft with torsional stiffness q is a torque $T\cos\omega_f t$,

$$I\frac{d^2\theta}{dt^2} + c\frac{d\theta}{dt} + q\theta = T\cos\omega_f t \tag{44}$$

and the steady state deflection θ_s is

$$\theta_s = \frac{T}{\sqrt{[c^2\omega_f^2 + (q - I\omega_f^2)^2]}}\cos(\omega_f t - \phi) \tag{45}$$

$$\phi = \tan^{-1}\left(\frac{c\omega_f}{q - I\omega_f^2}\right) \tag{46}$$

or where $\omega_n^2 = q/I$ and $2\zeta\omega_n = c/I$,

$$\frac{d^2\theta}{dt^2} + 2\zeta\omega_n\frac{d\theta}{dt} + \omega_n^2\theta = \frac{T}{I}\cos\omega_f t \tag{47}$$

and the steady state deflection θ_s is given by

$$\theta_s = \frac{T/I}{\sqrt{[4\zeta^2\omega_n^2\omega_f^2 + (\omega_n^2 - \omega_f^2)^2]}}\cos(\omega_f t - \phi) \tag{48}$$

$$\phi = \tan^{-1}\left(\frac{2\zeta\omega_n\omega_f}{\omega_n^2 - \omega_f^2}\right) \tag{49}$$

In the above discussion it has been assumed that the forcing disturbance was directly applied to the mass. If the disturbance is applied to the support from which the spring is supported then with a disturbance which results in a periodic displacement of $y = Y \cos \omega_f t$, the restoring force is $k(x - y)$ and so, in a similar way to the earlier discussion of the undamped motion,

$$m \frac{d^2x}{dt^2} + c \frac{dx}{dt} + k(x - y) = 0$$

$$\frac{d^2x}{dt^2} + \frac{c}{m} \frac{dx}{dt} + \frac{k}{m} x = \frac{ky}{m} \cos \omega_f t \qquad [50]$$

$$\frac{d^2x}{dt^2} + 2\zeta\omega_n \frac{dx}{dt} + \omega_n^2 x = \omega_n^2 Y \cos \omega_f t \qquad [51]$$

where $\omega_n^2 = k/m$ and $2\zeta\omega_n = c/m$. The resulting steady state deflection x_s is given by

$$x_s = \frac{\omega_n^2 Y}{\sqrt{[4\zeta^2\omega_n^2\omega_f^2 + (\omega_n^2 - \omega_f^2)^2]}} \cos(\omega_f t - \phi) \qquad [52]$$

$$\phi = \tan^{-1}\left(\frac{2\zeta\omega_n\omega_f}{\omega_n^2 - \omega_f^2}\right) \qquad [53]$$

Example

A mass of 2.0 kg is suspended from a vertical spring of stiffness 15 kN/m and subject to viscous damping of 5 N s/m. What is the amplitude of the forced oscillations produced when a periodic force of amplitude 25 N and angular frequency 100 rad/s acts on the mass? What is the maximum force transmitted to the support of the spring?

Using equation [41],

$$A = \frac{F}{\sqrt{[c^2\omega_f^2 + (k - m\omega_f^2)^2]}}$$

$$= \frac{25}{\sqrt{[5^2 \times 100^2 + (15 \times 10^3 - 2.0 \times 100^2)^2]}} = 4.98 \, mm$$

The dynamic force transmitted to the support is the vector sum of the spring stiffness force and the damping force and is given by equation [43] as

$$\text{max. force} = A \sqrt{[k^2 + c^2\omega_f^2]}$$
$$= 4.98 \times 10^{-3} \sqrt{[(15 \times 10^3)^2 + (5^2 \times 100^2)]}$$
$$= 74.7 \, N$$

The static force transmitted is $mg = 2.0 \times 9.8 = 19.6$ N and so the total force is 94.3 N.

Example

A machine of mass 500 kg is mounted on flexible supports which under the weight of the machine give a static deflection of 6.0 mm. The supports can be considered to give a damping force which is 20% of the critical value and which is proportional to the velocity. When the machine is running at 10 rev/s, oscillations at this frequency occur with an amplitude of 5.0 mm. What is the maximum value of the disturbing force produced by the machine?

A damping force of 20% of the critical means a damping factor ζ of 0.20. The static deflection of 6.0 mm under a weight of 500×9.8 N means a stiffness k of $500 \times 9.8/0.006 = 817$ kN/m. Thus $\omega_n = \sqrt{(k/m)} = 40$ rad/s. Thus, using equation [37], the amplitude is

$$A = \frac{F/m}{\sqrt{[4\zeta^2\omega_n^2\omega_f^2 + (\omega_n^2 - \omega_f^2)^2]}}$$

$$F = 0.005 \times 500 \sqrt{[4 \times 0.20^2 \times 40^2 \times (2\pi \times 10)^2 + (40^2 - 20^2\pi^2)^2]}$$

$$= 6.4\,\text{kN}$$

Problems

(1) A mass suspended from a spring is subject to viscous damping and oscillates with a damped frequency of 1.5 Hz with an amplitude which decreases by 20% in one complete oscillation. What is the damping factor and the frequency of the undamped oscillations?

(2) A mass of 10 kg is suspended from a vertical spring of stiffness 5 kN/m. Viscous damping reduces the amplitude of the oscillation by 25% in one complete oscillation. What is (a) the undamped frequency, (b) the damped frequency and (c) the damping ratio?

(3) Describe the form of the output variation with time when systems with a damping ratio of (a) 0, (b) 0.5, (c) 1.0, (d) 1.5 are set in oscillation.

(4) A mass of 2.0 kg is suspended from a vertical spring of stiffness 60 N/m. When the mass is displaced and released it is found that two successive amplitudes of the oscillation are 150 mm and 85 mm. What is the frequency of undamped motion, the damping ratio and the damping coefficient?

(5) A mass of 5.0 kg is suspended from a vertical spring and after being pulled down and released gives damped oscillations. The time taken for 10 complete oscillations is found to be 4.0 s and the ratio of the first to the sixth amplitude of the oscillation is 2.0 (note: 5 complete oscillations later). What is the stiffness of the spring and the damping coefficient?

(6) A machine of mass 70 kg is mounted on three identical springs,

each having a stiffness of 9 kN/m, with dash pot damping. If the amplitude of the oscillations of the machine decreases from 30 mm to 5.0 mm in two complete oscillations, what is the damping coefficient and the frequency of the damped motion?

(7) A mass of 10 kg is suspended from a vertical spring of stiffness 2.0 kN/m and has dash pot damping. When the mass is pulled down and released the oscillation amplitude is reduced from 100 mm to 10 mm in four complete oscillations. What is the damping coefficient and the damped oscillation frequency?

(8) A mass of 20 kg is suspended from a vertical spring of stiffness 1 kN/m. What is the damping coefficient required to give (a) critical damping, (b) a damping factor of 0.4?

(9) A rotor with a moment of inertia of 25 kg m^2 is attached to one end of a shaft of diameter 50 mm, length 2.0 m and modulus of rigidity 80 GPa, the other end of the shaft being rigidly fixed. The system is fitted with a torsional damper which exerts a damping torque proportional to the angular velocity and which results in successive amplitudes of torsional oscillation being reduced by a factor of 0.30. What is the damping factor and the frequencies of the damped and natural oscillations?

(10) A mass of 10 kg is supported by a vertical spring of stiffness 5 kN/m and is acted on by a periodic force of amplitude 20 N and frequency 4.0 Hz. What is the steady state amplitude of the forced vibration?

(11) A mass is supported by a vertical spring, with no damping, and acted on by a sinusoidal force of constant amplitude but variable frequency. At 10 Hz the amplitude of the oscillation of the mass is found to be 20 mm and this continuously decreases to 2.0 mm as the frequency is increased to 20 Hz. What is the resonant frequency of the system?

(12) An undamped oscillating system, which can be considered to effectively be a mass suspended from a vertical spring, is acted on by a sinusoidal force of constant amplitude but variable frequency. At 3.0 Hz the amplitude is found to be 50 mm and at 9.0 Hz it is 25 mm. The resonant frequency is known to lie between these two frequencies, what is it?

(13) An engine of mass 150 kg is supported on four springs, each of stiffness 25 kN/m. Due to the engine not being correctly balanced, when it runs at 15 rev/s a periodic force with a maximum value of 300 N acts on the system. What will be the maximum value of the force on the foundations on which the springs rest due to the engine running if it can be assumed that the engine only vibrates in the vertical direction?

(14) A machine mounted on the floor of a workshop produces a static

deflection of the floor 2.0 mm under the machine. Because the machine is not balanced it produces a vertical periodic force with the same frequency as that which the driving shaft of the machine rotates. What frequency of the driving shaft will lead to resonance? Assume that the floor is elastic and there is no damping.

(15) An instrument is rigidly attached to a platform which is supported by four springs, each having a stiffness 800 N/m. The floor to which the springs are attached is subject to a periodic displacement of amplitude 10 mm and angular frequency 8 rad/s. The total mass of the instrument plus platform is 20 kg and the instrument is constrained to move only vertically. What is the maximum vertical steady state displacement of the platform from the equilibrium position?

(16) A mass is suspended from a spring and when the spring support is stationary gives a frequency of oscillation of 1.5 Hz. What is the steady state amplitude when the spring support is subject to a periodic displacement of amplitude 30 mm and frequency 1.2 Hz assuming there is no damping?

(17) A mass of 3.0 kg is suspended from a vertical spring and produces a static deflection of 60 mm. If the system is subject to viscous damping of 36 N s/m, what will be the steady state amplitude of the oscillations produced when the mass is acted on by a periodic force of amplitude 10 N and frequency 1.5 Hz?

(18) A machine of mass 500 kg is mounted on flexible supports which contract by 6.0 mm under the weight. When the motor is running at 500 rev/min a forcing disturbance of the same frequency, and with an amplitude of 5.0 mm, is set up. If the system has a damping factor of 0.15, what is the maximum value of the disturbing force?

(19) A mass of 2.0 kg is suspended from a vertical spring of stiffness 15 kN/m. If the system is subject to viscous damping such that the damping force is 8 N at a velocity of 1 m/s, what will be the amplitude of the mass when it is acted on by a disturbing force of amplitude 20 N and frequency 100 Hz?

(20) A mass of 100 kg is suspended from a vertical spring of stiffness 10 kN/m, the other end of the spring being attached to a support which vibrates. If the system is subject to viscous damping such that the damping force is 200 N at a velocity of 1 m/s, what will be the amplitude of the mass when the support vibrates vertically with an amplitude of 3.0 mm and a frequency of 20 Hz?

(21) What is the maximum force transmitted to the ground through the mounting of a machine if when the speed of rotation of the machine is 30 rev/s the amplitude of vertical motion of the machine is 0.06 mm? The machine has a mass of 1000 kg and the mounting has a stiffness of 12 MN/m and a damping coefficient of 40 kN s/m.

(22)   A motor of mass 30 kg is supported by a flexible mounting of stiffness 800 N/m and damping ratio 0.20. The rotor of the motor is unbalanced so that the effect is as a mass of 4 kg located 60 mm from the axis of rotation. What is the amplitude of oscillation when the rotor is turning at a rate of 10 rad/s? (Hint: the periodic forcing force is the centripetal force due to the unbalanced rotor.)

Chapter 18
Vibrations of multi-degree systems

18.1   Degrees of freedom

Figure 18.1(a) shows a system which has one degree of freedom, and figure 18.1(b) shows a system with two degrees of freedom. When a system requires just one co-ordinate, i.e. only x_1, to describe its motion it is said to have *one degree of freedom*; when it uses two coordinates, i.e. x_1 and x_2, to describe its motion it has *two degrees of freedom*. The *degrees of freedom* of a system are the number of coordinates needed to describe its motion.

18.1.1   One degree of freedom system

Examples of systems with one degree of freedom have been considered in Chapters 16 and 17, e.g. a mass suspended from a vertical spring (as in figure 16.2) or a simple pendulum (section 16.2.1). In general, a useful model of a one degree system involves a mass and some means of providing a restoring force, e.g. a spring. Figure 18.1(a) can be considered to be an idealised model of such a system. The natural oscillations of the system are considered here without any consideration of damping.

When the mass m_1 is given a displacement x_1 then the spring is extended. If the spring has the spring stiffness k, then the restoring force acting on the mass is $-kx_1$, the minus sign being because the force is in the opposite direction to that direction in which x_1 increases. Thus the acceleration a resulting from this force is $-kx_1/m$. As $a = \mathrm{d}^2x_1/\mathrm{d}t^2$, we can write a differential equation for the displacement as

$$m_1 \frac{\mathrm{d}^2 x_1}{\mathrm{d}t^2} + kx_1 = 0 \qquad [1]$$

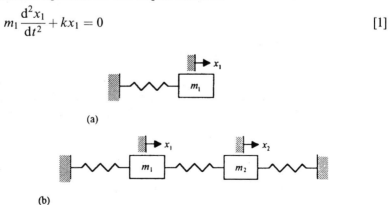

(a)

(b)

Fig. 18.1 (a) One, (b) two degrees of freedom.

This second-order differential equation has a solution of the form (see section 17.1):

$$x_1 = a \cos \omega_n t + b \sin \omega_n t \qquad [2]$$

where a and b are constant and ω_n is the natural angular frequency. This equation can also be written in the form

$$x_1 = A \sin (\omega_n t + \phi) \qquad [3]$$

The values of a, b and ϕ will depend on the initial conditions when the mass is first displaced from its equilibrium position. If $x_1 = 0$ when $t = 0$ then equation [3] gives $x_1 = A \sin \omega_n t$.

18.1.2   Two degrees of freedom system

Examples of systems with two degrees of freedom are quite common in engineering, e.g. two rotors connected by a flexible single shaft, or systems which can often be approximated to two degree systems, e.g. a car suspension. In general, a useful model of a two degree system involves two masses and springs to supply restoring forces as shown in figure 18.2. The natural oscillations of the system are considered here without any consideration of damping.

Consider only the forces acting on the mass m_1. If we assume that the three springs each have the same spring constant k, then when m_1 is displaced by x_1 and m_2 by x_2, the spring to the right of m_1 is extended and so gives a restoring force $-kx_1$ and the spring to the left of m_1 is compressed by $(x_1 - x_2)$ and so the restoring force due to it is $-k(x_1 - x_2)$. The total restoring force acting on the mass is thus $-kx_1 - k(x_1 - x_2)$. The differential equation describing the motion of m_1 is then

$$m_1 \frac{d^2 x_1}{dt^2} + k(2x_1 - x_2) = 0 \qquad [4]$$

Likewise for mass m_2, the spring to its left is extended by $(x_1 - x_2)$ and the spring to its right is compressed by x_2. Thus the total restoring force on m_2 is $k(x_1 - x_2) - kx_2$. Hence

$$m_2 \frac{d^2 x_2}{dt^2} + k(2x_2 - x_1) = 0 \qquad [5]$$

As equations [4] and [5] indicate, the two displacements are interlinked.

Assume that the masses m_1 and m_2 execute simple harmonic motions with the same frequency ω and different amplitudes A_1 and A_2, i.e.

$$x_1 = A_1 \sin \omega t \quad \text{and} \quad x_2 = A_2 \sin \omega t \qquad [6]$$

Substituting these into the differential equations [4] and [5] gives

$$-m_1 A_1 \omega^2 \sin \omega t + k(2A_1 \sin \omega t - A_2 \sin \omega t) = 0$$

$$-m_2 A_2 \omega^2 \sin \omega t + k(2A_2 \sin \omega t - A_1 \sin \omega t) = 0$$

which simplify to:

$$[-m_1 A_1 \omega^2 + k(2A_1 - A_2)] \sin \omega t = 0$$

$$[-m_2 A_2 \omega^2 + k(2A_2 - A_1)] \sin \omega t = 0$$

These equations must be true for all values of t. Hence we must have

$$-m_1 A_1 \omega^2 + k(2A_1 - A_2) = 0 \qquad [7]$$

$$-m_2 A_2 \omega^2 + k(2A_2 - A_1) = 0 \qquad [8]$$

Equations [7] and [8] can be written as

$$\frac{A_1}{A_2} = \frac{-k}{m_1 \omega^2 - 2k} = \frac{m_2 \omega^2 - 2k}{-k} \qquad [9]$$

and so

$$m_1 m_2 \omega^4 - 2k\omega^2(m_1 + m_2) + 3k^2 = 0 \qquad [10]$$

This is termed the *frequency equation* and its solution gives two values for ω.
Suppose we have $m_1 = m_2 = m$, then equation [10] becomes:

$$\omega^4 - 4(k/m)\omega^2 + 3(k/m)^2 = 0$$

$$[\omega^2 - 3(k/m)][\omega^2 - (k/m)] = 0$$

Thus the two solutions are $\omega_1 = \sqrt{(k/m)}$ and $\omega_2 = \sqrt{[3(k/m)]}$. These are
the two natural frequencies of the system. Substituting these values into
equation [9] gives $A_1 = A_2$ and $A_1 = -A_2$. Thus the displacements of the
two masses with the first solution are given by equation [6] as

$$x_1 = A \sin \omega_1 t \quad \text{and} \quad x_2 = A \sin \omega_1 t \qquad [11]$$

i.e., two in-phase oscillations of the same amplitude. For the second
solution

$$x_1 = A \sin \omega_2 t \quad \text{and} \quad x_2 = -A \sin \omega_2 t = A \sin (\omega_2 t + \pi) \qquad [12]$$

i.e., two oscillations of the same amplitude but out of phase by π. Equations
[11] and [12] define what are termed the *normal modes* of oscillation of the
system. We can display the modes graphically as in figure 18.2.

For a two-degree system, there are two normal modes of oscillation,
each with its own natural frequency. Thus, for the above example, if we
have an initial displacements of $x_1 = +1$ and $x_2 = +1$ for the two masses
and then release them, they will oscillate with a purely sinusoidal motion

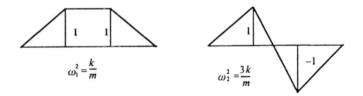

Fig. 18.2 Normal modes of the system with two equal masses.

with the frequency ω_1. They will oscillate in the first mode. If, however, the initial displacements are $x_1 = +1$ and $x_2 = -1$, they will oscillate with a purely sinusoidal motion with the frequency ω_2. They will oscillate in the second mode.

If the coupled masses are set in motion with initial displacements differing from those of the normal modes, we cannot start with the assumption of a solution of the form given by equation [6], i.e.

$$x_1 = A_1 \sin \omega t \quad \text{and} \quad x_2 = A_2 \sin \omega t$$

but need to assume a solution which is the superposition of two oscillations, one of which is in the first mode and the other in the second mode

$$x_1 = A_1 \sin(\omega_1 t + \phi_1) + B_1 \sin(\omega_2 t + \phi_2)$$
$$x_2 = A_1 \sin(\omega_1 t + \phi_1) + B_2 \sin(\omega_2 t + \phi_2)$$

[13]

Thus for the above example when we have an amplitude ratio for the first mode of $+1$ we have $A_1 = A_2 = A$ and with the amplitude ratio of -1 for the second mode, $B_1 = -B$ and $B_2 = B$. If we had displacements at $t = 0$ of $x_1 = 5$ and $x_2 = 0$ then

$$5 = A \sin \phi_1 - B \sin \phi_2 \quad \text{and} \quad 0 = A \sin \phi_1 + B \sin \phi_2$$

Hence we obtain $A \sin \phi_1 = 2.5$ and $B \sin \phi_2 = -2.5$. If we also had the initial condition that at $t = 0$ both masses had zero velocity, then if we differentiate equations [13]

$$0 = \omega_1 A \cos \phi_1 - \omega_2 B \cos \phi_2 \quad \text{and} \quad 0 = \omega_1 A \cos \phi_1 + \omega_2 B \cos \phi_2$$

Thus $\phi_1 = \phi_2 = 90°$. Hence $A = 2.5$ and $B = -2.5$. The oscillation is thus

$$x_1 = 2.5 \cos \omega_1 t - 2.5 \cos \omega_2 t$$
$$x_2 = 2.5 \cos \omega_1 t - 2.5 \cos \omega_2 t$$

Example

For the masses and spring system shown in figure 18.3, determine the normal modes of the system.

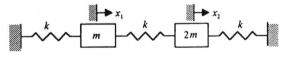

Fig. 18.3 Example.

The differential equations describing the displacement of each mass are

$$m \frac{d^2 x_1}{dt^2} + k(2x_1 - x_2) = 0$$

$$2m \frac{d^2 x_2}{dt^2} + k(2x_2 - x_1) = 0$$

Assuming solutions of the form

$$x_1 = A_1 \sin \omega t \quad \text{and} \quad x_2 = A_2 \sin \omega t$$

and then substituting these into the differential equations gives

$$-mA_1\omega^2 \sin \omega t + k(2A_1 \sin \omega t - A_2 \sin \omega t) = 0$$
$$-2mA_2\omega^2 \sin \omega t + k(2A_1 \sin \omega t - A_1 \sin \omega t) = 0$$

Hence we must have

$$-mA_1\omega^2 + k(2A_1 - A_2) = 0$$
$$-2mA_2\omega^2 + k(2A_2 - A_1) = 0$$

and so

$$\frac{A_1}{A_2} = \frac{-k}{m\omega^2 - 2k} = \frac{2m\omega^2 - 2k}{-k}$$

Thus

$$2\omega^4 - 6(k/m)\omega^2 + 3(k/m)^2 = 0$$

and so, using the equation for the roots of a quadratic equation

$$\omega^2 = \frac{6(k/m) \pm \sqrt{(36 - 24)(k/m)^2}}{4} = (1.5 \pm 0.866)(k/m)$$

Thus the two solutions are $\omega_1 = \sqrt{[0.634(k/m)]}$ and $\omega_2 = \sqrt{[2.366(k/m)]}$. These are the two natural frequencies of the system. Substituting these values into equation [9] gives $A_1 = 0.731A_2$ and $A_1 = -2.73A_2$. Thus the displacements of the two masses with the first solution are given by

$$x_1 = 0.731A \sin \omega_1 t \quad \text{and} \quad x_2 = A \sin \omega_1 t$$

and for the second solution as

$$x_1 = -2.73A \sin \omega_2 t = 2.73A \sin(\omega_2 t + \pi) \quad \text{and} \quad x_2 = A \sin \omega_2 t$$

We can display the modes graphically as in figure 18.4.

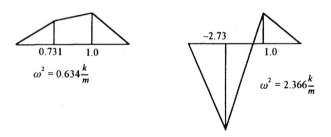

Fig. 18.4 Example.

18.2   Coupled pendulums

Consider the system shown in figure 18.5. Two simple, equal length, pendulums are coupled together by means of a light spring of stiffness k. The spring is unstrained when the pendulums are hanging in the vertical positions.

The spring is stretched by an amount which, for small angles of pendulum deflection, is $a(\theta_1 - \theta_2)$. For the first pendulum, the restoring force exerted by the spring will be $-ka(\theta_1 - \theta_2)$ and the restoring force exerted by gravity will be, for small angles, $-mg\theta_1$. Counter clockwise angular displacements are taken as positive.

Taking moments about the point of suspension of the first pendulum gives, for small oscillations

$$\text{restoring torque} = -mgL\theta_1 - ka^2(\theta_1 - \theta_2)$$

As $T = I\alpha$, where the moment of inertia I about the pivot point for the pendulum is mL^2 and the angular acceleration $\alpha = \mathrm{d}^2\theta_1/\mathrm{d}t^2$

$$mL^2 \frac{\mathrm{d}^2\theta_1}{\mathrm{d}t^2} = -mgL\theta_1 - ka^2(\theta_1 - \theta_2) \qquad [14]$$

Likewise for the second pendulum

$$mL^2 \frac{\mathrm{d}^2\theta_2}{\mathrm{d}t^2} = -mgL\theta_2 + ka^2(\theta_1 - \theta_2) \qquad [15]$$

If we assume solutions of the form

$$\theta_1 = A_1 \cos\omega t \quad \text{and} \quad \theta_2 = A_2 \cos\omega t \qquad [16]$$

and substitute in equations [14] and [15], we obtain

$$-mL^2 A_1 \omega^2 = -mgLA_1 - ka^2(A_1 - A_2)$$

$$-mL^2 A_2 \omega^2 = -mgLA_2 + ka^2(A_1 - A_2)$$

The two equations can be written as

$$\frac{A_1}{A_2} = \frac{ka^2}{-\omega^2 mL^2 + mgL + ka^2} = \frac{-\omega^2 mL^2 + mgL + ka^2}{ka^2}$$

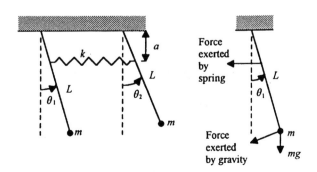

Fig. 18.5 Coupled pendulums.

Hence we obtain the frequency equation

$$\omega^4 - 2\left(\frac{g}{L} + \frac{ka^2}{mL^2}\right)\omega^2 + \left(\frac{g^2}{L^2} + 2\frac{ka^2g}{mL^3}\right) = 0$$

and so the solution of this quadratic equation gives

$$\omega^2 = \frac{g}{L} + \frac{ka^2}{mL^2} \pm \frac{ka^2}{mL^2}$$

Hence the two natural frequencies of the system are

$$\omega_1 = \sqrt{\frac{g}{L}} \quad \text{and} \quad \omega_2 = \sqrt{\frac{g}{L} + 2\frac{ka^2}{mL^2}} \qquad [17]$$

These give amplitude ratios of 1.0 and −1.0.

With the natural frequency ω_1, we have the same natural frequency as would be given by the pendulums oscillating without any coupling (see section 16.2.1) and the two move in phase with equal amplitudes. The coupling spring is then not extended or compressed during the motion. With the natural frequency ω_2, we have the two pendulums oscillating in opposition to each other with equal amplitudes. The spring is thus alternately elongated and compressed and this modifies the natural frequency, so increasing it.

Equation [16] assumes initial conditions where, at $t = 0$, we have $\theta_1 = A_1$ and $\theta_2 = A_2$. If we have at $t=0$, say, $\theta_1 = A$ and $\theta_2 = 0$ and both pendulums with zero velocity, then we can, as with equation [13] earlier in the chapter, assume solutions of the form

$$\theta_1 = A_1 \sin(\omega_1 t + \phi_1) + B_1 \sin(\omega_2 t + \phi_2)$$

$$\theta_2 = A_2 \sin(\omega_1 t + \phi_1) + B_2 \sin(\omega_2 t + \phi_2)$$

and solve for the given initial conditions. These give

$$\theta_1 = \tfrac{1}{2}A \cos\omega_1 t + \tfrac{1}{2}A \cos\omega_2 t$$

$$\theta_2 = \tfrac{1}{2}A \cos\omega_1 t - \tfrac{1}{2}A \cos\omega_2 t \qquad [18]$$

These equations can be rewritten as

$$\theta_1 = A \cos\left(\frac{\omega_1 - \omega_2}{2}\right)t \cos\left(\frac{\omega_1 + \omega_2}{2}\right)t$$

$$\theta_2 = -A \sin\left(\frac{\omega_1 - \omega_2}{2}\right)t \sin\left(\frac{\omega_1 + \omega_2}{2}\right)t \qquad [19]$$

If the coupling spring is weak then the two frequencies ω_1 and ω_2 will be nearly equal. Thus θ_1 will behave like $\cos\tfrac{1}{2}(\omega_1 + \omega_2)t$ with an amplitude which varies slowly with time and θ_2 will behave like $\sin\tfrac{1}{2}(\omega_1 + \omega_2)t$ with an amplitude which varies slowly with time, both showing beats.

18.3   Two rotor system

Consider the vibration of two rotors connected by a flexible shaft (figure 18.6). This could be, for example, a motor connected to an electric generator. The system has two degrees of freedom.

Let θ_1 and θ_2 be the angular displacements of the two rotors when measured in a counter-clockwise direction from the unstrained position of the shaft. The twist in the shaft when there are displacements will be $(\theta_1 - \theta_2)$. The restoring torque on rotor 1 will be $-k(\theta_1 - \theta_2)$, where k is the torsional stiffness, and that on rotor 2 will be $k(\theta_1 - \theta_2)$. Thus

$$I_1 \frac{d^2\theta_1}{dt} = -k(\theta_1 - \theta_2) \qquad [20]$$

$$I_2 \frac{d^2\theta_2}{dt} = k(\theta_1 - \theta_2) \qquad [21]$$

For the normal mode vibrations we will assume solutions of the form

$$\theta_1 = A_1 \cos \omega t \quad \text{and} \quad \theta_2 = A_2 \cos \omega t \qquad [22]$$

Substituting these into equations [20] and [21] gives

$$-I_1 A_1 \omega^2 \cos \omega t = -k(A_1 \cos \omega t - A_2 \cos \omega t)$$

$$-I_2 A_2 \omega^2 \cos \omega t = k(A_1 \cos \omega t - A_2 \cos \omega t)$$

These can be written as

$$\frac{A_1}{A_2} = \frac{k}{k - \omega^2 I_1} = \frac{k - \omega^2 I_2}{k} \qquad [23]$$

Equation [23] gives the frequency equation as

$$\omega^2 [I_1 I_2 \omega^2 - k(I_1 + I_2)] = 0 \qquad [24]$$

Hence the two natural frequencies are

$$\omega_1 = \sqrt{\frac{k(I_1 + I_2)}{I_1 I_2}} \quad \text{and} \quad \omega_2 = 0 \qquad [25]$$

These give, when substituted in equation [23], amplitude ratios of $A_1/A_2 = -I_1/I_2$ and $A_1/A_2 = +1.0$. With $\omega_2 = 0$ the amplitude ratio of $+1.0$ indicates that there is no twisting of the shaft and the two rotors rotate together.

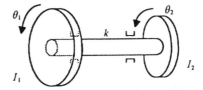

Fig. 18.6 Two rotors connected by a flexible shaft.

Although two coordinates have been used to specify the motion, there is only one finite natural frequency. This is because there is no second shaft. The presence of a second shaft leads to two finite natural frequencies.

18.3.1 Geared systems

Very often two rotors, such as a motor-generator system, are not mounted on a single shaft but on separate shafts connected by gearing (figure 18.7). In the analysis that follows we will assume that the gears and shafts have negligible moments of inertia.

For the first rotor we can write

$$I_1 \frac{d^2\theta_1}{dt^2} + k_1(\theta_1 - \theta_g) = 0 \tag{26}$$

and for the second rotor

$$I_2 \frac{d^2\theta_2}{dt^2} + k_2\left(\theta_2 - \frac{\theta_g}{n}\right) = 0 \tag{27}$$

For the gears we must have, for equilibrium

$$n_1 \times k_1(\theta_1 - \theta_g) = -k_2\left(\theta_2 - \frac{\theta_g}{n}\right) \tag{28}$$

Using this to eliminate θ_g from equations [26] and [27] gives

$$I_1 \frac{d^2\theta_1}{dt^2} + k^*\left(\theta_1 - \frac{\theta_2}{n}\right) = 0 \tag{29}$$

$$\frac{I_2}{n^2} \frac{d^2(\theta_2/n)}{dt^2} + k^*\left(\theta_1 - \frac{\theta_2}{n}\right) = 0 \tag{30}$$

with

$$\frac{1}{k^*} = \frac{1}{k_1} + \frac{1}{(k_2/n^2)} \tag{31}$$

Comparing equations [29] and [30] with the equations for the two rotor system (equations [20] and [21]), it is apparent that the geared system is equivalent to an ungeared two rotor system with moments of inertia I_1 and

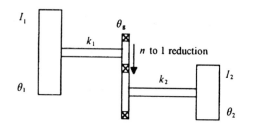

Fig. 18.7 Geared system.

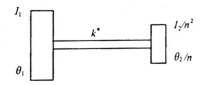

Fig. 18.8 Equivalent system.

I_2/n^2 and a shaft of torsional stiffness k^*, the rotation of rotor 1 being defined by the variable θ_1 and the rotation of rotor 2 by θ_2/n (figure 18.8). This then leads to the natural frequencies of the system being

$$\omega_1 = \sqrt{\left(\frac{k^*}{I_1} + \frac{k^*}{I_2/n^2}\right)} \quad \text{and} \quad \omega_2 = 0 \tag{32}$$

Example

A motor having moving parts with a moment of inertia of $10\,\mathrm{kg\,m^2}$ is used to drive a rotor of moment of inertia $50\,\mathrm{kg\,m^2}$ through a 2.5 to 1 reduction gear. The motor drive shaft has a torsional stiffness of $800\,\mathrm{kN\,m/rad}$ and the rotor drive shaft a torsional stiffness of $80\,\mathrm{kN\,m/rad}$. Assuming the moments of inertia of the shafts and gears are negligible, determine the natural frequencies of oscillation of the system.

The equivalent two rotor system will have a shaft of torsional stiffness given by equation [31] as

$$\frac{1}{k^*} = \frac{1}{800 \times 10^3} + \frac{2.5^2}{80 \times 10^3}$$

Hence $k^* = 12.6\,\mathrm{kN/rad}$. Hence, using equation [32]

$$\omega_1 = \sqrt{\frac{12.6 \times 10^3}{10} + \frac{12.6 \times 10^3 \times 2.5^2}{50}} = 53.2\,\mathrm{rad/s}$$

and

$$\omega_2 = 0$$

18.4 Forced vibrations

As an illustration of the effect of forced vibrations, consider the two degree mass–spring system shown in figure 18.9 when a harmonic force $F_1 \sin \omega t$ is applied to the mass m_1. We have

$$m_1 \frac{\mathrm{d}^2 x_1}{\mathrm{d}t^2} + k_2(x_1 - x_2) + k_1 x_1 = F_1 \sin \omega t \tag{33}$$

$$m_2 \frac{\mathrm{d}^2 x_2}{\mathrm{d}t^2} + k_2(x_2 - x_1) = 0 \tag{34}$$

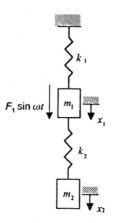

Fig. 18.9 Spring–mass system with forcing vibration.

Assume that the masses m_1 and m_2 execute simple harmonic motions with the same frequency ω and different amplitudes A_1 and A_2, i.e.

$$x_1 = A_1 \sin \omega t \quad \text{and} \quad x_2 = A_2 \sin \omega t \qquad [35]$$

Substituting these into the differential equations [33] and [34] gives

$$(k_1 + k_2 - m_1 \omega^2)A_1 - k_2 A_2 = F_1 \qquad [36]$$

$$-k_2 A_1 + (k_2 - m_2 \omega^2)A_2 = 0 \qquad [37]$$

Let ω_1 be the natural frequency of $\sqrt{(k_1/m_1)}$ and ω_2 the natural frequency of $\sqrt{(k_2/m_2)}$. The steady deflection of m_1 due to the steady force F_1 is F_1/k_1. If we denote this steady deflection by A_0, then equation [36] becomes

$$\left[1 + \frac{k_2}{k_1} - \left(\frac{\omega}{\omega_1}\right)^2\right]A_1 - \left[\frac{k_2}{k_1}\right]A_2 = A_0$$

and equation [37] becomes

$$-A_1 + \left[1 - \left(\frac{\omega}{\omega_2}\right)^2\right]A_2 = 0$$

Hence the amplitudes of the two masses in terms of A_0 are

$$\frac{A_1}{A_0} = \frac{\left[1 - \left(\dfrac{\omega}{\omega_2}\right)^2\right]}{\left[1 + \dfrac{k_2}{k_1} - \left(\dfrac{\omega}{\omega_1}\right)^2\right]\left[1 - \left(\dfrac{\omega}{\omega_2}\right)^2\right] - \dfrac{k_2}{k_1}} \qquad [38]$$

$$\frac{A_2}{A_0} = \frac{1}{\left[1 + \dfrac{k_2}{k_1} - \left(\dfrac{\omega}{\omega_1}\right)^2\right]\left[1 - \left(\dfrac{\omega}{\omega_2}\right)^2\right] - \dfrac{k_2}{k_1}} \qquad [39]$$

Equation [38] indicates that the amplitude A_1 of mass m_1 becomes zero when the exciting frequency ω is equal to the natural frequency ω_2. For this frequency the amplitude A_2 of m_2 is $-(k_1/k_2)A_0 = -F_1/k_2$. These conditions give the principle of the *vibration absorber*, the mass m_2 and its coupling being chosen to tune its natural frequency to that of the forcing frequency.

The amplitude of m_1 will become infinite when the denominator of equation [38] has a zero value. Because this is a quadratic equation there will be two frequencies at which this occurs, these being termed the *resonant frequencies*.

18.5 Systems with many degrees of freedom

With a one degree system we have one differential equation describing the motion; with a two degree system we have two differential equations. With an n degree system we will have n differential equations. Thus a system involving three rotors on a flexible shaft will give three differential equations and four rotors will give four differential equations.

Problems

(1)   For the masses and spring system shown in figure 18.10, determine the natural frequencies and normal modes.

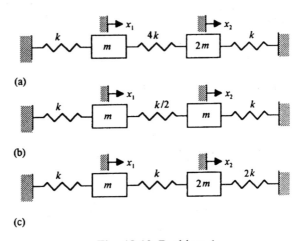

(a)

(b)

(c)

Fig. 18.10 Problem 1.

(2)   For the system described in figure 18.1(b) with both masses being 1 kg and all the springs having spring constants of 81 N/m, determine the two natural frequencies.

(3) For the system described in figure 18.1(b) with both masses being 1 kg and all the springs having spring constants of 81 N/m, determine how the displacements of the two masses vary with time given that at $t=0$ we have $x_1=4$, $x_2=0$, $dx_1/dt=0$ and $dx_2/dt=0$.

(4) For the masses and spring system shown in figure 18.11, determine the natural frequencies and normal modes.

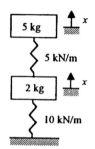

Fig. 18.11 Problem 4.

(5) Determine the frequency equation for the natural frequencies of vibration for the torsional system shown in figure 18.12.

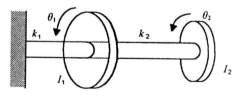

Fig. 18.12 Problem 5.

(6) Determine the frequency equation for the natural frequencies of vibration for the torsional system shown in figure 18.13.

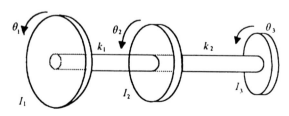

Fig. 18.13 Problem 6.

(7)   Determine the natural frequencies of the two rotor torsional system shown in figure 18.14. (Hint: see section 6.3.)

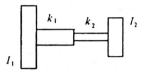

Fig. 18.14 Problem 7.

Chapter 19
Matrix methods of structural analysis

19.1 Matrix displacement method

This chapter is concerned with an introduction to the matrix displacement method of analysis of structures and forms an elementary introduction to the concept of the finite element method.

A pin-jointed plane frame (as for example that in figure 1.16) comprises a series of physically identifiable and distinct members that are interconnected at their extremities. The term *element* is used for a distinct member and the term *nodes* for their points of attachment to other parts of the structure. Consider first the representation of the behaviour of a single element by matrices and then their use in describing the behaviour of a structure, hence the determination of nodal displacements and forces in individual members.

19.1.1 A rod element

Consider a pin-jointed tie member for which we can assume that Hooke's law is obeyed. We can think of this as being equivalent to a spring of, say, stiffness k between nodes 1 and 2 (figure 19.1). As a convention, we will take the external force and displacement at a particular node to be denoted by F and u with a suffix indicating to which node they refer. Forces and displacements are taken to be positive when in the positive x increasing direction and negative when in the opposite direction.

For the spring shown in figure 19.1, if we consider node 1 we have

$$F_1 = k(u_1 - u_2) = ku_1 - ku_2 \tag{1}$$

and for node 2

$$F_2 = k(u_2 - u_1) = -ku_1 + ku_2 \tag{2}$$

We can use matrices to describe equations [1] and [2]

$$\begin{bmatrix} F_1 \\ F_2 \end{bmatrix} = \begin{bmatrix} k & -k \\ -k & k \end{bmatrix} \begin{bmatrix} u_1 \\ u_2 \end{bmatrix} \tag{3}$$

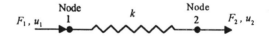

Fig. 19.1 Simple spring element.

The square matrix of the stiffnesses is termed the *stiffness matrix* **K**. The column matrix of the displacements **u** is termed the *nodal displacement vector* and the column matrix of the forces **f** the *element force vector*. The convention is used of representing matrices as bold, non-italic, symbols with column matrices being lower case and other matrices upper case letters. Thus equation [3] can be written as

$$\mathbf{f} = \mathbf{Ku} \tag{4}$$

For a simple rod element of uniform cross-sectional area A and length L with axial forces and displacements, assuming stress/strain = the modulus of elasticity E, the forces necessary for the displacements u_1 and u_2 are

$$F_1 = AE(u_1 - u_2)/L = (AE/L)u_1 - (AE/L)u_2 \tag{5}$$

$$F_2 = AE(u_2 - u_1)/L = -(AE/L)u_1 + (AE/L)u_2 \tag{6}$$

Writing equations [5] and [6] in matrix form gives

$$\begin{bmatrix} F_1 \\ F_2 \end{bmatrix} = \begin{bmatrix} AE/L & -AE/L \\ -AE/L & AE/L \end{bmatrix} \begin{bmatrix} u_1 \\ u_2 \end{bmatrix} \tag{7}$$

We can take the AE/L factor out of the stiffness matrix to give

$$\begin{bmatrix} F_1 \\ F_2 \end{bmatrix} = \frac{AE}{L} \begin{bmatrix} 1 & -1 \\ -1 & 1 \end{bmatrix} \begin{bmatrix} u_1 \\ u_2 \end{bmatrix} \tag{8}$$

19.1.2 In-line elements

Consider two in-line spring elements (figure 19.2(a)). Suppose we now imagine the two elements in isolation, making an imaginary cut at node 2 (figure 19.2(b)). We have then for element 1

$$F_1 = k_1(u_1 - u_2) = k_1 u_1 - k_1 u_2 \tag{9}$$

$$F_{12} = k_1(u_2 - u_1) = -k_1 u_1 + k_1 u_2 \tag{10}$$

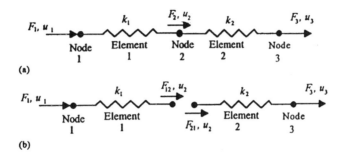

Fig. 19.2 Two elements in line.

and hence

$$\begin{bmatrix} F_1 \\ F_{12} \end{bmatrix} = \begin{bmatrix} k_1 & -k_1 \\ -k_1 & k_1 \end{bmatrix} \begin{bmatrix} u_1 \\ u_2 \end{bmatrix} \tag{11}$$

For element 2 we have

$$F_{21} = k_2(u_2 - u_3) = k_2 u_2 - k_2 u_3 \tag{12}$$

$$F_3 = k_2(u_3 - u_2) = -k_2 u_2 + k_2 u_3 \tag{13}$$

and hence

$$\begin{bmatrix} F_{21} \\ F_3 \end{bmatrix} = \begin{bmatrix} k_2 & -k_2 \\ -k_2 & k_2 \end{bmatrix} \begin{bmatrix} u_2 \\ u_3 \end{bmatrix} \tag{14}$$

As we must have $F_2 = F_{12} + F_{21}$, equations [9], [10], [12] and [13] can be combined to give

$$F_1 = k_1 u_1 - k_1 u_2$$

$$F_2 = F_{12} + F_{21} = -k_1 u_1 + k_1 u_2 + k_2 u_2 - k_2 u_3$$

$$= -k_1 u_1 + (k_1 + k_2)u_2 - k_2 u_3$$

$$F_3 = k_2(u_3 - u_2) = -k_2 u_2 + k_2 u_3$$

and hence

$$\begin{bmatrix} F_1 \\ F_2 \\ F_3 \end{bmatrix} = \begin{bmatrix} F_1 \\ F_{12} + F_{21} \\ F_3 \end{bmatrix} = \begin{bmatrix} k_1 & -k_1 & 0 \\ -k_1 & k_1 + k_2 & -k_2 \\ 0 & -k_2 & k_2 \end{bmatrix} \begin{bmatrix} u_1 \\ u_2 \\ u_3 \end{bmatrix} \tag{15}$$

This is just the addition of equations [11] and [14] and gives the *total* or *global stiffness matrix* of the structure. We can adopt the same procedure for any number of in-line elements, i.e. the global stiffness matrix is that obtained when we sum the force-displacement matrix equations for each element.

Example

Determine the stiffness matrix of the two in-line rods in figure 19.3.

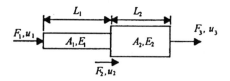

Fig. 19.3 Example.

The stiffness of rod 1 is A_1E_1/L_1 and that of rod 2 is A_2E_2/L_2. The situation is identical to that described in figure 19.2 and results in the global stiffness matrix of equation [15]. Thus the global stiffness of two in-line rods is

$$\begin{bmatrix} \dfrac{A_1E_1}{L_1} & -\dfrac{A_1E_1}{L_1} & 0 \\[2ex] -\dfrac{A_1E_1}{L_1} & \dfrac{A_1E_1}{L_1}+\dfrac{A_2E_2}{L_2} & -\dfrac{A_2E_2}{L_2} \\[2ex] 0 & -\dfrac{A_2E_2}{L_2} & \dfrac{A_2E_2}{L_2} \end{bmatrix}$$

19.2   Local and global coordinates

The elements considered so far in this chapter have been in the same straight line. However, in the analysis of frameworks this will generally not be the case for all elements, some rod elements being at different angles to one another. We have found out how to express the forces and displacements for each element in relation to a coordinate system which has the x-axis along the axis of the element and a y-axis at right angles to it. This is termed the *local coordinate system*. We need, however, to express the forces and deformations for each element in terms of the coordinate system of the framework as a whole, this being termed the *global coordinate system*.

Consider a spring element, between nodes 1 and 2, which has its local coordinate system at some angle θ to the global coordinate system (figure 19.4). The forces and displacements at each node can be resolved into components along the spring axis and at right angles to it. If we assume that the element is pin-jointed at the nodes, there will only be forces along the spring axis. Thus, for the four equations relating forces and displacements

$$\begin{bmatrix} F_{x1} \\ F_{y1} \\ F_{x2} \\ F_{y2} \end{bmatrix} = \begin{bmatrix} k & 0 & -k & 0 \\ 0 & 0 & 0 & 0 \\ -k & 0 & k & 0 \\ 0 & 0 & 0 & 0 \end{bmatrix} \begin{bmatrix} u_{x1} \\ u_{y1} \\ u_{x2} \\ u_{y2} \end{bmatrix} \qquad [16]$$

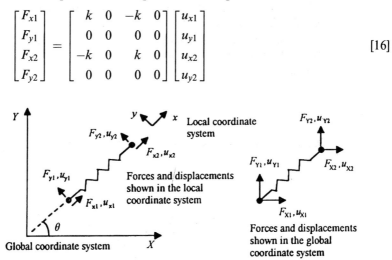

Fig. 19.4 Coordinate systems.

We can write this as

$$\mathbf{f_L} = \mathbf{K_e u_L} \tag{17}$$

where $\mathbf{f_L}$ is the matrix referring to the forces in the local coordinate system, $\mathbf{u_L}$ the matrix for the displacements in the local coordinate system and $\mathbf{K_e}$ is the element stiffness matrix.

We can relate displacements and forces in the local coordinate system to displacements and forces in the global coordinate system. Thus, if for node 1 we have displacements of u_{X1} and u_{Y1} in the global coordinate system, these give displacements in the local coordinate system of

$$u_{x1} = u_{X1} \cos \theta + u_{Y1} \sin \theta$$

$$u_{y1} = u_{Y1} \sin \theta + u_{Y1} \cos \theta$$

and, likewise for node 2 with displacements of u_{X2} and u_{Y2} in the global coordinate system

$$u_{x2} = u_{X2} \cos \theta + u_{Y2} \sin \theta$$

$$u_{y2} = u_{Y2} \sin \theta + u_{Y2} \cos \theta$$

These equations can be expressed as

$$\begin{bmatrix} u_{x1} \\ u_{y1} \\ u_{x2} \\ u_{y2} \end{bmatrix} = \begin{bmatrix} c & s & 0 & 0 \\ -s & c & 0 & 0 \\ 0 & 0 & c & s \\ 0 & 0 & -s & c \end{bmatrix} \begin{bmatrix} u_{X1} \\ u_{Y1} \\ u_{X2} \\ u_{Y2} \end{bmatrix} \tag{18}$$

where $c = \cos \theta$ and $s = \sin \theta$. This matrix equation can be written as

$$\mathbf{u_L} = \mathbf{T u_G} \tag{19}$$

where $\mathbf{T}$ is the transformation matrix relating local and global coordinates, $\mathbf{u_L}$ and $\mathbf{u_G}$ being the displacements in the local and global coordinate systems.

Similarly for the forces of F_{X1} and F_{Y1} in the global coordinate system, these give displacements for node 1 in the local coordinate system of

$$F_{x1} = F_{X1} \cos \theta + F_{Y1} \sin \theta$$

$$F_{y1} = F_{Y1} \sin \theta + F_{Y1} \cos \theta$$

and for node 2 with forces of F_{X2} and F_{Y2} in the global coordinate system

$$F_{x2} = F_{X2} \cos \theta + F_{Y2} \sin \theta$$

$$F_{y2} = F_{Y2} \sin \theta + F_{Y2} \cos \theta$$

These equations can be expressed as

$$\begin{bmatrix} F_{x1} \\ F_{y1} \\ F_{x2} \\ F_{y2} \end{bmatrix} = \begin{bmatrix} c & s & 0 & 0 \\ -s & c & 0 & 0 \\ 0 & 0 & c & s \\ 0 & 0 & -s & c \end{bmatrix} \begin{bmatrix} F_{X1} \\ F_{Y1} \\ F_{X2} \\ F_{Y2} \end{bmatrix} \qquad [20]$$

where $c = \cos\theta$ and $s = \sin\theta$. This matrix equation can be written as

$$\mathbf{f_L} = \mathbf{Tf_G} \qquad [21]$$

where $\mathbf{T}$ is the transformation matrix relating local and global coordinates, $\mathbf{f_L}$ and $\mathbf{f_G}$ being the forces in the local and global coordinate systems.
   Equation [17] can thus be written in terms of global coordinates as

$$\mathbf{Tf_G} = \mathbf{K_e Tu_G}$$

Multiplying each side of the equation by the inverse of the transformation matrix $\mathbf{T^{-1}}$ gives

$$\mathbf{T^{-1}Tf_G} = \mathbf{T^{-1}K_e Tu_G}$$

But $\mathbf{T^{-1}T} = \mathbf{I}$, the unit matrix, and so

$$\mathbf{f_G} = \mathbf{T^{-1}K_e Tu_G} = (\mathbf{T^{-1}K_e T})\mathbf{u_G} = \mathbf{K_G u_G} \qquad [22]$$

where $\mathbf{K_G}$ is the stiffness matrix in global coordinates. The transpose of a matrix is obtained by interchanging its columns with its rows and it turns out that the inverse transformation matrix $\mathbf{T^{-1}}$ is the transposed transformation matrix $\mathbf{T^T}$. Thus

$$\mathbf{K_G} = \mathbf{T^T K_e T} \qquad [23]$$

and so

$$\mathbf{K_G} = \begin{bmatrix} c & -s & 0 & 0 \\ s & c & 0 & 0 \\ 0 & 0 & c & -s \\ 0 & 0 & s & c \end{bmatrix} \begin{bmatrix} k & 0 & -k & 0 \\ 0 & 0 & 0 & 0 \\ -k & 0 & k & 0 \\ 0 & 0 & 0 & 0 \end{bmatrix} \begin{bmatrix} c & s & 0 & 0 \\ -s & c & 0 & 0 \\ 0 & 0 & c & s \\ 0 & 0 & -s & c \end{bmatrix}$$

$$= k \begin{bmatrix} c^2 & cs & -c^2 & -cs \\ cs & s^2 & -cs & -s^2 \\ -c^2 & -cs & c^2 & cs \\ -cs & -s^2 & cs & s^2 \end{bmatrix} \qquad [24]$$

Example

Determine the forces and displacements at each node for the structure shown in figure 19.5 if for each member we have $AE = 200\,\text{MN}$.

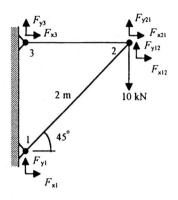

Fig. 19.5 Example.

For the horizontal rod, i.e. the rod between nodes 2 and 3, we have $k = AE/L = 200/(2\cos 45°) = 100/\sqrt{2}\,\text{MN/m}$ and since, for this member, $\cos\theta = 1$ and $\sin\theta = 0$, the global stiffness matrix is given by equation [24] as

$$\mathbf{K_G} = \frac{100}{\sqrt{2}}\begin{bmatrix} 1 & 0 & -1 & 0 \\ 0 & 0 & 0 & 0 \\ -1 & 0 & 1 & 0 \\ 0 & 0 & 0 & 0 \end{bmatrix}\text{MN/m}$$

Equation [22] thus gives for the force–displacement relationship

$$\begin{bmatrix} F_{x21} \\ F_{y21} \\ F_{x3} \\ F_{y3} \end{bmatrix} = \frac{100}{\sqrt{2}}\begin{bmatrix} 1 & 0 & -1 & 0 \\ 0 & 0 & 0 & 0 \\ -1 & 0 & 1 & 0 \\ 0 & 0 & 0 & 0 \end{bmatrix}\begin{bmatrix} u_{x2} \\ u_{y2} \\ u_{x3} \\ u_{y3} \end{bmatrix}\text{MN}$$

For the other element, $AE/L = 100\,\text{MN/m}$, $\sin\theta = 1/\sqrt{2}$ and $\cos\theta = 1/\sqrt{2}$. Thus the stiffness matrix is given by equation [24] as

$$\mathbf{K_G} = 100\begin{bmatrix} \frac{1}{2} & \frac{1}{2} & -\frac{1}{2} & -\frac{1}{2} \\ \frac{1}{2} & \frac{1}{2} & -\frac{1}{2} & -\frac{1}{2} \\ -\frac{1}{2} & -\frac{1}{2} & \frac{1}{2} & \frac{1}{2} \\ -\frac{1}{2} & -\frac{1}{2} & \frac{1}{2} & \frac{1}{2} \end{bmatrix}\text{MN/m}$$

Equation [22] thus gives for the force–displacement relationship

$$\begin{bmatrix} F_{x1} \\ F_{y1} \\ F_{x12} \\ F_{y12} \end{bmatrix} = 100\begin{bmatrix} \frac{1}{2} & \frac{1}{2} & -\frac{1}{2} & -\frac{1}{2} \\ \frac{1}{2} & \frac{1}{2} & -\frac{1}{2} & -\frac{1}{2} \\ -\frac{1}{2} & -\frac{1}{2} & \frac{1}{2} & \frac{1}{2} \\ -\frac{1}{2} & -\frac{1}{2} & \frac{1}{2} & \frac{1}{2} \end{bmatrix}\begin{bmatrix} u_{x1} \\ u_{y1} \\ u_{x2} \\ u_{y2} \end{bmatrix}\text{MN}$$

As $F_{x2} = F_{x12} + F_{x21}$ and $F_{y2} = F_{y12} + F_{y21}$, combining the above force–displacement relationships gives

$$
\begin{bmatrix} F_{x1} \\ F_{y1} \\ F_{x2} \\ F_{y2} \\ F_{x3} \\ F_{y3} \end{bmatrix} = 100
\begin{bmatrix}
\frac{1}{2} & \frac{1}{2} & -\frac{1}{2} & -\frac{1}{2} & 0 & 0 \\
\frac{1}{2} & \frac{1}{2} & -\frac{1}{2} & -\frac{1}{2} & 0 & 0 \\
-\frac{1}{2} & -\frac{1}{2} & \frac{1}{2}+\frac{1}{\sqrt{2}} & \frac{1}{2} & -\frac{1}{\sqrt{2}} & 0 \\
-\frac{1}{2} & -\frac{1}{2} & \frac{1}{2} & \frac{1}{2} & 0 & 0 \\
0 & 0 & -\frac{1}{\sqrt{2}} & 0 & \frac{1}{\sqrt{2}} & 0 \\
0 & 0 & 0 & 0 & 0 & 0
\end{bmatrix}
\begin{bmatrix} u_{x1} \\ u_{y1} \\ u_{x2} \\ u_{y2} \\ u_{x3} \\ u_{y3} \end{bmatrix} \text{MN}
$$

We can now enter the boundary conditions for the structure. At the fixed supports we have zero displacements and so $u_{x1} = u_{y1} = u_{x3} = u_{y3} = 0$. At node 2 we have $F_{x2} = 0$ and $F_{y2} = -0.01$ MN. Hence

$$
\begin{bmatrix} F_{x1} \\ F_{y1} \\ 0 \\ -0.01 \\ F_{x3} \\ F_{y3} \end{bmatrix} = 100
\begin{bmatrix}
0.5 & 0.5 & -0.5 & -0.5 & 0 & 0 \\
0.5 & 0.5 & -0.5 & -0.5 & 0 & 0 \\
-0.5 & -0.5 & 1.21 & 0.5 & -0.71 & 0 \\
-0.5 & -0.5 & 0.5 & 0.5 & 0 & 0 \\
0 & 0 & -0.71 & 0 & 0.71 & 0 \\
0 & 0 & 0 & 0 & 0 & 0
\end{bmatrix}
\begin{bmatrix} 0 \\ 0 \\ u_{x2} \\ u_{y2} \\ 0 \\ 0 \end{bmatrix} \text{MN}
$$

Thus:

$$F_{x1} = 100(-0.5u_{x2} - 0.5u_{y2})$$

$$F_{y1} = 100(-0.5u_{x2} - 0.5u_{y2})$$

$$0 = 100(1.21u_{x2} + 0.5u_{y2})$$

$$-0.01 = 100(0.5u_{x2} + 0.5u_{y2})$$

$$F_{x3} = 100(-0.71u_{x2})$$

$$F_{y3} = 0$$

Hence we obtain $u_{x2} = 0.14\,\text{mm}$ and $u_{y2} = -0.34\,\text{mm}$, $F_{x1} = 10\,\text{kN}$, $F_{y1} = 10\,\text{kN}$ and $F_{x3} = -10\,\text{kN}$.

19.3   Finite element method

The finite element method is based on the concept that a body can be divided into regions, the so-called elements, for which the behaviour can be described in a simpler manner than that of the body as a whole. When the behaviour of the elements has been formulated, the elements are then patched together according to some rules and hence the behaviour of the body as a whole is predicted. This method is not restricted to structural frameworks but can be applied to such problems as determining the temperature distribution resulting from heat flow, the analysis of vibrations, the elastic behaviour of plates and shells, etc., indeed any problem which can be modelled as an assemblage of simple elements.

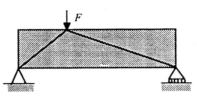

Fig. 19.6 Modelling a two-dimensional shape.

For example, we might model a plane two-dimensional shape by an array of plane elements which are joined together at node points. Figure 19.6 shows a simple situation where a bar is loaded off-centre. The choice of element shape and size in such an analysis is dictated by the geometry of the shape and the accuracy required; the more elements we have the greater the number of computations. The method enables us to determine, in the same way as outlined for frameworks, the distribution of forces and displacements within the shape.

Problems

(1)   Determine the stiffness matrix for three in-line springs with stiffnesses k_1, k_2 and k_3.

(2)   Three dissimilar materials are joined together and placed between two rigid supports as shown in figure 19.7. For material 1 we have $AE/L = 100$ kN/mm, for material 2 we have $AE/L = 200$ kN/mm and for material 3 we have $AE/L = 150$ kN/mm. Determine the displacements of the interfaces when the forces of 100 kN and 50 kN are applied as shown.

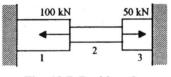

Fig. 19.7 Problem 2.

(3)   Using the matrix displacement method, determine the displacements at each of the nodes of the pin-jointed frameworks shown in figure 19.8. Assume that $AE = 200$ MN for each of the members.

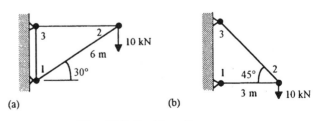

Fig. 19.8 Problem 3.

Answers to problems

Chapter 1

    (1)   213 N at 35.3° from normal

    (2)   184 N at 58.3° to AB

    (3)   52.5 N, 76.7 N

    (4)   60 N, 104 N

    (5)   32 N

    (6)   4.57 m

    (7)   206 mm

    (8) $4r/3\pi$

    (9) $x = 49.5$ mm, $y = 28.0$ mm

  (10) $y = 150$ mm, $x = 33.9$ mm

  (11) 2180 N m

  (12) (a) 51.8 N, 73.0 N, (b) 179.3 N, 146.3 N

  (13) 466 N at 24.4° to the beam, 332 N vertically

  (14) 7.1 kN

  (15) 2.3 kN

  (16) 225 kN, 135 kN

  (17) 276 N

  (18) 208 N, 120 N

  (19) $R_x = 0$, $R_y = 200$ N, $M = 400$ N m

  (20) 660 N, 613 N at 24.1° to the bar

  (21) $R_A = 3.3$ kN, $R_{Fx} = 0$, $R_{Fy} = 1.7$ kN, $F_{AB} = 4.7$ kN, $F_{AC} = 3.3$ kN, $F_{BD} = 3.3$ kN, $F_{CD} = 2.4$ kN, $F_{CE} = 1.7$ kN, $F_{ED} = 0$, $F_{EF} = 1.7$ kN, $F_{DF} = 2.4$ kN. Ties are BC, CD, AC, CE, EF; struts AB, BD, DE

  (22) $R_{Ay} = 10$ kN, $R_{Cy} = 10$ kN, $F_{AB} = 14.1$ kN, $F_{BC} = 14.1$ kN, $F_{AC} = 10$ kN, struts are AB and BC, with AC a tie.

  (23) 2.9 kN

  (24) 11.5 kN

  (25) (a) 6.9 kN, (b) 10 kN

  (26) (a) 2.4 kN, (b) 0.5 kN

Chapter 2

    (1)   204 m

    (2)   4.6 MPa, 2.2 MPa

    (3)   2.4 m

    (4)   (a) 10 MPa, (b) 13 MPa

    (5)   94.2 kN

    (6)   491 kN

    (7)   1.0 MN

(8) 2.6 mm
(9) 0.78 mm
(10) 0.91 mm
(11) 1.2 mm
(12) Weight acts at centre of gravity, i.e. half bar length
(13) See earlier in chapter for a similar example
(14) 2.3 kN
(15) 1.7 kN
(16) 0.37 mm
(17) 7.2 mm
(18) 2.35 MN
(19) 26
(20) 3.84 MPa, 38.4 MPa
(21) 80 MPa, 10 MPa
(22) 109 MPa, 70 MPa
(23) 208 MPa, 208 MPa
(24) 49 MPa, 33 MPa
(25) 6.8 MPa, 8.7 MPa
(26) 62 MPa, 189 MPa
(27) 197 MPa, 101 MPa
(28) 0.18 mm, $-3.6\,\mu$m, $-1.8\,\mu$m
(29) 0.50 mm, $-7.5\,\mu$m
(30) $1.9 \times 10^{-3}\,\text{m}^3$

Chapter 3

(1) (a) $+2.5\,\text{kN}\,\text{m}$, $-2.5\,\text{kN}$ (b) $+5.0\,\text{kN}\,\text{m}$, $-2.5\,\text{kN}$ or $+2.5\,\text{kN}$
(c) $+2.5\,\text{kN}\,\text{m}$, $+2.5\,\text{kN}$
(2) (a) $-6.0\,\text{kN}\,\text{m}$, $+12\,\text{kN}$, (b) $-12.0\,\text{kN}\,\text{m}$, $+12\,\text{kN}$, (c) $-24.0\,\text{kN}\,\text{m}$,
$+12\,\text{kN}$
(3) $-\frac{1}{2}wx^2$, wx
(4) $+106.5\,\text{kN/m}$, $-25.5\,\text{kN}$
(5) See figure A.1.
(6) AB $V = R_A$, $M = R_A x$; BC $V = R_A - F_1$, $M = R_A x - F_1(x - a)$;
CD $V = R_A - F_1 - F_2$, $M = R_A x - F_1(x - a) - F_2(x - b)$
(7) (a) 90 kN m at fixed end, (b) 60 kN at fixed end
(8) (a) 15 kN m, (b) 15 kN
(9) (a) 25 kN, 62.5 kN m; (b) 25 kN, 31.25 kN m; (c) 83.3 kN, 80.2 kN m
(10) 1.67 kN, 20 kN m
(11) 78.75 kN m, 5.5 m from A
(12) 46.7 kN m, 1 m from A
(13) $0.207L$ from each end
(14) 750 kN
(15) 0.60 m and 3.0 m from A
(16) 1.35 m and 2.73 m from A
(17) 797 MPa

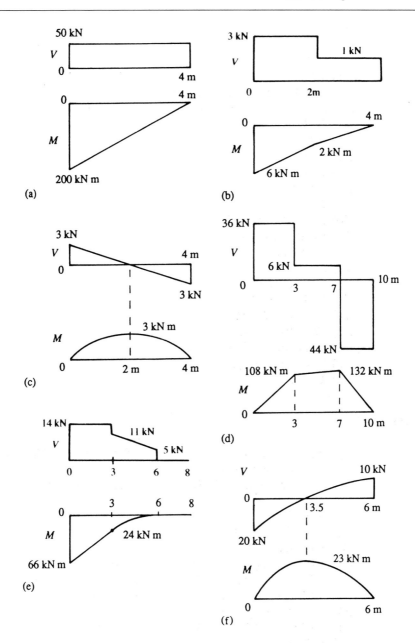

Fig. A.1 Problem 5 Chapter 3.

(18) 3.999 m
(19) 797 MPa
(20) 150 MPa, 7.87 N m
(21) (a) 3.0 MPa, (b) 1.0 MPa
(22) 322 mm

(23) 64 MPa
(24) 171 MPa
(25) 80 MPa
(26) (a) $4.7 \times 10^6 \, \text{mm}^4$, $8.9 \times 10^6 \, \text{mm}^4$, (b) $10.7 \times 10^6 \, \text{mm}^4$, $4.1 \times 10^6 \, \text{mm}^4$,
       (c) $28.0 \times 10^6 \, \text{mm}^4$, $1.51 \times 10^8 \, \text{mm}^4$, (d) $1.95 \times 10^8 \, \text{mm}^4$, $1.95 \times 10^6 \, \text{mm}^4$
(27) 5.49 m
(28) 15.7 kN m
(29) 506 N m
(30) $2.5 \times 10^{-5} \, \text{m}^3$
(31) 40.3 kN/m
(32) (a) $2.25 \times 10^{-3} \, \text{m}^3$, (b) 18 kN m
(33) 353 mm
(34) 20.5 m
(35) 200 MPa tensile, 184 MPa compressive
(36) 400 kN
(37) +250.5 MPa, −270.9 MPa
(38) ≤250/8 mm
(39) 0, −4.0 MPa
(40) +156 kPa, −219 kPa at opposite corners
(41) 3.0 MPa
(42) 4.2 MPa, 3.6 MPa
(43) 3.7 MPa
(44) 14.9 kN m
(45) 5.6 kN m
(46) 5.7 MPa, 119 MPa
(47) 103 kN m
(48) $1.74 \times 10^{-3} \, \text{m}^2$
(49) $1216 \, \text{mm}^2$
(50) 25.9 kN
(51) $8.24 \times 10^5 \, \text{mm}^3$
(52) 400 kN m
(53) As given in the problem
(54) 1.33 kN
(55) 35.9 kN/m
(56) (a) 1.5 kN, (b) 1.05 mm

Chapter 4

(1) 6.0 mm, 0.030 rad
(2) 0.0067 rad, 10 mm
(3) 4.3 mm
(4) 3.0 mm, 4.9 mm
(5) 2.7 mm
(6) 0.55 mm
(7) $y = (w_0 x^2 / 120 LEI)(10L^3 - 10L^2 x + 5Lx^2 - x^3)$
(8) $0 \leq x \leq 1.5L$ $y = -(Fx/12EI)(L^2 - x^2)$;
     $L \leq x \leq 1.5L$ $y = (F/12EI)(3L - x)(L - 2x)$

(9) $0 \le x \le a$ $y = (wx^2/24EI)(6a^2 - 4ax + x^2)$;
   $a \le x \le L$ $y = (wa^3/24EI)(4x - a)$
(10) $(FL^3/3EL) + wa^3 (4L - a)/24EI$
(11) 0.69 mm
(12) $0 \le x \le a$ $y = (Fx/6EI)(3aL - 3a^2 - x^2)$;
   $a \le x \le (a + b)$ $y = (Fa/6EI)(3Lx - 3x^2 - a^2)$
(13) $2FL^3/9EI$
(14) $3FL^2/64EI$
(15) $wL^4/8EI$
(16) $14FL^3/3EI$
(17) $3FL^3/256EI$
(18) $y = -10^3/EI(90x^3/6 - 20x^4/24 + 20\{x - 4\}^4/24 - 120\{x - 6\}^3/6 - 540x)$,
   14.1 mm
(19) $y = (1/EI)wbx^2(3L + 3a - 2x)/12 + w\{x - a\}/24$
(20) $y = (1/EI) (Fbx) (L^2 - b^2 - x^2)/6L + F\{x - a\}^3/6$
(21) $y = (1/EI) Fx(3aL - 3a^2 - x^2)/6 + F\{x - a\}^3/6 + F\{x - L + a\}^3/6$
(22) 2.33 mm
(23) 14.0 mm
(24) 21.4 mm
(25) 3.75 m, 14.8 mm
(26) 2.58 m, 1.4 mm
(27) 54.8 mm at end B
(28) $M_1 = (Fa^2/L^2)(L - a)^2$, $M_2 = (Fa^2/L^2)(L - a)$, $y = Fa^3(L - a)^3/3EIL^3$
(29) $wL^4/768EI$
(30) Reactions each 50 kN, fixed end moments each 31.9 kN m, deflection 0.70 mm
(31) Reactions each 50 kN, fixed end moments each 62.5 kN m, deflection 1.63 mm
(32) $3wa/8 + wb + 3wb^2/4a$
(33) 2.5F, 1.5F, 0.5FL
(34) 4F/3, 2F/3, FL

Chapter 5

(1)   1.72 kN
(2)   58.3 mm
(3)   425 kN
(4)   6.8 MN
(5)   59 kN
(6)   7.74 MN
(7) $\pi^2 I/\alpha AL^2$
(8)   15.5 kN
(9)   1.65 kN
(10) 103 MPa
(11) (a) 6.85 MN, (b) 27.42 MN, (c) 1.71 MN, (d) 14.03 MN
(12) 7.42 kN/m, 15.2 kN/m

(13) 104
(14) 1.04 m
(15) See problem
(16) 30 mm
(17) 464 kN
(18) 91.6 mm by 183.2 mm
(19) 1.2 MN

Chapter 6

(1) 41 MPa, 0.040 rad
(2) 1.7 kN m, 0.033 rad
(3) 1.8 kN
(4) 23.8 kN m
(5) 120 mm
(6) Hollow shaft has 64.2% mass of solid shaft
(7) 7.9 N m
(8) 4.8 MPa
(9) 0.28 MPa, 0.35 MPa
(10) 4.7 MPa
(11) 150 mm
(12) 21.1 MPa, 33.8 MPa
(13) 2.36 MW
(14) 14.8 rev/s
(15) 60 MPa, 0.01 rad
(16) $+0.29$ kN m, -1.29 kN m
(17) 14.8 MPa
(18) 365 N m, -565 N m
(19) $17L/32$
(20) 34.3 MPa, 51.5 MPa
(21) 23.6 MPa, 40.7 MPa, 0.014 rad
(22) $d_a/(d_a + d_b)$
(23) 673 N m in BC
(24) 53.3 mm
(25) 3.77 MPa, 86.1 MPa
(26) 20.1 mm, 40.2 mm
(27) 45 MPa, 13 MPa

Chapter 7

(1) 14.6 J
(2) 1.26×10^5 J
(3) 12.9 J
(4) $5F^2L/16EA$
(5) Strain energy increased by factor 3

(6) $\pi\rho^2 g^2 d^2 h^3/360E$

(7) $5F^2/\sqrt{2}EA$

(8) $\sqrt{2}F^2/EA$

(9) 0.32 mm, 12.6 MPa

(10) $x = x_{st} + x_{st}\sqrt{(1 + 2L/x_{st})}$, 0.78 m

(11) 17 mm

(12) 16 mm

(13) (a) 40.7 MPa, (b) $2 \times 40.7 = 81.4$ MPa, (c) 681 MPa

(14) 170 J

(15) $5F^2L^3/162EI$, $5FL^3/162EI$

(16) $U = (F^2L^3/96E)(1/I_1 + 1/I_2)$, $\delta = (FL^3/96E)(1/I_1 + 1/I_2)$

(17) $w^2L^5/240EI$

(18) 10 mm

(19) $y = y_{st} + \sqrt{(y_{st}^2 + 2hy_{st})}$

(20) $\sigma = \sigma_{st} + \sigma_{st}\sqrt{[1 + (2h/y_{st})]}$

(21) $FR^3\pi/4EI$

(22) $\pi FR^3/2EI$

(23) $Fb^3/3EI + Fb^2a/EI$

(24) $wL^3/8EI$

(25) $FR^3(0.75\pi - 2)/EI$, $FR^3/2EI$

(26) $FL^3/27EI$

(27) $3FL^3/EI + 2FL^3/GJ$

(28) 8.0 mm

(29) 2.7 mm, 1.7 mm

(30) 69 mm

Chapter 8

(1) (a) 18 MPa, 9 MPa, (b) 12 MPa, 12 MPa

(2) 15.6 MPa, 27.1 MPa, 31.3 MPa

(3) 40 MPa

(4) 25 MPa, 43.3 MPa

(5) (a) $\sigma_\theta = 40$ MPa, $\tau_\theta = 10$ MPa, (b) $\sigma_\theta = 60$ MPa, $\tau_\theta = 20$ MPa, (c) $\sigma_\theta = 61.0$ MPa, $\tau_\theta = 23.7$ MPa, (d) $\sigma_\theta = 43.5$ MPa, $\tau_\theta = 19.3$ MPa, (e) $\sigma_\theta = 43.5$ MPa, $\tau_\theta = -71.3$ MPa

(6) (a) 105.9 MPa at 13.3°, -5.9 MPa at 103.3°, 55.9 MPa at 58.3° and 148.3°, (b) 94.1 MPa at 25.1°, 15.9 MPa at 115.1°, 39.1 MPa at 70.1° and 160.1°, (c) 81.6 MPa at 125.7°, 18.4 MPa at 215.7°, 31.6 MPa at 170.7° and 260.7°, (d) -37.6 MPa at 75.7°, 87.6 MPa at 165.7°, -62.6 MPa at 120.7° and 210.7°, (e) 116 MPa at 66.3°, -46.4 MPa at 136.4°.

(7) 65.3 MPa, 42.7 MPa

(8) -23.0 MPa, 7.0 MPa, 15.0 MPa

(9) -191 MPa, 89 MPa, 140 MPa

(10) 94 kN

(11) -12 MPa, $+58$ MPa

(12) 54.9 MPa
(13) 0.65 MPa, -71 MPa, 36 MPa
(14) 20.7 MPa, -2.9 MPa, 11.8 MPa
(15) See the answers for problem 5.
(16) See the answers for problem 6.
(17) 76.5 MPa, 54.9 MPa
(18) (a) 465×10^{-6}, 55×10^{-6}, (b) 210×10^{-6}, -390×10^{-6}, (c) 230×10^{-6}, -130×10^{-6}
(19) 462×10^{-6}, 38×10^{-6}, 104 MPa, 39 MPa
(20) 803×10^{-6}, -103×10^{-6}, 168 MPa, 30 MPa
(21) 405×10^{-6}, 228×10^{-6}, 104 MPa, 77 MPa
(22) 579×10^{-6}, -179×10^{-6}, 115 MPa, -1.2 MPa
(23) See the answer to problem 18
(24) 108 mm
(25) (a) 9.5 mm, (b) 9.0 mm
(26) (a) 42.8 mm, (b) 39.1 mm
(27) 3.1
(28) 3.3
(29) 17.2 GPa
(30) 0.27, 149 GPa

Chapter 9

(1) 60 MPa, 30 MPa
(2) 15 mm
(3) 98 MPa
(4) 1.06 kPa
(5) 16.7 mm
(6) 800 kPa
(7) 628.5 kPa
(8) 3.75 mm
(9) 2.5 MPa
(10) 6.8×10^{-5} m
(11) 0.063%
(12) 10.6×10^{-3} m^3
(13) 14.4×10^{-3} m^3
(14) 20 MPa, 10 MPa
(15) 37.2 mm
(16) 0.077 mm
(17) 24 mm
(18) 61 MPa
(19) 124 MPa, 84 MPa
(20) ratio $= \frac{1}{2}(k^2 + 1)$
(21) 2.38×10^{-6} m
(22) 80 MPa
(23) 127 MPa
(24) 1.78×10^{-5} m

Chapter 10

(1) $-4\,\mathrm{m}$, $-18.7\,\mathrm{m}$
(2) $0.28\,\mathrm{m/s}$
(3) $9\,\mathrm{m/s}$, $6\,\mathrm{m/s}^2$
(4) $8\,\mathrm{m/s}$, $4\,\mathrm{m/s}^2$
(5) $31.5\,\mathrm{s}$
(6) $7.5\,\mathrm{s}$, $112.5\,\mathrm{m}$
(7) $7.5\,\mathrm{s}$
(8) $3.1\,\mathrm{s}$
(9) $1.65\,\mathrm{s}$
(10) $15.4\,\mathrm{kN}$
(11) $2.35\,\mathrm{kN}$
(12) $1.96\,\mathrm{m/s}^2$
(13) $5.89\,\mathrm{m/s}^2$
(14) $30.6\,\mathrm{N}$
(15) $(1/k)\ln(1+u/g)$
(16) $v_\mathrm{t} = \sqrt{(g/k)}$
(17) $\tan\theta$
(18) $3.0\,\mathrm{m/s}$
(19) $14.1\,\mathrm{m/s}$ at $45°$
(20) $1.28\,\mathrm{m}$, $8.83\,\mathrm{m}$
(21) $30.9°$ or $59.1°$
(22) $14\,\mathrm{s}$
(23) $2u^2\sin(\theta+\phi)\cos\theta/g\cos^2\phi$
(24) $188\,\mathrm{m/s}$
(25) $10.8\,\mathrm{m/s}$
(26) $11.5°$
(27) $v = \sqrt{(gdr/2h)}$
(28) $2.63\,\mathrm{m/s}$
(29) $21.2\,\mathrm{m/s}$ at $20.7°$ to vertical
(30) $500\,\mathrm{km/h}$ at $53.1°$ east of south
(31) $3.1\,\mathrm{rad/s}$
(32) $0.0785\,\mathrm{rad/s}^2$, $36.6\,\mathrm{rad}$
(33) $91.6°$
(34) $2.62\,\mathrm{m/s}$
(35) $21.3\,\mathrm{m/s}$
(36)   (a) $0.32\,\mathrm{rad/s}^2$, (b) 0.95
(37) $1.5\,\mathrm{rad/s}$
(38) 7.96
(39) $11.2\,\mathrm{kN\,m}$
(40) $188\,\mathrm{N\,m}$
(41) $2.81\,\mathrm{m/s}^2$
(42) $3.1\,\mathrm{rad/s}^2$
(43) $102\,\mathrm{N}$
(44) $0.187\,\mathrm{kg\,m}^2$
(45) $336\,\mathrm{mm}$
(46) $436\,\mathrm{mm}$

(47) 0° and 180°, 31.6 N
(48) about 296 N
(49) 71.1 N, 165.8 N
(50) 48 kN
(51) 100 N
(52) 1.6 N m
(53) 6.9 rev/s
(54) 75 kW
(55) 108 W
(56) 6.52 kW
(57) $1.20\sqrt{(gh)}$
(58) $L^2\omega^2/6g$
(59) 3.2
(60) (a) 14 rad/s^2, (b) 3.5 N, (c) 44 J

Chapter 11

(1) 4 rad/s
(2) (a) 3.2 m/s, (b) 12.7 m/s
(3) 5.1 m/s at 90° to DC
(4) 8 m/s
(5) 6.2 rad/s, 300 rad/s^2
(6) 2.1 m/s at 76.7°, 12.5 m/s^2 at 35.5°
(7) 33.4 rad/s, 175 rad/s^2
(8) 45 m/s^2 at 30°
(9) 302 rad/s^2
(10) 25 m/s^2
(11) $v = (r\omega p/L)\cos\theta$, $a = -(r\omega^2 p/L)\sin\theta$
(12) 0.30 m/s, 1.95 m/s^2
(13) 333 N m
(14) 260 N m
(15) 4.28 kN
(16) 2.85 kN, 0.27 kN
(17) 8.82 kN, 1.91 kN, 2.52 kN m
(18) 2.69 kN m,
(19) 167 N m
(20) 609 N m

Chapter 12

(1) 0.0030 or 0.30%
(2) 1989 kg m^2
(3) 239 kg m^2
(4) 4109 kg m^2
(5) 7.07 kg m^2
(6) 10.9 kg m^2

(7)   200 W, 64.8 kg m^2
(8)   1.4 kW, 0.115 kg m^2
(9)   75 N m
(10) 6.81 kg m^2
(11) 0.277 kg m^2

Chapter 13

(1)   125 mm, 375 mm
(2)   32, 128, 4.0
(3)   15, 30, 15, 30
(4)   30, 94, 15, 47, 124 mm
(5)   16, 32, 27, 21, 21, 27, 16, 32
(6)   15, 30, 26, 19, 21, 24, 15, 30
(7)   20, 3.0
(8)   4.25 rev/s in same direction
(9)   − 700 rev/min
(10) − 762.5 rev/min
(11) 438 rev/min
(12) 923 rev/min
(13) −36.8 rev/min
(14) 46.5 N m
(15) − 477.5 N m, 397.9 N m
(16) 605 N m
(17) 80 N m
(18) 121 N m
(19) 1563 N m
(20) 15.5
(21) 13.1
(22) 1.51 m/s^2, 1.13 m/s^2, 0.74 m/s^2, 0.51 m/s^2, 7.48
(23) (a) 7.57 s, (b) 8.76 s
(24) 13.0

Chapter 14

(1)   1437 N, 53.8 kPa
(2)   11.26 kN, 36.6 kW
(3)   7.31 kN
(4)   16.3 kW
(5)   8.8 s
(6)   21.0 kW
(7)   13.9 N m
(8)   2.71 s
(9)   193.4 W
(10) 18.5 W
(11) 5

(12) 2.17 kW
(13) 215 W
(14) 5.65 W
(15) 32.2 W
(16) (a) 236 W, (b) 13.1 W
(17) 168.52°, 191.48°, 3.31 kW
(18) 4.91 kW
(19) 850.7 N, 349.3 N, 2.36 kW
(20) 67.6 kW, 45.2 m/s
(21) 18.8 m/s, 8.81 kW
(22) 19.1 m/s
(23) 3
(24) 4

Chapter 15

(1)   603 N
(2)   125.7 kN m
(3)   377 N m
(4)   300 N m
(5)   126 N m
(6)   409 N m, raise nose and depress tail
(7) $392 + 197$ N downwards, $392 - 197$ N downwards
(8)   105 N
(9)   262 N m, front load increases and rear decreases

Chapter 16

(1)   (a) 1.9 m/s, (b) 0.27 m/s, (c) 17.8 m/s^2, (d) 8.9 m/s^2, (e) 35.5 N, (f) 17.8 N
(2)   0.0167 s
(3)   62 mm, 1.2 m/s
(4)   63 N, 2.2 m/s
(5)   1.6 Hz
(6)   0.72 Hz
(7)   (a) 12.3 Hz, (b) 25.2 Hz
(8) $f = (1/2\pi) \sqrt{[kb^2/m_1 (a+b)^2 + m_2 c^2]}$
(9)   0.36 Hz
(10) 8.1 Hz
(11) 203 Hz
(12) 369 Hz
(13) 0.94 m, 9.4 Hz
(14) 2.23 s
(15) $2\pi \sqrt{(d/g)}$

(16) Torque $= 3(mg\phi/3) = mgr\theta/L = I\alpha = \frac{1}{2}mr^2\alpha$, hence $T = 2\pi\sqrt{(L/2g)}$

(17) $306\,\mathrm{kg\,m^2}$

(18) 4.3 Hz

(19) 13 Hz

(20) 1.1 Hz

(21) 32 Hz

(22) 70 Hz

(23) $2.759\sqrt{(EI/FL^3)}$

(24) 2.94 Hz

(25) 1.66 Hz

(26) 21.1 Hz

(27) 41.4 Hz

(28) 11.2 Hz

(29) 43.3 rev/s, 0.0028 mm

(30) 31 rev/s

(31) 0.038 mm

(32) 0.58 mm

(33) 8.5 rev/s to 12.7 rev/s

Chapter 17

(1) 0.25, 1.55 Hz

(2) (a) $f = (1/2\pi)\sqrt{(k/m)} = 3.6\,\mathrm{Hz}$, (b) 3.5 Hz, (c) 0.22

(3) (a) continuous undamped oscillations, (b) underdamped oscillations decaying with time, (c) critical damping, (d) overdamped with no oscillations

(4) 120 Hz, 0.090, 0.99 N s/m

(5) 1.2 kN/m, 3.5 N s/m

(6) 485 N s/m, 3.51 Hz

(7) 258 N s/m, 2.24 Hz

(8) (a) 283 N s/m, (b) 113 N s/m

(9) 0.19, 4.89 Hz, 4.98 Hz

(10) 0.015 m

(11) 8.2 Hz

(12) 4.4 Hz

(13) 24.3 N

(14) 11.1 Hz

(15) 16.7 mm

(16) 83.3 mm

(17) 24.6 mm

(18) 3.2 N

(19) 3.95 mm

(20) 5.18 mm

(21) 10.5 kN

(22) 10.5 mm

Chapter 18

(1) (a) $\omega_1 = \sqrt{(0.66k/m)}$, 0.92, 1.0; $\omega_2 = \sqrt{(6.84k/m)}$, -2.17, 1.0,
   (b) $\omega_1 = \sqrt{(k/m)}$, 1.0, 1.0; $\omega_2 = \sqrt{(2k/m)}$, 1.0, -1.0, (c) $\omega_1 = \sqrt{(k/m)}$,
   1.0, 1.0; $\omega_2 = \sqrt{(5k/2m)}$, 1.0, -2.0

(2) 9 rad/s, 15.6 rad/s

(3) $x_1 = 2\cos\omega_1 t + 2\cos\omega_2 t$, $x_2 = 2\cos\omega_2 t - 2\cos\omega_2 t$, $\omega_1 = 9$ rad/s, $\omega_2 = 15.6$ rad/s

(4) 26.1 rad/s, 85.5 rad/s, ratios 0.86, -0.46

(5) $\omega_n^4 - \left(\dfrac{k_1 + k_2}{I_1} + \dfrac{k_2}{I_2}\right)\omega_n^2 + \dfrac{k_1 k_2}{I_1 I_2} = 0$

(6) $\omega_n^4 - \left[k_1\left(\dfrac{1}{I_1} + \dfrac{1}{I_2}\right) + k_2\left(\dfrac{1}{I_1} + \dfrac{1}{I_3}\right)\right]\omega_n^2 + \dfrac{k_1 k_2}{I_1 I_2 I_3}(I_1 + I_2 + I_3) = 0$

(7) 0, $\sqrt{\dfrac{k_1 k_2 (I_1 + I_2)}{(k_1 + k_2) I_1 I_2}}$

Chapter 19

(1) $\begin{bmatrix} k_1 & -k_1 & 0 & 0 \\ -k_1 & k_1 + k_2 & -k_2 & 0 \\ 0 & -k_2 & k_2 + k_3 & -k_3 \\ 0 & 0 & -k_3 & k_3 \end{bmatrix}$

(2) -0.385 mm, -0.077 mm

(3) (a) $u_{x1} = u_{x3} = u_{y3} = 0$, $u_{y1} = -0.15$ mm, $u_{x2} = 0.49$ mm, $u_{y2} = -2.12$ mm,
   (b) $u_{x1} = u_{y1} = u_{x3} = u_{y3} = 0$, $u_{x2} = -0.15$ mm, $u_{y2} = -0.57$ mm

Index